Biochemical toxicology

a practical

TITLES PUBLISHED IN
THE
PRACTICAL APPROACH
SERIES

Biochemical toxicology

a practical approach

Edited by

K Snell
Division of Toxicology, Department of Biochemistry,
University of Surrey, Guildford, Surrey GU2 5XH, UK

B Mullock
Department of Clinical Biochemistry, University of Cambridge,
Addenbrookes Hospital, Hills Road, Cambridge CB2 2QR, UK

IRL PRESS
OXFORD · WASHINGTON DC

IRL Press Limited
P.O. Box 1,
Eynsham,
Oxford OX8 1JJ,
England

First published February 1987
First reprinting July 1987

British Library Cataloguing in Publication Data

Biochemical toxicology: a practical approach.
  1. Toxicology   2.   Biological chemistry
  I.   Snell, Keith   II. Mullock, B.
  615.9

ISBN 0-947946-67-5 (hardbound)
ISBN 0-947946-52-7 (softbound)

Printed by Information Printing Ltd., Oxford, England.

Preface

Toxicology is a multi-disciplinary science dealing with the adverse effects of chemical and other agents on living systems and includes major contributions from biochemistry, pharmacology and pathology. Professional toxicologists are involved in establishing the safe limits of chemical substances intended for use as drugs, pesticides, food additives, cosmetics, and industrial chemicals. Research in toxicology is aimed at elucidating the nature and molecular mechanisms of toxic interactions. Such knowledge not only provides a basis for the practice of toxicology and allows the rational prediction of potential toxic hazards but also has been, and is, of great importance in elucidating the metabolic biochemistry of living organisms. Our primary aim in this book has been to provide detailed practical protocols and descriptions of methods which will allow biochemists to enter the fascinating area of toxicological research and will allow toxicologists to apply biochemical techniques and approaches to their studies. We believe that the book will prove indispensable to the novice in providing access to the all-important 'tricks of the trade' which are so often omitted from methods descriptions in research papers. However, we also believe that it will be valuable to experienced toxicologists in guiding them through the range of biochemical approaches which may be applied. The levels of biological complexity to which these methods are applied range from biochemical macromolecules, through subcellular preparations, to the whole animal. Of course, within the limitations of space it is not possible to cover every biochemical technique or biological preparation which can be used in toxicological research. However, we have aimed to include most of the more significant and fundamental practical approaches that are used in this area, and even some which have only recently been developed. The rationale for the choice of topics is provided in the Foreword and the only major topics which have been consciously omitted are *Mutagenicity Testing*, which is covered by another volume in this series edited by S.Venitt and J.M.Parry, and *Carcinogenicity Testing*, which is covered in a book edited by A.D.Dayan and R.W.Brimblecombe (MTP Press, Lancaster, 1978).

It is unfortunate that we have to introduce some notes of regret into this preface. However, it is with sadness that we have to record the untimely death of Professor Eric D.Wills, one of the first of our contributors to complete his chapter. We regret that he never saw the completed version of this book, nor indeed of his own textbook on the *Biochemical Basis of Medicine*. Both provide testimony to his clarity of expression and his erudition and will surely be fitting memorials. Our other note of regret is the long time span between the submissions of the first contributors and the last contributors to this book. The latter are in no way responsible for the publication delay since they responded at short notice to replace certain contributors who withdrew their commitment at a late stage. Although we accept full editorial responsibility for the result of this publication schedule, we do not believe that any contribution is diminished because of it. Indeed, we wish to express our thanks to all the authors for the quality of their contributions and for their forbearance. We are grateful to the staff at IRL Press and, in particular, to Eva Gooding for her unfailing patience and encouragement.

Finally, we wish to dedicate this book in honour of Professor Dennis Parke, Head of Department and Professor of Biochemistry at the University of Surrey, a post he has held with distinction for the past twenty years. He is, of course, a world-renowned scientist in toxicology and biochemical pharmacology, but we wish to recognise here the significant and pioneering achievements he has made in promoting the science of toxicology in the United Kingdom through his research and his teaching.

Keith Snell and Barbara Mullock

Contributors

D.J.Benford
Robens Institute of Industrial and Environmental Health and Safety, University of Surrey, Guildford, Surrey GU2 5XH, UK

N.A.Brown
MRC Experimental Embryology and Teratology Unit, Woodmansterne Road, Carshalton, Surrey SM5 4EF, UK

K.Cain
MRC Toxicology Unit, Woodmansterne Road, Carshalton, Surrey SM5 4EF, UK

M.E.Coakley
MRC Experimental Embryology and Teratology Unit, Woodmansterne Road, Carshalton, Surrey SM5 4EF, UK

M.Dobrota
Robens Institute of Industrial and Environmental Health and Safety, University of Surrey, Guildford, Surrey GU2 5XH, UK

L.G.Dring
Drug Metabolism Department, Upjohn Ltd., Fleming Way, Crawley, Sussex RH10 2NJ, UK

S.J.Freeman
Smith Kline and French Laboratories Ltd., Welwyn Garden City, Herts AL7 1EX, UK

R.C.Garner
Cancer Research Unit, University of York, Heslington, York YO1 5DD, UK

S.A.Hubbard
Health and Safety Executive, Magdalen House, Stanley Precinct, Bootle, Merseyside L20 3QZ, UK

R.S.Jones
Division of Toxicology, Department of Biochemistry, University of Surrey, Guildford, Surrey GU2 5XH, UK

B.G.Lake
British Industrial Biological Research Association, Woodmansterne Road, Carshalton, Surrey SM5 4DS, UK

C.N.Martin
Cancer Research Unit, University of York, Heslington, York YO1 5DD, UK

J.V.Møller
Institute of Medical Biochemistry, University of Aarhus, DK-8000 Aarhus C, Denmark

M.I.Sheikh
Institute of Medical Biochemistry, University of Aarhus, DK-8000 Aarhus C, Denmark

D.N.Skilleter
MRC Toxicology Unit, Woodmansterne Road, Carshalton, Surrey SM5 4EF, UK

K.Snell
Division of Toxicology, Department of Biochemistry, University of Surrey, Guildford, Surrey GU2 5XH, UK

E.D.Wills*
Department of Biochemistry, Medical College of St. Bartholomew's Hospital, Charterhouse Square, London EC1M 6BQ, UK

*Deceased

Contents

7. PREPARATION AND USE OF RENAL AND INTESTINAL PLASMA MEMBRANE VESICLES FOR TOXICOLOGICAL STUDIES

M. Iqbal Sheikh and Jesper V. Møller

8. PREPARATION AND CHARACTERISATION OF MICROSOMAL FRACTIONS FOR STUDIES OF XENOBIOTIC METABOLISM

Brian G. Lake

Abbreviations

BHT	2,6-ditert-butyl-p-methylphenol
BSP	bromosulphophthalein
CHO cells	Chinese hamster ovary cells
CIP	chloroform-isoamyl alcohol-phenol
CP	cyclophosphamide
DCPIP	2,4-dichloro-phenol-indophenol
DMSO	dimethylsulphoxide
EDTA	ethylenediamine tetraacetic acid
EGTA	ethyleneglycobis(β-aminoethyl)ether tetraacetic acid
g.l.c.	gas-liquid chromatography
GLDH	glutamate dehydrogenase
GOT	aspartate transaminase
GPT	alanine transaminase
HBSS	Hank's balanced salt solution
Hepes	N-2-hydroxyethylpiperazine-N$'$-2-ethanesulphonic acid
h.p.l.c.	high-performance liquid chromatography
LAP	leucine aminopeptidase
LDH	lactate dehydrogenase
MEM	minimal essential medium
ODS	octadecyl silicate
PBS	phosphate-buffered saline
PDH	pyruvate dehydrogenase
PMSF	phenylmethylsulphonyl fluoride
PVP	polyvinyl pyrrollidone
RCR	respiratory control ratio
RSA	relative specific activity
S9	post-mitochondrial supernatant
SDH	sorbitol dehydrogenase
SDS	sodium dodecyl sulphate
TBA	thiobarbituric acid
TCA	trichloroacetic acid
t.l.c.	thin-layer chromatography
TMPD	tetramethyl-p-phenaline diamine
WME	William's medium E

Foreword

KEITH SNELL

The science of toxicology is not a 'pure' discipline; it is a unification of a number of scientific disciplines (e.g. biochemistry, pharmacology and pathology) orientated towards the common goal of the identification, quantification and mechanistic explanation of adverse interactions between a chemical substance and a living organism or biological system. It is concerned with investigations of toxicity; but toxicity is a subjective term since, for example, an anti-bacterial agent is clearly toxic to the organism it is directed against but hopefully not to the organism to which it is administered. Even within a single organism toxicity depends on dose; hence acetylsalicylic acid (aspirin) can have beneficial analgesic properties in humans at low doses, but can be a gastric irritant and ulcerogenic agent at high doses or after chronic administration. It is the goal of toxicity testing to define such parameters as dose-response characteristics and species selectivity and to refine these parameters (along with others) into a quantitative risk-benefit analysis that allows the value judgement of safety to be applied to a chemical substance which may have potential economic or therapeutic use for man. The approaches and methodologies of toxicity testing procedures are not dealt with in this book, but are covered elsewhere (1−4), and in general textbooks of toxicology (5−8). However, the development of appropriate toxicity testing procedures is critically dependent on fundamental studies on the molecular mechanisms of toxic effects which is the province of Biochemical Toxicology. Only with this basic knowledge is it possible to devise meaningful approaches to the detection of toxicity or indeed to make rational predictions about the nature of the toxic response which might determine the type of testing protocol to be employed. The present book focuses on the application of biochemical methods to investigations of mechanisms of toxicity. Such investigations aim to define the molecular targets of toxic interactions, so as to provide a biochemical explanation for the overt toxicity manifested in the whole organism as well as the basis for the selectivity of toxic actions.

Apart from the differential cellular sensitivity conferred by the presence of critical molecular targets, another major determinant of selectivity can be the generation of the ultimate toxicant chemical species at the susceptible site of toxic interaction. Investigations of this latter aspect involve considerations of pharmacokinetics and of xenobiotic metabolism. Thus the measurement of the parent chemical compound and its metabolic products in body fluids is an essential tool and the relevant techniques are considered in Chapter 1. Since the liver is the most biochemically active organ in the metabolism of xenobiotics, the activation or detoxification of chemical substances is frequently assessed using liver-derived systems. The most physiological of these is the intact perfused liver (Chapter 2). The use of liver cells, either freshly isolated or in primary culture (Chapter 3), has the advantage that a single cell type (hepatocyte) is being studied and that different experimental conditions can be employed with a preparation from a single animal. Even more defined is the liver microsomal subcellular fraction (Chapter 8), where the reactions and enzymes of many of the pathways of xenobiotic metabolism can be studied in isolation from many other intracellular biochemical pathways. A major attribute of this preparation is the cytochrome P-450 mixed function oxidase (monooxygenase) enzyme system which carries out the bioactivation of many toxic chemicals (9, 10). For this reason a crude microsomal preparation with supplementations is often included in biological preparations which otherwise have a limited capacity for the bioactivation of chemical toxicants (Chapters 3 − 5). Liver perfusion and cell culture techniques are also useful in defining and elucidating the cellular responses to toxic insult, free from the potential ambiguities of interpretation inherent in studies at the whole animal level (Chapters 2 and 3). Similarly, the post-implantation embryo culture system (Chapter 4) affords a useful tool for the study of teratological mechanisms without the ambiguities that might arise from maternal-conceptus interactions *in vivo*.

Ultimately, the sensitivity of a biological system to toxic insult is defined and characterised by the presence of critical molecular targets. Of these macromolecules, proteins possess highly specific functional characteristics and are difficult to consider in a generalised fashion; each must be studied individually. For other cellular macromolecules such as nucleic acids and lipids, it is feasible to study a more generalised interaction with the toxic chemical. In the case of nucleic acids, DNA is a critical toxicological target, through the covalent binding of reactive chemical toxicants, because of the known associations between chemical modification of DNA and mutagenicity and carcinogenicity (Chapter 5). For lipids, the most significant chemical damage comes from peroxidative attack (Chapter 6) and the consequent disturbances of structural integrity and functioning of biological membranes.

In many cases, the prime interest for the biochemist in linking the toxicant-target interaction to cellular damage, is the consequence for the normal functioning of the target molecule. A valuable approach in elucidating the mechanism of an agent at this level is to study the functional properties of the subcellular organelle in which the target macromolecule is located. More usually it is the cellular response to the toxicant which implicates a particular subcellular process, and then the demonstration of a direct effect on the isolated subcellular organelle will provide clues to the identity of the ultimate target molecule. These approaches are detailed in this book for subcellular fractions derived from the plasma membrane (Chapter 7), the endoplasmic reticulum (Chapter 8), mitochondria (Chapter 9), and lysosomes and peroxisomes (Chapter 10).

With the methodological details provided, it should be possible for biochemists to apply their skills to problems of toxicological interest. The principle aim of this book is to encourage such approaches and provide the practical means to follow them.

REFERENCES

1. Gorrod,J.W., ed. (1981) *Testing for Toxicity*, Taylor and Francis Ltd., London.
2. World Health Organization (1978) *Environmental Health Criteria 6; Principles and Methods for Evaluating the Toxicity of Chemicals*, W.H.O., Lyon.
3. Brown,V.K.H. (1980) *Acute Toxicity in Theory and Practice*, Wiley and Sons Ltd., Chichester.
4. Witschi,H.R., ed. (1980) *The Scientific Basis of Toxicity Assessment*, Elsevier, Amsterdam.
5. Timbrell,J.A. (1982) *Principles of Biochemical Toxicology*, Taylor and Francis Ltd., London.
6. Hodgson,E. and Guthrie,F.E., eds. (1980) *Introduction to Biochemical Toxicology*, Elsevier, New York.
7. Casarett,L.J. and Doull,J., eds. (1980) *Toxicology*, 2nd edition, Macmillan, New York.
8. Loomis,T.A. (1978) *Essentials of Toxicology*, 3rd edition, Lea and Febiger, Philadelphia.
9. Parke,D.V. (1974) *The Biochemistry of Foreign Compounds*, Pergamon Press, Oxford.
10. Gibson,G.G. and Skett,P. (1986) *An Introduction to Drug Metabolism*, Chapman and Hall, London.

CHAPTER 1

Methods for Studying Metabolism and Distribution *In Vivo* of Radiolabelled Drugs

L.GRAHAM DRING

1. INTRODUCTION

The methods used for studying drug metabolism and distribution have advanced greatly in the last 20 years, largely because of the increasing requirement for drug safety evaluation after the thalidomide disaster. That metabolic data would become part of the regulatory requirements for drug registration became apparent on the publication of the 'Goldenthal letter' (1) in the USA: 'Although at present, we are not insisting that metabolic data be submitted while the drug is under investigational exemption, we will expect to see information of this type in most New Drug Applications for new entities in the coming year'.

As a result, regulatory bodies have followed suit in many countries, thus in the UK the major requirements of the DHSS are set out in Notes on Applications for Clinical Trial Certificates (2). These notes recommend that metabolism studies include the following.

(i) *Plasma levels.* Single dose, peak drug levels and calculation of half-life in species used in toxicology. Chronic drug administration to identify any accumulation and to test for enzyme induction.

(ii) *Distribution.* To include plasma levels and autoradiography or quantitative studies, of major organs and the pregnant animal.

(iii) *Excretion.* Total urine and faeces collection. Times should allow reasonably complete recovery. Evidence for enterohepatic recycling.

(iv) *Metabolites.* Identification or separation conducted as far as is technically reasonable.

These are really the minimum requirements and most pharmaceutical companies when studying new chemical entities will have the basic pharmacology and biopharmaceutical information related to the drug and will be actively supporting the toxicity studies and pharmaceutical development. After judicious choice of formulation, route of administration and other biopharmaceutical factors, the animal experiments would comprise bioavailability/dose proportionality studies at the dose levels used in the toxicology studies. This can often give the toxicologist a valuable insight into the behaviour of the drug in the organism at increasing dose levels. It is also possible to generate much information using pregnant animals and the foetus which could help in the interpretation of the peri- and post-natal toxicology. Tissue distribution studies not only help the toxicologist, who may find that the accumulation of a drug in a particular organ goes far in explaining the toxicity to that organ, but also are of importance when assessing the feasibility of human studies with the radiolabelled drug.

Many powerful techniques are now at our disposal including radioisotope labelling, metabolite separation by many different forms of chromatography and identification of metabolites by varied physicochemical techniques such as mass spectrometry and nuclear magnetic resonance spectrometry. The complexity of the subject is directly proportional to the number of drugs which are known and it would require at least an entire book to give practical instructions for all eventualities. Consequently, generalised instructions are presented together with specific examples used as illustration.

2. RADIOLABELLED DRUGS

Radiolabelled forms of the drug being studied are conventionally used to facilitate the work, since, although radioisotopes are expensive, the time saved by their utilisation fully justifies their cost. The isotopes most commonly used are ^{14}C, ^{3}H, ^{35}S and ^{32}P. The preferred isotope is ^{14}C, although it is expensive. Often the ease of incorporation of tritium into a molecule coupled with its cheapness outweigh the problems frequently encountered with this isotope. The major problem with tritium is the possibility of loss of the isotope from the molecule by metabolism or chemical exchange, but judicious choice of the position of labelling can often circumvent this problem. Per unit of radioactivity ^{3}H is some 20 times cheaper than ^{14}C. Again, with ^{14}C the position of labelling is of great importance. A good example of the effect of position of labelling can be seen in the metabolism of phenacetin (3). The two major routes of metabolism (*Figure 1*) are its conversion to either paracetamol or phenetidine. Labelling with [^{14}C]acetyl would be suitable as long as no estimate of the phenetidine produced was required as this would no longer be labelled. An estimate of the paracetamol would however be possible. Equally the degree of de-acetylation would be easy to measure as this would be directly proportional to the $^{14}CO_2$ produced by the metabolism of the acetyl group. Labelling of the ethoxy group would enable the measurement of the phenetidine, but not the paracetamol. Ring labelling would enable the concurrent measurement of both phenetidine and paracetamol, but it is unfortunately several orders of magnitude more expensive to synthesise than the other forms. The radiochemical purity of a compound

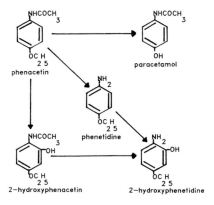

Figure 1. The major metabolic pathways of phenacetin.

4

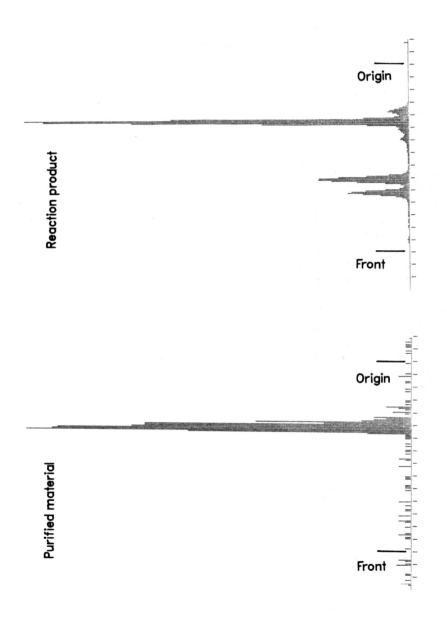

Figure 2. The use of t.l.c. to follow the purification of a ³H-labelled semisynthetic antibiotic.

(4) is an often overlooked but important point to be remembered. Many workers have inadvertently examined the metabolism of a compound and its impurities! Usually t.l.c. in two or three systems and possibly an h.p.l.c. system is sufficient to prove radiochemical purity, which should be greater than 98%. The use of t.l.c. to follow the purification of a radiolabelled drug is shown in *Figure 2*.

After synthesis, steps should be taken to minimise decomposition (5) by, e.g., storage at low temperature and dissolution in a suitable solvent to reduce autoradiolysis. With ^{14}C-labelled compounds, it is usually sufficient to store the solid at $-70°C$ or in liquid nitrogen since these have relatively low specific activities. On the other hand, high specific activity ^{3}H-labelled compounds are usually best stored in a solvent, e.g., ethanol at $+4°C$.

3. ANIMAL STUDIES

The laboratory animals most commonly used are the mouse, rat, rabbit, dog and various non-human primates.

3.1 **Dosing**

Animals may be dosed by any route appropriate in the human population and should at least include the proposed route of administration in man e.g., orally, intraperitoneally, intramuscularly, subcutaneously, intravenously or by inhalation. Preparation of a radiochemical dose has its own peculiar problems which depend upon the dose form to be used. The actual dose in mg/kg body weight to be used will depend upon a number of factors:

(i)   the toxicity;
(ii)   the projected human dose;
(iii)   if the compound is other than a drug, e.g. industrial chemical, then the expected exposure level.

The two major dose forms are a solution, which should preferably be isotonic if it is to be given parenterally, or a solid. Any solid dosage form should be completely homogeneous. If the radiolabelled form of the latter has to be diluted with the non-radioactive form, this must be achieved by recrystallisation and not by a method such as simple mechanical mixing of the two solids. A recommended technique is that of Sword and Waddell (6). The dose levels of radioactivity in *Table 1* can be regarded as a guide.

Table 1. Suggested Doses of Radioactivity in a Number of Animal Species.

Species	Body weight	^{14}C μCi(Bq)	^{3}H μCi(Bq)
Mouse	20 g	4 (148 kBq)	10 (370 kBq)
Rat	200 g	20 (740 kBq)	50 (1.85 MBq)
Dog	10 kg	50 (1.85 MBq)	100 (3.70 MBq)

Studies in man require about 50 μCi of ^{14}C and proportionately more for ^{3}H. It must be borne in mind that permission for such human studies is required in most countries from the appropriate regulatory agency.

Figure 3. A metabolism cage for the collection or urine, faeces and expired air from small animals.

3.2 Collection of Biological Samples

After dosing with the radiolabelled drug, the animals are placed in metabolism cages suitable for the collection of urine, faeces and expired air. Such an apparatus for small mammals is shown in *Figure 3* and is of all-glass construction (Metabowl by Jencons, Leighton Buzzard, Beds, UK). Similar cages of plastic construction are also available but these should not be used for the collection of excreta for metabolite isolation since the plasticisers cause extensive contamination problems. The gas absorption towers may most appropriately be filled with a suitable trapping agent (methoxyethanol-ethanolamine 8:2 v/v for trapping $^{14}CO_2$) and the expired air is passed through these. The urine may be collected conveniently in containers cooled in either ice or solid carbon dioxide to minimise the bacterial/chemical degradation of the drug and metabolites. Blood samples

7

can be taken at the desired time intervals from a vein, placed in heparinised tubes and a portion centrifuged to harvest the plasma, which may be frozen at $-20°C$ until it can be analysed.

4. RADIOCHEMICAL ANALYSIS

Low energy beta particle-emitting isotopes are those most frequently used in drug metabolism studies for a number of reasons, including ease of incorporation into molecules, safety and short path length of the beta particles which leads to high resolution in autoradiography. These particles are now almost always detected and counted by means of liquid scintillation counters. The use of these will not be discussed here as there are excellent treatises on the subject (7). The preparation of samples for liquid scintillation counting is the subject of one of the excellent monographs produced by Amersham International, Amersham, UK (8). The following procedures are those most often adopted for such biological samples.

4.1 Estimation of Total Radioactivity in Excreta, Blood, Plasma and Tissues

The most frequently used method is that of liquid scintillation counting with an emulsion-type system containing Triton X-100, toluene and a liquid scintillator cocktail. The system can exist as either a clear fluid, as two phases or as a gel, depending upon the proportions of aqueous phase and counting medium (*Figure 4*). As might be anticipated, the clear fluid is the most efficient way of counting samples, the gel is slightly less efficient and the two-phase system should be avoided at all costs since the efficiency is both poor and variable. The cocktail can be purchased ready for use, e.g., Optiphase 'X', LKB Instruments Ltd., South Croydon, Surrey, UK, or, alternatively, it can be made in the laboratory, e.g., Patterson and Greene who described (9) the preparation

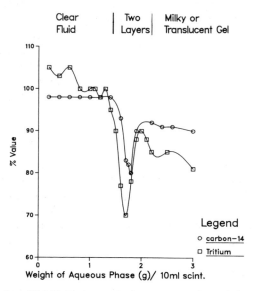

Figure 4. Emulsion counting of $^{14}C(U)$-D-glucose and tritiated water using toluene/Triton X-100 (2:1 v/v). After Turner (15).

of a system containing 0.4% PPO and 0.1% POPOP in toluene/Triton X-100, 2:1 (v/v).

4.1.1 *Urine*

This is usually counted as a gel in 20 ml low background scintillation vials. The proportions of scintillant and aqueous phase must be measured exactly otherwise an unstable biphasic system may result. If fairly active samples are being counted, proportionately smaller samples may be counted in 5 ml plastic minivials with suitable quantities of scintillant. Where large numbers of samples are involved this technique can lead to substantial cost savings.

4.1.2 *Plasma*

This can be counted as a gel, like urine, in either glass or plastic minivials.

4.1.3 *Blood*

The strong red colour of whole blood significantly reduces the efficiency of counting and must therefore be removed. This can be done by either bleaching or combustion (see below). To prepare a sample by bleaching:

(i)   Mix whole blood (100 μl) with water (1 ml) in order to haemolyse the red cells, butan-1-ol (100 μl) to reduce frothing and 600 μl of 100 volume H_2O_2.

(ii)   Leave overnight.

(iii) Digest the precipitated protein in the now decolorised sample with a suitable tissue solubiliser, e.g., Fisosolve (3 ml, Fisons plc, Loughborough, UK) by heating at 60°C in a water bath for 1 h.

(iv)   After cooling the solution, add a suitable scintillant, e.g., Fisofluor 3 (15 ml, Fisons plc). The mixture is then ready for counting.

4.1.4 *Solid Biological Samples*

These can be tissues, faeces or blood which can be considered to be in the same category for the purposes of the procedure for processing. After homogenisation, divide the samples into portions suitable for combustion (~ 300 mg) and burn in a sample oxidiser e.g. the Packard Tri-Carb 306 Sample Oxidiser, Packard Instrument Co. Ltd., Caversham, Berks, UK (see *Figure 5*). The ^{14}C is converted to $^{14}CO_2$ which can be trapped in a suitable reagent (usually ethanolamine-based) and mixed with scintillant. This is an automated process. The ^3H in the samples is converted to ^3H_2O which is condensed and mixed with a suitable scintillant (again automatically).

4.2 **Differential Analyses for Unchanged Drug and Metabolites**

The biological fluids most frequently analysed for unchanged drug and metabolites are urine, bile and plasma. Faeces are analysed somewhat less frequently. The major analytical methods (for non-radioactive drug) are h.p.l.c., g.l.c., g.l.c.-m.s., t.l.c. densitometry and radioimmunoassay. The actual method of detection after separation of the compounds will depend upon the chemical nature of the compound. The use of a radiolabel greatly facilitates the quantitation and detection of a drug. In conjunction the radioactive and non-radioactive methods complement each other. No general hard and fast rules can be given, so the general principles will be illustrated by a few examples.

4.2.1 *Quantitation of the Metabolites of [^{14}C]Feprazone in Rat Urine*

The example given is that of a 0 − 24 h urine sample from a rat dosed orally with the

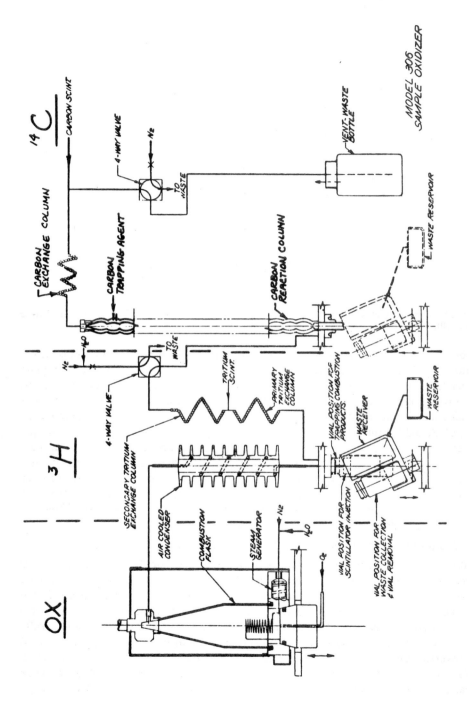

Figure 5. Diagram of the Packard Tri-Carb 306 Sample Oxidiser. Reprinted by permission of Packard Instrument Co. Ltd.

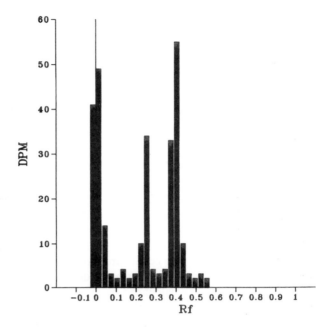

Figure 6. A histogram of the t.l.c. separation of the urine of a rat dosed with [¹⁴C]feprazone. After Donetti *et al.* (10).

labelled drug (10). Urine samples usually contain sufficient radioactivity to be chromatographed directly. In this case, 10 μl of neat urine was spotted onto the t.l.c. plate (silica Gel G F254, E.Merck A.G., Darmstadt, FRG), dried with the aid of a stream of nitrogen and developed (CHCl₃/EtOH, 9:1 v/v). After drying, the silica gel was divided into 5 mm bands which were scraped off, suspended as a gel (3 ml Optiphase 'X', LKB, and 1.5 ml of water) and counted in a scintillation counter. The use of nitrogen for drying the applied spot is recommended since many compounds decompose on silica if a solution of the compound is dried down with a stream of hot air. After counting, a histogram is constructed (*Figure 6*) and peaks can be quantified as a percentage of the dose originally given to the animal. The identity of each peak can be ascertained in separate experiments and by reference to standard compounds which have been run on the same plate.

4.2.2 *Quantitation of the Amine Metabolites of Tiflorex*

In this example (11) solvent extraction was used to concentrate the metabolites in either urine or plasma. This was followed by derivatisation and reverse phase t.l.c., thus: 1 ml of the biological sample was adjusted to pH 12 with 1 M NaOH, extracted with toluene (5 ml) and centrifuged (10 min, 1000 g, 4°C). An aliquot of this was transferred to a silanised vial and treated with trichloracetyl chloride in toluene (10 μl 1:1000 v/v). The mixture was heated (1 h, 80°C), evaporated to 100 μl with N₂ and analysed by reverse phase t.l.c., (silanised Kieselgel 65-F254, thickness 0.25 mm, E.Merck A.G.). The radioactive areas corresponding to the unchanged drug and metabolites were located on the chromatogram plates by autoradiography (Section 4.2.3). The silica gel radioactive areas were scraped off and counted as a gel (see Section 4.2.1). It is then a relatively simple procedure to calculate the proportion of each metabolite in terms of a percentage of the dose given.

Variations on the above procedures can be used for both urine and plasma samples. Thus a very simple but effective method of deproteinising plasma is to add three volumes of methanol to one of plasma, mix and centrifuge. This is effective for most drugs and metabolites, obviates the need for any strong acids or alkalis which might destroy labile metabolites and gives an aqueous methanol extract which can be used for chromatography.

4.2.3 *Location of Radioactive Areas on Chromatograms*

(i) *T.l.c.* By far the most sensitive but the most time-consuming method is to scrape narrow bands (say 5 mm) of silica off the plate, and after adding scintillant, count them in the scintillation counter and construct a histogram of the number of counts against the distance from the end of the plate (*Figure 6*).

A far easier but less sensitive method is to use a plate scanner with a windowless Geiger counter as the detecting head. The counts are fed to a microcomputer, which then plots a histogram of much the same type as that produced by scraping and counting. The area of each peak can also be calculated by the microcomputer. One major advantage of the system is the automatic functioning and the ability to scan a number of chromatography runs on the same plate (ESI-Panax Scanner, Rotheroe and Mitchell, South Ruislip, Middlesex, UK, *Figure 7*). A more recent but more expensive innovation is the linear analyser. This has a modified Geiger tube which can scan the whole length of a chromatography run at the same time, can therefore accumulate many more counts in a given time and is therefore much faster to use (Laboratorium Professor Dr. Berthold, D-7547 Wildbad-1,FRG).

A third method of detecting radioactivity on t.l.c. plates is by means of autoradiography. Again this is not particularly sensitive, but on the other hand, it does have the advantage of resolution and also gives the ability to look at the whole area of a two-dimensional radiochromatogram and is carried out as follows.

(i)   Mark the eluted t.l.c. plate with radioactive ink in three corners and transfer to the darkroom. All further operations must be carried out using illumination provided by the dark green safelight in the darkroom only.

(ii) Sandwich a sheet of Kodak X-Omat AR X-ray film (Kodak Ltd., Hemel Hempstead, Herts, UK) (10 × 8 in, suitable for 20 × 20 cm t.l.c. plate) between the t.l.c. plate and a sheet of plywood.

(iii) Clamp by means of two large paper clips and place in a light-tight plastic bag which is sealed by means of Scotch tape.

(iv) Store the bag in a −80°C deep freeze to increase the sensitivity of the process.

After a suitable time for exposure (which will depend upon the amount of radioactivity in each spot on the chromatogram) develop the film.

(i)   Developer. To 100 ml of Kodak DX-80 developer, add 300 ml of water and place the solution in a developing dish (12 × 10 in).

(ii) Fixer. To 100 ml of Kodak FX-40 fixer add 300 ml of water and place the solution in a fixing dish (12 × 12 in). The temperature of both solutions should be

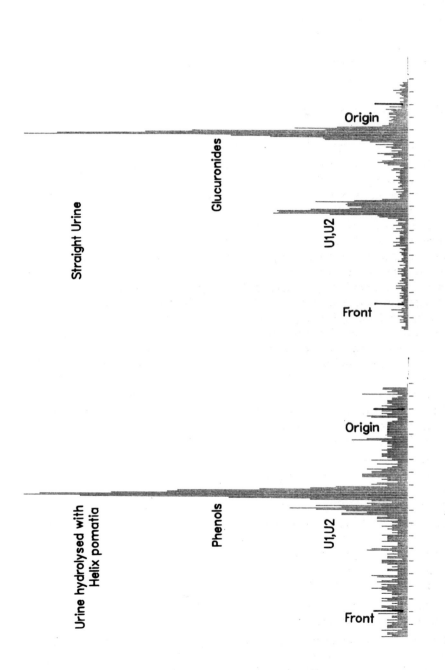

Figure 7. A radiochromatogram scan [see Section 5.3.(i)] of a t.l.c. separation (butanol/acetic acid/water, 10:2:4 v/v on Silica Gel G) of straight urine and hydrolysed urine (6–24 h) from a male rabbit dosed with [^{14}C]tazadolene (4 mg/kg; 4 μCi/kg). *Helix pomatia* refers to a beta-glucuronidase/sulphatase preparation from the intestinal juice of the snail *Helix pomatia* [see Section 5.3.(i)]. U1 and U2 are, as yet unidentified metabolites.

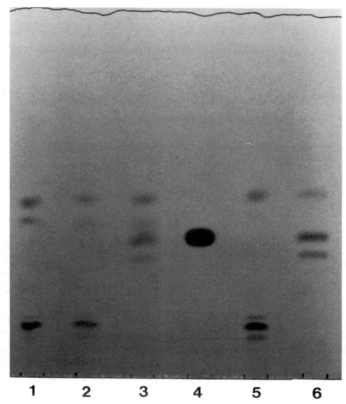

Figure 8. An autoradiogram of the t.l.c. separation of urine from male rabbits dosed with [¹⁴C]tazadolene. **1**. 40 mg/kg, 0−24 h. **2**. Incubation control. **3**. As **1** but incubated overnight with beta-glucuronidase/sulphatase preparation [see Section 5.3(i)] **4**. [¹⁴C]tazadolene. **5**. 4 mg/kg, 0−24 h. **6**. As **5**, but incubated overnight with beta-glucuronidase/sulphatase preparation. See Section 5.3(i).

approximately room temperature (20°C) but should be accurately measured so that the development time can be determined from the manufacturer's tables.

(iii) Unpack the film using safelight illumination only and place it quickly in the developer.

(iv) Agitate the dish for about 10 sec in every minute for 6 min (or the manufacturer's recommended time if the temperature is greater or less than 20°C).

(v) Remove the film from the developer and wash in water for 2 sec to remove excess developer.

(vi) Place the film in the fixer with agitation again for 10 sec in every minute for 15 min.

(vii) Finally, wash the film for 1 h in running tap water, attach film clips, and hang in an airy place to dry.

An example of such an autoradiogram is shown in *Figure 8*. The radioactive areas can then be easily located on the t.l.c. plate by sandwiching a carbon paper between the film and the plate and lightly circling round the spots on the film with a pencil.

On removing the carbon paper and film, the radioactive areas will be circled with carbon.

This method in itself is quite adequate for ^{14}C and ^{35}S which produce beta particles of sufficiently high energy to expose the film efficiently. However, the beta particles emitted by ^3H are of a lower energy and, in order to visualise radioactive areas, exposure times of months are often required. Two ways to overcome this problem are:

(a)   by use of a film specially designed for shorter exposure times (LKB Ultrofilm, LKB Instruments Ltd.);

(b)   by use of a fluorographic enhancer. This is sprayed onto the t.l.c. plate and acts as an amplifying agent by emitting several photons for every beta particle with which it interacts. In this way, exposure times can be reduced to 1/10th of the time which would previously have been required. A useful form of this enhancer is EN^{3}HANCE (NEN, Dreieich, FRG).

(ii) *H.p.l.c.* There are two principal methods whereby drugs and metabolites may be located and quantified as they elute from the h.p.l.c. column.

(a)   For relatively large quantities of radioactivity a scintillation detector can be used. This is essentially a scintillation counter with a flow-through detector. The effluent flows over solid scintillant which emits photons when excited by the beta particles from the radiochemical. These photons are detected by photomultipliers in exactly the same way as a conventional liquid scintillation counter. The system is reasonably good with ^{14}C but is not suitable for ^3H due to the low efficiency with that isotope. The detector may be coupled with a microcomputer which plots a graph as the solute elutes in exactly the same way as a conventional u.v. h.p.l.c. system. It is also capable of calculating the areas under selected peaks to enable quantification of those peaks. One of the most frequently used instruments is the Berthold LB 503 radioactivity monitor for h.p.l.c. (Laboratorium Professor Dr. Berthold, D-7547 Wildbad 1, FRG).

(b)   Where the levels of radioactivity are below the limits of detection for the scintillation flow cell an alternative method is that of sample collection. Usually $1-5$ ml aliquots are collected serially across a chromatography run and, after adding scintillant, are then counted in a conventional liquid scintillation counter. Although this is probably one of the most sensitive methods available to us, it is both time-consuming and demanding of scintillation counter time. A histogram can then be plotted of d.p.m. against elution volume in order that the compounds can be located and quantified.

5. METABOLITE ISOLATION AND STRUCTURAL DETERMINATION

The ease of isolation and identification of metabolites is again dependent upon the chemical nature of the compound and there are no hard and fast rules for this. Urine is most conveniently used as the source of metabolites (i) because most drugs and metabolites are excreted by this route, and (ii) because it is much easier technically and kinetically to isolate metabolites from urine than faeces. Usually a crude extraction procedure is used in the first step, to remove inorganic salts such as phosphate. This first step could be a solvent partition, a solid phase extraction with, for example, XAD-2 resin or an ion-exchange resin followed by one or possibly more separation steps such as t.l.c., h.p.l.c. or g.l.c. The two most useful analytical methods in current use for the structural determination of the isolated metabolites are mass spectrometry (m.s.) and nuclear magnetic resonance spectrometry (n.m.r.) for either protons or ^{13}C. The use of n.m.r. and i.r. to analyse the small quantities of metabolites which are finally isolated has been made possible by the relatively recent introduction of Fourier transform analysis. Another important consideration is the availability of authentic potential metabolites. The structure of these can often be arrived at by consideration of precedents with similar compounds. Frequently, reaction intermediates in the synthesis of the drug can be of great use. Since there are so many permutations and combinations for such studies, examples will be given from the literature by way of illustration.

5.1 Tiflorex Metabolites (11)

The $0 - 48$ h urine (30 ml) from three rats dosed with the labelled compound (10 mg/kg; 250 μCi/kg) was made alkaline (pH 12) with 10 M NaOH and extracted with toluene (2 $\times$ 60 ml). The organic phase was removed, dried over sodium sulphate and, after filtration, evaporated to dryness. The sample was taken up into a small volume of toluene and subjected to g.l.c/m.s. The g.l.c. was performed on a Hewlett-Packard Model 5703 gas chromatograph equipped with a glass column (2 m $\times$ 4 mm i.d.) packed with OV-17 (3% w/w) on Gas-Chrom Q (80 – 100 mesh). Helium was the carrier gas (40 ml/min) and the oven temperature was 150°C. The column effluent was transferred to VG-Micromass model 70 – 70 mass spectrometer *via* a jet separator. The ionisation beam had an energy of 70 eV. *Figure 9* shows the total ion current trace of the urine extract and the mass spectra of these metabolites and the chemically synthesised metabolites are shown in *Table 2*.

5.2 Feprazone Metabolites (12)

The ^{14}C-labelled drug (12) was suspended in water containing carboxymethyl cellulose (0.1% w/v) and Tween 80 (1% v/v) and administered orally to nine female Wistar albino rats (60 mg/kg; 0.65 μCi per animal). The urines were collected for 24 h, pooled, adjusted to pH 2 with 10 M HCl, diluted to 100 ml and saturated with NaCl. This was then extracted with dichloromethane (2 $\times$ 200 ml) which was back-extracted with 2 M NaOH (50 and 20 ml). The aqueous extract was acidified to pH 2 with 10 M HCl, saturated with NaCl and again extracted with dichloromethane (2 $\times$ 200 ml). The organic layer was dried using phase separating paper (Whatman 1 PS) and sodium sulphate and subjected to t.l.c. in CHCl$_3$/EtOH, 9:1 (v/v) on silica gel GF254, 1 mm thick (E.Merck A.G., Darmstadt, FRG). The major metabolites were located at R_f 0.26 and 0.43 by means of a chromatogram scanner and also by virtue of their quenching of the fluorescence of the plates under u.v. light. The bands were scraped off the plate and eluted with methanol. The eluates were taken to dryness and dissolved in a few drops of dichloromethane.

The band of R_f 0.43 was amenable to g.l.c./m.s. using an OV-101 chromatography column at 240°C and introducing the column eluate into a Varian MAT CH-5 mass spectrometer *via* a Bieman-Watson molecular separator. The compound had a retention time of 8.5 min and a spectrum identical to that of synthetic *trans*-3'-hydroxy feprazone.

On the other hand, the band of R_f 0.26 was not amenable to g.l.c. and had to be

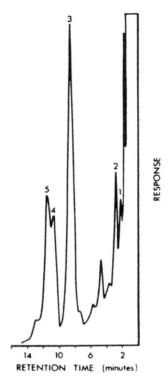

Figure 9. The total ion current trace of the g.l.c./m.s. of an extract of urine from rats dosed with [^{14}C]tiflorex. Numbered peaks are those not seen in the control urine. See *Table 2* for the mass spectra of these and of chemically-synthesised metabolites. After Mas-Chamberlin *et al.* (11).

Table 2. The Mass Spectra of Metabolites of Tiflorex.

Substance	Fragments (relative intensities)
	m/z (%)
Peak 1	236(100), 237(14.3), 238(14.5)
Nortiflorex	236(100), 237(12.2), 238(5.1)
Peak 2	72(45.1), 264(100), 265(14.5), 266(16.0)
Tiflorex	72(75.1), 264(100), 265(15.0), 266(5.6)
Peak 3	236(100), 252(29.7), 268(36.5)
Nortiflorex sulfone	236(28.6), 268(100), 269(12.0), 270(5.7)
Peak 4	72(75), 296(100), 297(15.0), 298(16.7)
Peak 5	72(70.4), 264(100), 280(63.5)
Tiflorex sulfoxide	72(80), 264(100), 280(40.5)
Tiflorex sulfone	72(52.7), 296(77.6), 297(11), 298(4.0)

The major significant ions are reported. For technical reasons, the ion source had to be changed during the analysis and the temperature of the source could not be fully stabilised. This explains the differences between some fragment intensities. After Mas-Chamberlin *et al.* (11).

17

further purified in the same t.l.c. system but on an analytical plate 250 μm thick. Direct insertion spectrometry of this metabolite gave a spectrum consistent with the structure for *p*-hydroxy-*trans*-3'-hydroxy feprazone.

5.3 **Drug Conjugates**

Many drugs and their metabolites are excreted as conjugates with carbohydrates, amino acids and, less frequently, other compounds. Usually in order to identify them they must be isolated and characterised as above. However, with two groups, the glucuronides and sulphates, enzymic methods are available for identifying the conjugating agent and at the same time liberating the aglycone for further analysis. This is accomplished as follows.

(i)   Mix urine (0.2 – 0.5 ml) with an equal volume of 0.2 M acetate buffer (pH 5.0) containing either bovine liver β-glucuronidase (2500 units/ml, Glucurase, Sigma London Chemical Co. Ltd., Poole, Dorset, UK) or Type H-2 β-glucuronidase (2000 units/ml, crude extract ex. *Helix pomatia*, also containing arylsulphatase activity, Sigma.)

(ii)   Incubate each mixture at 37°C overnight together with suitable controls, namely the mixtures lacking enzyme, mixtures with deactivated enzyme (100°C in a water bath for 10 min), and two authentic conjugates, phenolphthalein glucuronic acid and *p*-nitrophenyl sulphate.

(iii)   Before further separation, remove the excess protein from the enzyme mixture by the addition of an equal volume of methanol and centrifugation. It must be borne in mind that different substrates cause the pH optimum to move (if this is suspected a pH range from 4 to 7 should be tried), and that some glucuronides will not hydrolyse. Authentic glucuronides are often difficult to make and isolate which makes good quantitation difficult.

6. WHOLE-BODY AUTORADIOGRAPHY

This method can be considered to complement quantitative tissue analysis since, whereas the latter can give levels of drug and/or metabolite per organ in terms of percentage dose or microgram equivalents, autoradiography can, in addition, give the cellular location and indeed the sub-cellular location if tritium is the radioisotope used, due to the short pathlength of the beta particles emitted by this isotope. Also, autoradiography can be used to investigate all tissues within the body, not only the ones which can be dissected out. Thus the localisation in the many small glands in the experimental animal's body e.g. the pituitary, can be investigated.

6.1 **Animals**

The physical size of the sledge microtome limits the size of the animal to be used. This is usually limited to a maximum of 0.5 kg and therefore it is only possible in practice to use mice, rats and other small mammals, e.g., the marmoset. In order to keep the exposure time of the autoradiogram to a reasonable period, relatively large doses of radioactivity must be used — 100 μCi/kg (3.7 MBq) for ^{14}C and 2 mCi/kg (74 MBq) for ^{3}H.

Section through left kidney

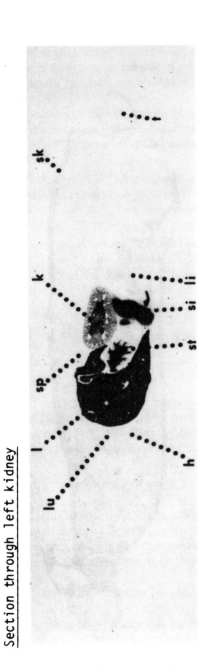

Section through mid-line

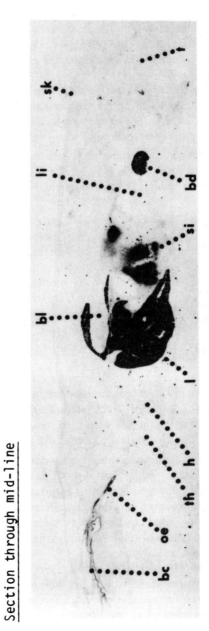

Figure 10. An autoradiogram of a rat sacrificed 0.5 h after being dosed orally with [^{14}C]tazadolene (4 mg/kg, p.o.). The density of exposure of the film is proportional to the amount of radioactivity.

lu = lungs; l = liver; sp = spleen; k = kidney; sk = skin; t = testis; bd = bladder; si = small intestine; st = stomach, h = heart; li = large intestine; bl = bile; th = thymus; oe = oesophagus; bc = buccal cavity.

6.2 Preparation of the Animal for Autoradiography

(i)   Kill mice and other small animals by CO_2 inhalation and larger animals by using a lethal dose of barbiturate.

(ii)   Immediately after the death of the animal place it in a press to prevent the twisting of the carcass during freezing. This press is a spring-loaded device which pushes two flat metal plates on the left and right hand sides of the animal (see reference 13).

(iii)   Place the press in a solid CO_2/hexane freezing mixture for about 30 min for the average sized rat and proportionately more or less depending upon the size of the animal. The animal can then be kept frozen in a deep freeze at $-20°C$ until required for sectioning.

The embedding medium used is carboxymethylcellulose and the cheapest and most convenient form of this is Polycell wallpaper paste (Polycell Products Ltd., Welwyn Garden City, Herts, UK). A rectangular mould, of dimensions $20 \times 7 \times 5$ cm, is used for rats.

(i)   Cool the mould to $-70°C$ in the freezing mixture and pour a small quantity of the embedding medium down the sides of the mould.

(ii)   Remove the rat from the freezer, soak with cold water and place in the mould right side down (the spleen, stomach and heart lie to the left side of the body and are thus more easily reached if they are towards the top of the mould).

(iii)   Pour the embedding medium into the mould until it is full.

(iv)   Place the mould in the freezing mixture until the embedding medium is frozen solid (~ 0.5 h for a 200 g rat).

6.3 Sectioning Procedure

Sections are made by using a microtome mounted in a cryostat, a technique first described by Ullberg (14). At present there are two manufacturers of these instruments (i) LKB (see above), and (ii) Bright Instrument Company Ltd., Huntingdon, UK. The exact method of using the cryostat/microtome will depend of course upon the make of instrument but the general procedure is the same with all.

(i)   Trim the frozen carcass embedded in the block down to the area of interest in a coarse fashion and discard these sections.

(ii)   When the area of interest is reached, stick a piece of adhesive tape of sufficient size to cover the whole section on the frozen section. Probably the most commonly used tape for this procedure is that manufactured by the 3M Company, the type 810 (matt backing) which adheres well to frozen tissue and does not crack at low temperature.

(iii)   Use the microtome to cut a section (usually 20 μm) under the tape.

(iv)   Lift the whole section off with the tape.

(v)   Take further sections at different levels through the animal to give adequate coverage of all the organs.

(vi)   Support the tape bearing each section on a frame which is usually suspended inside the cryostat for several days to freeze-dry the section before autoradiography in exactly the same way as t.l.c. plates (see Section 4.2.3).

A typical whole-body autoradiograph is shown in *Figure 10*. The whole subject is dealt with in detail in an excellent monograph by Curtis *et al.* (13) who also consider the interpretation of autoradiograms; a subject which falls outside the scope of this chapter.

7. ACKNOWLEDGEMENTS

Upjohn Ltd., UK, are thanked for their permission to reproduce *Figures 2, 7, 8* and *10*.

8. REFERENCES

1. Goldenthal,E.I. (1968) Office of New Drugs, Bureau of Medicine, Letter to Industry dated 15 July 1968.
2. D.H.S.S. (1977) *MAL-4. Notes on Applications for Clinical Trial Certificates (Medicines for Human Use)*.
3. Timbrell,J.A. (1982) *Principles of Biochemical Toxicology,* published by Taylor and Francis Ltd., London.
4. Catch,J.R. (1968) *Purity and Analysis of Labelled Compounds. Review 8,* published by Amersham International, Amersham, UK.
5. Evans,E.A. (1976) *Self-decomposition of Radiochemicals. Principles, Control, Observations and Effects. Review 16,* published by Amersham International, Amersham, UK.
6. Sword,I.P. and Waddell,A.W. eds. (1981) *Standard Operating Procedures. Analytical Chemistry and Metabolism,* published by MTP Press Ltd., Lancaster, UK.
7. Horrocks,D.L. and Peng,C-T., eds. (1971) *Organic Scintillators and Liquid Scintillation Counting,* published by Academic Press, NY.
8. Peng,C.-T. (1977) *Sample Preparation in Liquid Scintillation Counting,* published by Amersham International, Amersham, UK.
9. Patterson,M.S. and Greene,R.C. (1965) *Anal. Chem.,* **37**, 853.
10. Donetti,A., Dring,L.G., Hirom,P.C. and Williams,R.T. (1977) in *Mass Spectrometry in Drug Metabolism,* Frigerio,A. and Ghisalberti,E.L. (eds.), Plenum Publishing Corp., NY, p. 1-11.
11. Mas-Chamberlin,C., Gillet,G., Gomeni,R., Dring,L.G. and Morselli,P.L. (1981) *Drug Metab. Disposition,* **9**, 150.
12. Toft,P., Dring,L.G., Hirom,P.C., Williams,R.T., Donetti,A and Midgley,J.M. (1975) *Xenobiotica,* **5**, 729.
13. Curtis,C.G., Cross,S.A.M., McCulloch,R.J. and Powell,G.M. (1981) *Whole-Body Autoradiography,* published by Academic Press, London.
14. Ullberg,S. (1977) in *Science Tools,* Alvfeldt,O. (ed.), LKB-Producter AB, Sweden, p. 1-29.
15. Turner,J.C. (1967) *Sample Preparation for Liquid Scintillation Counting. Review 6,* published by Amersham International, Amersham, UK.

CHAPTER 2

Liver Perfusion Techniques in Toxicology

ROBIN S. JONES

1. INTRODUCTION

Although there are many sophisticated techniques available to study the effects of compounds on organ function and *vice versa*, the isolated organ preparation is unique. Not only does the tissue retain its architecture and intact vascular system, but also it is isolated completely from the influences of all other tissues and organs. Unlike isolated cell preparations, tissue homogenates, tissue slices and organelle fractions, the perfusion technique allows substrates to be delivered by the physiological route to cells whose absorption, transport, excretion and secretion mechanisms and biochemical pathways have not been subjected to chemical and mechanical abuse. In contrast to the *in vivo* situation, the technique allows very precise monitoring and control of the composition and flow-rate of the perfusing medium. Organ donors may be pre-treated in all manner of desired fashions prior to experiments, yet general or specific organ functions can then be followed precisely under strictly standardised conditions.

Any organ which has an easily accessible blood supply and which can be surgically isolated from the animal may be perfused, though with varying degrees of difficulty. There are reports in the literature of the liver (1,2), kidney (3 − 5), lung (6 − 8), heart (9 − 11), intestine (12,13), pancreas (14 − 16), brain (17 − 19), thyroid (20), mammary gland (21), placenta (22), ovaries (23), testis (24,25), anterior pituitary (26), stomach (27), adrenal gland (28), spleen (28), bone (28) and the uterus (28), of various animal species being perfused, of which, for the toxicologist, the liver is probably the most important. For this reason, and because it would be impossible to give in this one chapter details of the surgical and experimental procedures to isolate and perfuse each of these organs, the liver has been selected as a representative tissue.

Details will be provided of the apparatus required to perform isolated liver perfusion experiments. The composition of the perfusion medium will be discussed, and a thorough description of the surgical isolation procedure presented. After outlining methods of assessing normal function and viability of preparations, a few brief examples of its use in toxicology will be given.

2. METHODS AND MATERIALS

2.1 Apparatus and Equipment

2.1.1. *Liver Perfusion Apparatus (Recirculating System)*

A perfusion apparatus must ensure a continuous supply of well-mixed, oxygenated perfusate to the isolated liver, and maintain a constant temperature of 37°C. There should

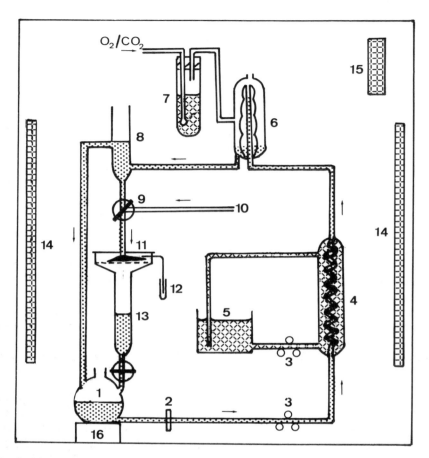

Figure 1. Diagram of a recirculating liver perfusion apparatus. **1**, reservoir; **2**, filter; **3**, peristaltic pump; **4**, heat exchanger; **5**, water bath (37°C); **6**, multibulb oxygenator; **7**, humidifier; **8**, constant head device; **9**, three-way stopcock and two-way metering stopcock; **10**, perfusion pump; **11**, liver; **12**, vial to collect bile (LP3 tube); **13**, liver platform and flow meter; **14**, heating bars; **15**, thermostatic control unit; **16**, magnetic stirrer. All glassware is connected by Portex pp. 320 polypropylene tubing and is housed in a constant temperature cabinet constructed from angle iron, laminate panels with polystyrene being used as insulation material. A fluorescent light is also fitted to the ceiling of the unit and ample fused 13-A power points are incorporated for the use of electric equipment inside the cabinet.

also be a means of adjusting both the pressure head of fluid above the liver and the flow-rate through the organ, as well as providing for the easy withdrawal of perfusate and bile samples. Further refinements such as the inclusion of oxygen and pH electrodes or spectrophotometers and fluorimeters can be made if necessary. Thus, the basic system will consist of little more than a constant temperature cabinet housing the glassware which contains, oxygenates, heats and directs the perfusate (see *Figure 1*)

The cabinet itself should have internal measurements of approximately 100 cm wide × 120 cm high and 65 cm deep, and can be constructed from an angle iron frame, panelled with laminate sheets and insulated with polystyrene blocks. An upward sliding Perspex door avoids the necessity for extra space required for outwardly opening hinged

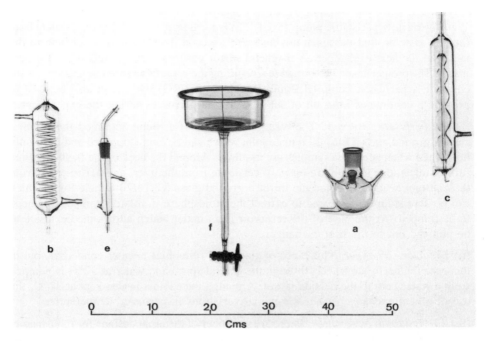

Figure 2. Some of the glassware for a recirculating liver perfusion apparatus. **a**, reservoir (100 ml); **b**, heat exchanger; **c**, multibulb oxygenator; **e**, constant head device; **f**, liver platform and flowmeter.

doors. In addition to a thermostatically-controlled heating system (Cressal Manufacturing Co. Ltd., Birmingham, UK), there should be ample fused 13-A power points for other electrical equipment such as the peristaltic pump and the magnetic stirrer. Thus the cabinet will need to be connected to the mains electricity supply with a heavy duty cord. (A ready-constructed perfusion chamber is available from MRA, Clearwater, FL, USA.)

The glassware (*Figure 2*) for a recirculating system consists of:

(i)   reservoir (50 ml or 100 ml);
(ii)   heat exchanger;
(iii) oxygenator;
(iv)   gas humidifier (Dreschel bottle);
(v)   constant head device;
(vi)   liver platform and flow meter;
(vii) a length of glass tubing (20 cm, i.d. 3.5 mm).

The glassware is supported by clamps on a tubular frame within the cabinet and is connected together by means of Portex polypropylene tubing (pp. 320, i.e. 2.69 mm from Portex Ltd., Hythe, Kent, UK) which is secured by means of elastic tubing sleeves. A luer 2-way metering stopcock and a luer 3-way Teflon stopcock (Biorad Laboratories, 32nd and Griffin Avenue, Richmond, CA 94804, USA) are also connected into the recirculating system, as is a filter containing disposable nylon mesh ('bolting cloth', mesh size 125 μm, John Staniar and Co. Ltd., Manchester, UK).

25

Although a variety of pump types has been used in perfusion experiments (29) a Watson-Marlow peristaltic pump (MHRE 3) is ideal as it can be used to circulate both the perfusate around the apparatus and also water at 37°C from a water bath to the jacket of the heat exchanger. A magnetic stirrer ensures adequate mixing of the perfusate. The pre-perfusion performed *in situ* during the operative procedure is accomplished using a perfusion pump (B.Braun, Melsungen, FRG) which can also be used to provide a continuous infusion of substrate to the perfusate during the experiment.

(i) *The perfusate reservoir.* The reservoir consists of a round-bottomed flask with a central ground glass ST 24/29 female joint which can be kept stoppered and need only be opened when perfusate samples are required. Around the neck of the flask are three further inlets, one allowing re-entry of perfusate from the liver, another re-entry from the constant head device and the third, a ground glass ST 14/20 female joint, which can be kept stoppered or used to extract the atmosphere if volatile components need to be trapped. At the base of the reservoir is an outlet which allows the perfusate to be pumped out to the heat exchanger.

(ii) *The heat exchanger.* This piece of glassware resembles a water condenser, but in this case the perfusate travels through the central spiral and water at 37°C is pumped from a water bath to the outside jacket. Although the cabinet is also kept at 37°C, the use of a heat exchanger ensures minimal variations in perfusate temperature.

(iii) *The multibulb oxygenator.* There are a number of different designs for oxygenators (28 − 32). The one used in this laboratory is a multibulb device (33), in which the perfusate travels up a central tube, and is then allowed to overflow down the outside surface of five bulbs of increasing diameters. In this way the perfusate forms a thin film which allows maximum gaseous exchange. Oxygen/carbon dioxide (95/5, v/v) is supplied to the artificial lung *via* the side arm and the now oxygenated medium leaves through an outlet at the bottom of the vessel.

(iv) *The gas humidifier.* A Dreschel bottle can be used to humidify the gas mixture before it enters the oxygenator. In the absence of such a device, the drying action of the gas would cause loss of water by evaporation and thus concentrate the perfusate. Also, if there is any possibility of the gas mixture not being sterile, it is advisable to use two Dreschel bottles in tandem, the first containing a 1% (w/v) solution of copper sulphate, and the other distilled water.

(v) *The constant head device.* It is usual to perfuse isolated livers under a constant head of fluid. This can be provided by a constant head device which receives the oxygenated perfusate from the oxygenator and then distributes it to the liver, with an overflow returning it directly to the reservoir. It is important that the flow of perfusate is sufficient to ensure that there is always an overflow, otherwise the head may vary.

The constant head device is connected to the isolated liver itself by means of a glass tube (the length of which can be varied to give an appropriate head of fluid), a two-way metering stopcock and a three-way Teflon stopcock, all connected in series. The side arm of the three-way stopcock is attached to the pre-perfusion system.

(vi) *The liver platform and the flow meter.* The isolated liver is rested on a perforated porcelain plate, which can be cut from a Büchner suction filter. This is seated in a container which allows the perfusate to flow down into a calibrated tube and which returns the perfusate to the reservoir. This lower part of the platform is constructed from the lower part of a burette, and therefore flow can be interrupted by turning the tap at its base. By measuring the time it takes to fill the burette (10 ml), the flow-rate through the liver can be measured. Any adjustment to the flow-rate which may be required can be accomplished by means of the two-way metering stopcock previously mentioned.

The upper portion of the platform has a small hole in its side to allow the bile cannula to be threaded through. It also has a ground glass rim which accommodates a ground glass cover. The cover itself has a groove cut into it along its diameter and to just beyond its centre point. This permits the lid to be slid around the hepatic portal cannula and keeps the liver mostly covered from the drying atmosphere of the cabinet.

2.1.2 *Surgical Instruments*

In order to perform the total hepatectomy of the rat it is essential to have a well maintained set of surgical instruments. After a little experience each individual will learn which instruments suit him/her best, but until that time the following may be found useful.

Iridectomy scissors, De Wecker's 11 cm (both points sharp)	1 pair
Plastic surgery scissors, fine blunt points, 11.5 cm	1 pair
Dressing scissors, straight vigo points, one sharp, one blunt, 15 cm	1 pair
Artery forceps, Spencer Wells, straight, box joint, 12.5 cm	4 pairs
Ligature holding forceps, Cancer Research pattern, curved 10 cm	2 pairs
Scalpel handle (Swann Morton No. 4)	1
Surgical blades (No. 22)	

(All the above are available from Holborn Surgical Instruments Co. Ltd., Dolphin Works, Margate Road, Broadstairs, Kent CT10 2QQ, UK.)

2.1.3 *Cannulae and Ligatures*

Polypropylene tubing (Portex Ltd., Hythe, Kent) is ideal for cannulating both the bile duct and the hepatic portal vein. It is available in a number of different sizes, but pp. 25 (i.d. 0.46 mm) is most suitable for the bile duct and pp. 60 (i.d. 0.79 mm) for the hepatic portal vein. To help the cannulae enter the vessels/ducts, it is useful to cut the ends obliquely with a scalpel. Non-capillary braided silk suture (3/0 BPC, Ethicon Ltd., from A.W.Staniforth, 15 Station Road, Penarth, UK) should be used for the ligatures. Ordinary cotton thread is totally unsuitable for this purpose as it expands when wet, causing ligatures to loosen and cannulae to become unsecure.

2.2 **Perfusion Media**

It is the function of the perfusion medium (perfusate) to maintain the isolated organ in a viable state for the duration of the experiment by providing an adequate supply of oxygen, nutrients and other substrates. These are carried in a solution which must be of correct electrolyte and protein composition, and thus display physiological osmotic and colloid osmotic potentials. The pH must also be physiological and adequately

Table 1. The Composition of TC Medium 199.

Ingredients per litre

L-Arginine	70 mg	Calcium Pantothenate	0.01 mg
L-Histidine	20 mg	Biotin	0.01 mg
L-Lysine	70 mg	Folic acid	0.01 mg
L-Tyrosine	40 mg	Choline	0.5 mg
DL-Tryptophan	20 mg	Inositol	0.05 mg
DL-Phenylalanine	50 mg	*p*-Aminobenzoic acid	0.05 mg
L-Cystine	20 mg	Vitamin A	0.1 mg
DL-Methionine	30 mg	Calciferol	0.1 mg
DL-Serine	50 mg	Menadione	0.01 mg
DL-Threonine	60 mg	α-Tocopherol phosphate	0.01 mg
DL-Leucine	120 mg	Ascorbic acid	0.05 mg
DL-Isoleucine	40 mg	Glutathione	0.05 mg
DL-Valine	50 mg	Cholesterol	0.2 mg
DL-Glutamic acid	150 mg	L-Glutamine	100 mg
DL-Aspartic acid	60 mg	Adenosinetriphosphate	1 mg
DL-Alanine	50 mg	Adenylic acid	0.2 mg
L-Proline	40 mg	Ribose	0.5 mg
L-Hydroxyproline	10 mg	Desoxyribose	0.5 mg
Glycine	50 mg	Bacto-dextrose	1 mg
L-Cysteine	0.1mg	Tween 80	5 mg
Adenine	10 mg	Sodium acetate	50 mg
Guanine	0.3 mg	Iron (as ferric nitrate)	0.1 mg
Xanthine	0.3 mg	Sodium chloride	8 g
Hypoxanthine	0.3 mg	Potassium chloride	0.4 g
Thymine	0.3 mg	Calcium chloride	0.14 g
Uracil	0.3 mg	Magnesium sulphate	0.2 g
Thiamine hydrochloride	0.01 mg	Disodium phosphate	0.06 g
Riboflavin	0.01 mg	Monopotassium phosphate	0.06 g
Pyridoxine hydrochloride	0.025 mg	Sodium bicarbonate	0.35 g
Pyridoxal hydrochloride	0.025 mg		
Niacin	0.025 mg		
Niacinamide	0.025 mg	Triple-distilled water	1000 ml

buffered.

All these properties could be provided by whole or defibrinated blood but because of the variable nature of these due to the presence of circulating hormones, vasoconstrictive agents and other compounds, it is now much more common to use an artificial medium, the composition of which can be carefully defined.

2.2.1 *Electrolyte Composition, pH and Buffering Capacity*

Krebs' Mammalian Ringer solutions (34), Krebs-Henseleit solution (35) or Tyrode's solution (36) are adequate for this purpose, but pre-prepared tissue culture media provide a ready source of nutritionally complete phosphate-bicarbonate buffering systems containing the principal electrolytes. Bicarbonate is essential to buffer the effects of carbon dioxide. One such medium which has been used successfully for isolated liver perfusion experiments is tissue culture medium 199, (37,38) the composition of which is given in *Table 1*. However, if such a medium is to be used, the presence of pH indicators, such as Phenol Red, must be avoided otherwise they may compete with experimental substrates for uptake, metabolism and excretion, and adversely affect the results of experiments. Besides containing amino acids, vitamins and salts, tissue culture medium 199 also contains dextrose (1 g/l) which acts as an energy source for the perfused organ. Glucose (5.0 mM), lactate (1.0 mM) and/or pyruvate (0.1 mM) are often used to supplement less complete media. Bile salts, such as sodium taurocholate (10 mM) can be added to ensure an adequate bile flow.

2.2.2 *Plasma Expanders*

Perfusion of an organ with even a perfectly balanced and buffered physiological electrolyte solution would result in the swelling of the tissue as a result of induced oedema. The addition to the perfusate of a plasma expander to confer the correct colloid osmotic potential will prevent this.

Bovine serum albumin (fraction V), at a concentration of between 2.5 and 5% w/v is most often used for this purpose. It is said that commercially prepared protein should first be defatted by treatment with ethyl acetate, and then dialysed to remove low molecular weight contaminants, before being added to the perfusate. However, in drug metabolism studies this seems to be unnecessary. As H^+ and HCO_3^- bind strongly to protein it is necessary to avoid adjusting the pH of a bicarbonate buffer with carbon dioxide until the protein has been added. Also, drugs under investigation may bind with the albumin, but as this is normal it should not be considered a disadvantage. On occasions when protein-binding is not desired a synthetic alternative can be used.

Ficol 70 is a synthetic co-polymer of sucrose and epichlorohydrin (mol. wt. $70\ 000 \pm 10\ 000$) which does not bind to drugs, yet a 4% w/v solution is able to confer on the perfusate the required colloid osmotic potential (39). Also, this multibranched molecule is rich in hydroxyl groups which makes it highly water soluble, and far less viscous than dextran, a polysaccharide which has been used for the same purpose (40). Dextran (mol. wt. 40 000 – 70 000) has also been known to cause anaphylactoid-like responses in rat livers (41), and can interfere with various biochemical assays. Hydroxyethyl starch (30 mg/ml) is another polysaccharide used occasionally as a plasma expander (42).

Polyvinyl pyrrollidone (PVP), a polymer of acetylene, ammonia and formaldehyde (mol. wt. 25 000 – 30 000) has been used as a plasma expander on occasions at a concentration of 3.5% w/v (43). However, it can be taken up by fluid phase pinocytosis into the liver cells where it accumulates in the lysosomes. *In vivo*, this results in increased hepatic lysosomal enzyme activity (44). Stroma-free haemoglobin (6% w/v) can also be used (45) and this has the advantage of being an oxygen-carrier as well as a plasma expander, but the problems of protein binding return.

2.2.3 *Oxygen Carriers*

In order to maintain the isolated tissue in a viable state it is essential to provide adequate oxygen. At 37°C and 760 mm Hg, water can carry up to 2.8% v/v oxygen, as compared with 46% for red blood cells. Thus for an aqueous solution to deliver amounts of oxygen similar to whole blood it would be necessary to perfuse the organ at rates 7 – 10 times normal. Although this may lead to vascular damage after prolonged perfusions, livers perfused with cell-free media have been shown to take up amounts of oxygen similar to those perfused with blood (46,47). A flow-rate of 4 ml/g liver/min is now generally adopted for oxygen-saturated cell-free perfusates (48).

If it is considered necessary to have an oxygen carrier present in the artificial medium, a number of alternatives are available. Most commonly, either homologous or heterologous red blood cells are used at haematocrit values of $25 - 35\%$. If homologous cells were to be used for isolated rat liver preparations 35 ml would be required for each 100 ml of perfusate, and thus a number of animals would have to be sacrificed to obtain sufficient blood. To avoid this, heterologous cells may be used without adverse effects. Large volumes of bovine or sheep blood may be collected from an abattoir, or smaller quantities may be obtained regularly from donors by vena puncture. Goats are good blood donors as their erythrocytes have a smaller diameter than those of rats, and therefore pass through the hepatic vascular bed more easily. Aged human cells from a blood bank are also convenient for use, but these cells are often fragile and haemolyse very rapidly when added to a recirculating system.

Once the blood has been collected into a heparinised vessel the cells must be separated and washed thoroughly. The blood (120 ml) is first centrifuged at about 2000 g for 10 min and the plasma and buffy coat removed by aspiration and rejected. The cell pellet is then gently resuspended in 2 volumes of heparinised physiological saline (500 units/100 ml of 0.9% NaCl), and the centrifugation repeated. The whole procedure is repeated twice with unheparinised saline, to ensure that the cells are free of all unwanted materials. After the final centrifugation, 35 ml of cells are added to 65 ml of the buffered nutrient solution, and the perfusate is ready for use. Stroma-free haemoglobin (6% w/v) may be used as an alternative to erythrocytes (45).

Unfortunately, there are disadvantages associated with the use of erythrocyte-containing perfusion media, which may preclude their use in certain experiments. For instance, the continual passage of blood through a peristaltic pump will inevitably lead to an increasing haemolysis of the cells, with the concurrent release of haemoglobin into the medium, which could interfere with absorbance or fluorescence measurements. Also haemolysis results in the leakage of transaminases (alanine and aspartate) into the medium, which devalues their use as indices of liver damage. The cells may also adsorb, absorb and metabolise substrates which are under investigation, though the use of suitable control experiments should overcome this problem. The use of aged blood cells from expired human blood reduces glucose utilisation by glycolysis in the cells to negligible levels, where this might be a disadvantage, without prejudicing oxygen-carrying functions.

In experiments where the use of red blood cells is contraindicated, yet where an extra oxygen-carrying capacity is required, it is possible to make use of a group of perfluoro compounds known as fluorocarbons (37,139). Compounds belonging to this family have a high capacity for transporting oxygen and carbon dioxide, have low toxicity and are not themselves metabolised. Their oxygen-carrying capacity is similar to that of red blood cells, so that a $35 - 40\%$ emulsion of a fluorocarbon in a buffered solution is a satisfactory substitute for blood. Perfluorotributylamine (FC 43, 3M Company, Düsseldorf, FRG), perfluorobutyltetrahydrofuran (FC 80, 3M Company, Düsseldorf, FRG) and perfluorodecalin (Flutec PP5, Aldrich Chemical Co., Ltd., Gillingham, UK) are fluorocarbons which have been used for this purpose. FC 43 maintains viability of perfused livers for up to 20 h (49) whereas FC 80 maintains serum-protein synthesis in perfused livers for up to 6 h (37). Flutec PP5 has no adverse effects on liver microsomal activities, bile formation, leakage of transaminases or on various other parameters (34). Fluosol 43, which is a mixture of fluorotributylamine (0.86 M) and pluronic F-68 (26 mg/ml; Wyandotte Chemical Corp., Wyandotte, MI, USA) also maintains normal hepatic function (42).

Unfortunately, fluorocarbons are themselves not without problems. For example, unless care is taken when preparing the emulsion to ensure a particle size of below 0.5 μM diameter, the particles will be taken up by the Kupffer cells with an associated increase in portal pressure and a decrease in perfusate flow (50). Sonification overcomes this problem. FC 80 reacts with amino acids, particularly glycine, present in the perfusate to form complexes (51). Even so, it has been used successfully in drug metabolism studies using rabbit livers (52). Fluorocarbons are unsuitable for work utilising lipophilic compounds, because such compounds tend to partition between the oxygen carrier and the aqueous phase. This has been demonstrated with steroids and FC 43, in which system the steroids partition out of the aqueous phase (53). (NOTE: FC 43 is now designated FC 47 by the 3M Company, whereas FC 80 is sometimes referred to as FX 80 or FC 75).

2.2.4 *Anti-coagulants*

Even when using a completely artificial perfusate, it is advisable to add anti-coagulant to prevent any possibility of clots forming in the microcirculation. Heparin and citrate are used for this purpose, but each should be used with caution, thought being given to the experiment in hand. For example, heparin interferes with fatty acyl lipase activity (54), whereas citrate cannot be used in experiments involving the study of intermediary metabolism.

2.2.5 *Antibiotics*

The complete perfusate is an ideal medium for bacterial growth, and for this reason it is advisable to add an antibiotic to control bacterial contamination. The chosen antibiotic should be broad spectrum and have minimum toxicity. Neomycin (50 μg/ml) is very often used because it is effective at concentrations as low as 15 μg/ml, and yet is not cytotoxic until concentrations of 3 mg/ml have been reached. Gentamycin (200 μg/ml) can also be used, but antibiotics such as chloramphenicol and tetracycline should be avoided as the error of margin is too small.

2.2.6 *Typical Perfusion Media*

A number of alternatives have been presented for each physical, chemical and biochemical parameter required of the perfusion medium, so that perfusates can be designed to suit individual needs. However, the medium detailed below has been used successfully in a number of drug metabolism and organic anion transport experiments, and could be used until personal requirements are clarified.

Initially, a medium, comprising 100 ml of tissue culture medium 199, without Phenol Red (Difco Laboratories, Detroit, MI, USA), containing 2.5 g bovine serum albumin, fraction V (Sigma Chemical Company, St. Louis, MO, USA), 500 units heparin (BDH Chemicals Ltd., Poole, Dorset, UK) and 75 mg gentamycin (Flow Laboratories Ltd.,

Irvine, Ayrshire, UK) is prepared, heated to 37°C, and the pH adjusted to pH 7.4 with O_2/CO_2 (95/5, v/v). 65 ml of this medium is then made up to 100 ml with goat erythrocytes (washed with isotonic saline) to give the perfusate. The remaining cell-free medium (35 ml) is used as the pre-perfusate.

Less complicated media can be based on either Krebs-Henseleit bicarbonate buffer solution (pH 7.4), or Tyrode's solution, the most common additions being glucose (5.0 − 12 mM), pyruvate (0.1 − 0.3 mM), lactate (1.0 − 3.0 mM), bovine serum albumin (2.0 − 5.0%), red blood cells (15 − 35%) and an antibiotic (e.g., oxatetracycline, 1.5 mg/100 ml; neomycin, 50 mg/100ml; gentamycin, 75 mg/100 ml). Defibrinated homologous or heterologous blood is sometimes used at a ratio of 1 to 2 with Krebs-Ringer bicarbonate buffer (pH 7.4) containing bovine serum albumin (4% w/v) and glucose (0.1% w/v).

2.3 Liver Donors

Although in some experiments there may be a specific need to use a liver from a particular animal species, the rat is the most usual liver donor. This is not only because of the amount of background data available for this species, but also because the animal is cheap, easily obtainable and is a convenient size for the surgical and subsequent procedures. If the weight of the animal to be used is not an important criterion, then from the surgical point of view 200 − 250 g is the ideal weight range. Rats much smaller than this have ducts and veins too fine to work with easily, whereas larger animals have too much fat deposited around organs and vessels, making identification and isolation of vessels difficult.

In addition to the rat, a variety of other animal species have been used, including the mouse, hamster, rabbit, ferret, cat, dog, sheep, pig and calf. The mouse, although an economical animal to work with, is far too small for the operative procedure to be completed successfully by anyone other than a skilled surgeon. This species also has a gall bladder which complicates the bile duct cannulation procedure. All other species are larger than the rat, but although the surgery may become less intricate, far greater volumes of perfusate will be required to maintain the isolated organ and all apparatus must be unmanageably large for the average laboratory.

2.4 Surgical Procedure

In the UK, before any experiments on living animals can be performed, permission must be obtained from the Home Office. Once such permission is obtained, a licence, with or without accompanying certificates, is issued which allows the experimenter to undertake certain specific procedures. The surgical procedure to isolate an intact liver from an anaesthetised animal for the purpose of liver perfusion experiments is no exception and anyone wishing to perform such experiments should seek the advice of their local Home Office Inspector before embarking on the programme.

Liver donors are anaesthetised with ether, halothane or a barbiturate, though the latter should be used with care as it can affect the activity of the hepatic drug-metabolising enzymes. Pentabarbital (Sagital, May & Baker, 60 mg/kg body weight) is a common veterinary anaesthetic which should be administered intraperitoneally at least 20 min before surgery is to begin. If complete anaesthesia is not attained after 30 min with the first dose, a repeat dosage of 20 mg/kg can be used, but it is essential not to overdose the animal nor to begin the surgical removal of the organ until the animal is totally anaesthetised.

When the perfusion apparatus has been properly assembled, and all reagents, surgical instruments and cannulae are ready for use, the isolation surgery may begin.

(i)   Restrain the animal on the operating platform and remove the fur on the anterior thorax and abdomen with small-animal, electric hair clippers. Cleanse this area with a swab of 70% v/v aqueous alcohol solution.

(ii)   After checking that the animal is properly anaesthetised, make a midline incision of the skin from the symphysis pubis to the suprasternal notch. In order to achieve a clean cut, a new scalpel blade should be used for each operation and the skin should be held tight by applying downward pressure on the abdomen on either side of the mid-line with the fingers of the non-cutting hand.

(iii)   Make a small incision with the scalpel in the peritoneum in the region of the umbilicus on the midline to avoid significant bleeding. Extend this to just above the xiphoid and to the symphysis pubis with a pair of plastic surgery scissors.

(iv)   As this is a non-recovery animal experiment, it is permissible and advisable to ease access to the working surgical area as much as possible. This is achieved by fixing two Spencer-Wells Artery forceps parallel to each other and at right angles to each cut surface of the peritoneum approximately halfway down the incision. By cutting with scissors between the forceps down through the peritoneum and skin towards the dorsal surface of the animal, flaps will be generated.

(v)   Open these flaps by repositioning the Spencer-Wells forceps adjacent to the four limbs of the animals, which results in the whole of the abdomen being exposed and made more accessible for surgery.

(vi)   Secure the clamps adjacent to the forelimbs in a position which applies slight

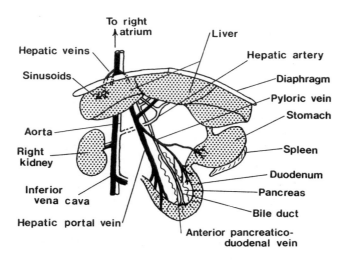

Figure 3. The abdominal cavity of the rat.

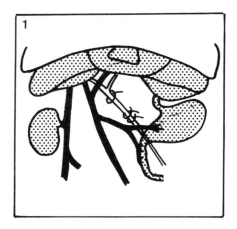

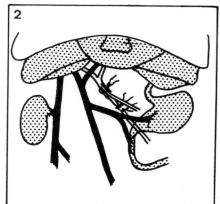

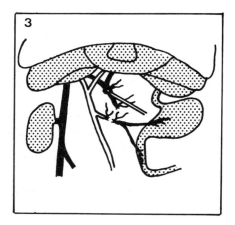

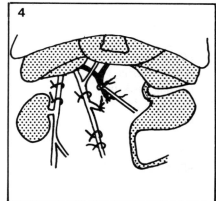

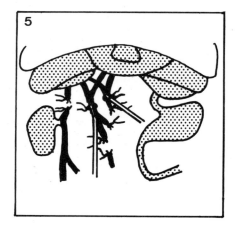

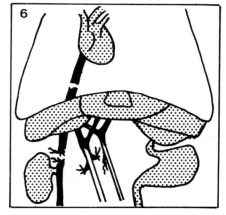

34

tension on the thorax, so that it is possible to raise and rest the lobes of the liver on the diaphragm.

(vii)   Once the gastrointestinal tract has been retracted to the animal's left, and swathed in a saline-soaked cotton wool swab, the porta hepatis is exposed and the hepatic portal vein, the bile duct and the inferior vena cava will be seen (*Figure 3*).

Cannulation of the bile duct. The bile duct will be visible as a thin silvery line embedded in pancreatic tissue.

(i)   Place curved ligature holding forceps, in the closed position, under the bile duct at a position midway between the liver and duodenum. This will involve piercing the pancreatic tissue with the forceps as close to the bile duct as possible and allowing the curved ends to repierce the adherent tissue on the opposite side of the duct.

(ii)   Allow the forceps to open gently, and insert between the arms another closed pair of similar forceps held in the other hand.

(iii) By careful manipulation of these two instruments, clear a section of the duct of most of its adhering tissue.

(iv)   Once this has been accomplished, place three lengths of surgical thread around the bile duct, one proximate to the liver and the others proximate to the duodenum (*Figure 4.1*).

(v)   After ensuring that the bile cannula tube is at hand, knot the duodenal ligatures causing back pressure distension of the duct.

(vi)   To avoid damage due to this induced cholestasis, make a small incision in the duct with a pair of De Wecker's iridectomy scissors midway between the upper two threads as quickly as possible, and insert the cannula towards the liver. It might prove helpful to hold the cannula in the right hand whilst at the same time applying a little tension to the duct by pulling the ends of the duodenal ligatures with the left hand. If the handle of a pair of forceps is placed under the bile

Figure 4. Diagramatic representation of the surgical removal of a liver from an anaesthetised rat for the purpose of isolated perfusion studies. **1**, (i) Locate and clear bile duct. (ii) Place a single ligature loosely around the duct ~1 cm away from the biliary tree. (iii) Place a double ligature loosely around the duct proximate to the duodenum. **2**, (i) Tighten the lower ligatures. (ii) Using iridectomy scissors, make an incision in the bile duct midway between the single and double ligatures. (iii) Insert the bile duct cannula and secure by tightening the single ligature. (iv) Again using the iridectomy scissors, sever the bile duct between the knots of the double ligature. **3**, (i) Locate, and carefully clear the hepatic portal vein and the pyloric vein. (ii) Place two ligatures tightly around the pyloric vein ~1 cm apart. (iii) Using iridectomy scissors, sever the pyloric vein between the ligatures. **4**, (i) Place a single ligature loosely around the hepatic portal vein immediately above the pyloric vein. (ii) Place a double ligature around the hepatic portal vein ~2 cm below the pyloric vein. (iii) Locate the abdominal inferior vena cava. (iv) Place two single ligatures loosely around the inferior vena cava immediately above the right renal vein. **5**, (i) Tighten the lower double ligature on the hepatic portal vein. (ii) Using iridectomy scissors make a small incision in the hepatic portal vein between the single and double ligatures. (iii) Insert the hepatic portal vein cannula and secure by tightening the upper ligature. (iv) Allow reflux of blood to fill cannula and then attach to the pre-perfusion pump and commence perfusion. (v) Sever the hepatic portal vein between the knots of the double ligature. (vi) Tighten the ligatures on the inferior vena cava. (vii) Sever the inferior vena cava between the ligatures. **6**, (i) Using heavy duty scissors cut open the rib cage along the sternum. (ii) Sever the thoracic inferior vena cava immediately above the diaphragm. (iii) Remove liver from the animal after severing the hepatic artery (not shown) and clearing all other connections with the animal.

duct not only is the incision more clearly visible but also there is something solid to push the cannula against.

(vii)   Push the cannula firmly into the duct, taking care that it does not reach the biliary tree. When in position tie it in place with the previously positioned thread. Bile will immediately flow into the cannula, and can be collected in a small collecting tube (e.g., LP3 tube) during the remainder of the operative procedure.

(viii) Once the flow has been established cut the bile duct between the lower two ligatures with the iridectomy scissors (*Figure 4.2*).

Cannulation of the hepatic portal vein. Before the hepatic portal vein is cleared and cannulated, the pyloric vein must be located and occluded.

(i)   Place two ligatures around this vein a few millimetres apart and tie immediately.

(ii)   Cut the vessel between the ligatures with the iridectomy scissors (*Figure 4.3*).

(iii)   Clear the portal vein itself, which runs parallel to the bile duct, of connective tissue in the manner previously described for the bile duct, taking extra care as this blood vessel is considerably more fragile than the duct.

(iv)   Choose a section immediately above and below the pyloric vein, taking care not to damage the hepatic artery.

(v)   When clear, place three ligatures loosely around the vessel, one immediately above the pyloric vein and the others about 1 cm below.

(vi)   Before proceeding with the cannulation, locate the inferior vena cava in the abdominal cavity, and place two ligatures loosely around it immediately above the right renal vein (*Figure 4.4*). Again, it is helpful to place a forceps handle beneath the hepatic portal vein between the upper two ligatures.

(vii)   Tie the lower ligatures on the hepatic portal ensuring that the vein does not twist in the process.

(viii) At this point, because blood would normally be flowing from the intestine to the liver, the vessel will collapse, making the cannulation process more difficult than had been the case for the bile duct. Using the iridectomy scissors, make a small incision between the two upper ligatures. If the hepatic artery has been undamaged during the surgical procedure so far, blood should flow from the incision. The integrity of the hepatic artery is essential, as it provides the liver with blood during the period when the flow through the hepatic portal vein is arrested.

(ix)   Apply slight tension to the vessel by pulling the tied lower ligatures with the left hand, and insert the cannula (previously filled with heparinised pre-perfusion medium) with the right, pushing it past the untied ligature but being careful not to reach the hepatic portal tree.

(x)   Once it is in position tie the ligature, which should result in a small flow of blood out of the cannula. If this does not occur apply slight suction to draw blood out of the cannula because it is essential not to have any air space in it. Otherwise air bubbles will enter the liver and cause blockage of part of the vascular system of the organ.

(xi)   Once secured, attach the blood-filled cannula *via* a three-way stopcock to the pre-perfusion system, and perfuse the liver immediately *in situ* with oxygenated erythrocyte-free medium at a rate of approximately 1 ml/min/g tissue.

(xii) Sever the portal vein between the lower two ligatures (*Figure 4.5*), tighten the ligatures around the abdominal region of the vena cava and cut the blood vessel between them.

(xiii) At this point, cut open the thorax with dressing scissors and sever the inferior vena cava to kill the animal and to release the inflowing perfusate (*Figure 4.6*). Within moments of beginning the pre-perfusion, the liver should become a uniform fleshy colour. Areas remaining pink are not being perfused, and if they fail to clear abort the experiment at this stage.

(xiv) Slow the rate of infusion of the pre-perfusate to about 0.5 ml/min/g to allow more time to complete the operation.

(xv)   It now remains only to remove the liver from the rat. This process should be undertaken quickly, but patiently, taking care to avoid damaging the liver and to avoid pulling out the cannulae. Ease the liver down from the diaphragm, return to its normal anatomical position and dissect out the diaphragm from the thorax. If it is left attached to the liver, it can be used as a convenient 'handle' to manipulate the liver whilst the dissection is completed.

(xvi) Thus, with the freed diaphragm being held with forceps in the left hand, free the liver by cutting with scissors all points of attachment to the rat. If possible, avoid damage to the gastrointestinal tract to prevent the bacterial contamination which would ensue. It is extremely difficult to describe this excision process adequately on paper, and it is left for the operator to find the technique which best suits her/him.

(xvii) Finally, when the liver is free, disect away the diaphragm, leaving the intact liver with perfusate being infused *via* the hepatic portal vein, and bile being collected *via* the cannula in the bile duct. Perfusate will leave the liver by the severed unligated end of the inferior vena cava.

(xviii) Place the liver on the perforated disc in such a manner that all the lobes, the portal vein and the cannulated bile duct are in their normal anatomic positions. In order to arrange the liver in this position, first suspend it by means of the hepatic portal cannula (which will allow any twists to resolve themselves) and then gently lower onto the disc. Care should be taken to ensure that there are no twists in the bile duct, which would prevent the free flow of bile.

An excellent description of a similar operative procedure is given by Miller (1), whereas procedures for the removal of other organs are described by Ross (55). Ross (55) also gives details for the cannulation of the hepatic artery for isolated rat liver preparations.

2.5 The Perfusion Experiment

2.5.1 *Preparing for the Experiment*

It is best to set up the apparatus, prepare the pre-perfusion medium (100 ml) and wash the erythrocytes on the day prior to the experiment. The apparatus can be tested for leaks by circulating around it physiological saline, which can be left in the reservoir until the day of the experiment and then used to wet the surfaces of the glassware and tubing prior to addition of the perfusion medium proper. The perfusate itself can be prepared on the day of the experiment by adding the washed red blood cells (35 ml) to the pre-perfusion medium (65 ml).

(i)   Once prepared, place the perfusate in the reservoir (now drained of saline) and recirculate to allow equilibration. At this stage, carefully dislodge any small air bubbles appearing.

(ii)   Draw the 35 ml of pre-perfusate into the barrel of the infusion pump, and connect this by a length of polypropylene tubing to the luer three-way stopcock, to which is attached the pp. 60 portal vein cannula. For each experiment it is advisable to use venous cannulae of standard length, otherwise awkward adjustments to the positioning of the apparatus in the cabinet may be required. Also the pp. 25 bile duct cannulae should be the same length for each set of experiments, otherwise an extra variable is introduced which will affect the lag time before metabolites appear in the collected bile samples.

(iii)   Whilst the apparatus is equilibrating, place the surgical instruments on the operating table in convenient positions. It is helpful to develop the habit of placing each instrument in a specific position on every occasion, so that during the surgical procedure a particular instrument can be found quickly. After use, the instrument should be returned to its usual position. It may also be helpful to place all cutting instruments to one side of the animal and all clamps and forceps to the other.

(iv)   Finally, before anaesthetising the rat, cut an adequate number of 15 cm lengths of surgical ligature and place in a convenient position. If they are stuck to the edge of the table by a length of adhesive tape they can be pulled loose and used as required.

2.5.2 *Performing the Experiment*

(i)   Connect the isolated liver into the perfusion system by holding the luer fitting of the three-way tap vertically adjacent to its intended position in the cabinet and, after ensuring that there is no air space present (by adding a few drops of saline with a Pasteur pipette) attach it to the male joint. Turn the three-way tap to complete the connection to the perfusion system, and to disconnect the pre-perfusion circuit.

Immediately the isolated liver is connected it should regain its uniform physiological pink colour (only if the perfusate contains erythrocytes). If a lobe fails to regain its colour, it is probably due to the constriction of the supplying branch of the portal vein and a slight adjustment of the position of the liver on the platform may rectify the situation. If areas near the edges of the liver fail to colour up then it is likely that air or cellular debris has entered the organ and is causing blockage of the vascular system. This cannot be rectified and the experiment is best aborted at this stage.

(ii)   As soon as one is satisfied that the liver is being completely perfused, measure the flow of perfusate through the organ and adjust to 1 ml/min/g liver. To avoid having to weigh the organ, assume that the liver represents 4% of the total body weight of the animal, so that a 250 g rat will have a liver weighing approximately 10 g.

(iii)   Carefully thread the bile duct cannula through the hole in the liver platform and place its end in a collecting vial (e.g., LP3 tube) and collect the bile. If bile fails to flow, then it may be because of a constriction in the duct, which can often be resolved by twisting the cannula slightly, first in one direction and then in the other, until the flow is restored.

It is usual to leave the liver to equilibrate for about 30 min before any experiment is begun. This time should be spent adjusting the perfusate flow and ensuring a patent bile flow as described above. Also, the perfusion apparatus should be checked carefully to ensure that it is functioning and that no leaks have appeared. Any frothing of the perfusate should be stopped as this may lead to the formation of a stable foam which will introduce air into the liver.

The tube collecting bile can be changed as frequently or as infrequently as necessary. If it is anticipated that a compound under investigation is going to be rapidly biliary excreted, and a sensitive assay is available, then tubes could be changed every 5 min. This permits pharmacokinetic analysis of biliary excretion. Otherwise, 15 or 30 min samples should be satisfactory.

Perfusate samples can be taken as often as necessary from the reservoir, but the volume withdrawn each time should be kept to an absolute minimum (<0.5 ml), otherwise the total volume of the perfusate would be quickly diminished. It is always a compromise between how often and what volumes of samples may be taken, and careful thought should be given to this problem before the experiment is begun. There is always the temptation to do a large number of viability tests, yet these must be limited when using a recirculating system for the reasons outlined. It is important to record accurately the volumes of aliquots removed, so that adjustments can be made when calculating the results (55).

Differences between the composition of influent and effluent perfusate can be determined by collecting aliquots from the constant head device and from the burette portion of the liver platform. Disposable syringes (1 ml) with an attached length of polypropylene tubing can be used for this purpose.

Substrates or other agents can be added to the perfusate at any time *via* the central ground-glass female joint of the reservoir. They may be added as a single bolus or as a continuous infusion from the pump previously used for the *in situ* pre-perfusion. Taurocholate (15 μmol/h) is sometimes added to the perfusion medium by infusion to maintain the level of bile salts and to ensure a good biliary flow. If the bile salt concentration of the perfusate becomes depleted, then the bile flow will decrease.

It is not unusual to maintain an isolated liver in a viable state for 6 h or more using a recirculating system such as the one described above, but most workers restrict the time to about 4 h. It is up to an individual to confirm the continued viability of the preparation.

2.6 Recirculating Versus Non-recirculating Systems

In 1855, Claude Bernard (56) attached one end of a hose pipe to a water tap, the other to the hepatic portal vein of an isolated pig liver, and showed that the effluent water contained glucose. This classical demonstration of the conversion of glycogen to glucose was the first recorded use of an isolated liver preparation as a research tool, and though crude, is an example of a non-recirculating system. It was not until the late forties and early fifties of this century, when the technique became accepted by physiologists and biochemists, that the perfusion medium became more sophisticated and a recirculating system was introduced.

Although Monsieur Bernard had no need to worry about the cost of his perfusion medium, it is now a very real concern for modern day scientists. The more sophisticated the medium is, the more expensive it becomes, and the less willing we are to perform non-recirculating perfusions, in which the perfusate passes through the liver only once, and is not used further except for analytical purposes. For this reason, where prolonged perfusions need to be performed, the recirculating system is the desired method. However, the composition of the recirculating perfusate will be defined only until it has passed through the liver and thereafter its exact composition will be unknown unless very thorough analyses are performed. Thus, if it is necessary to provide a perfusion medium, the composition of which must remain constant throughout the experiment, then a non-recirculating system must be used. In such systems it is usual to employ cell-free perfusates based on Krebs-Henseleit, Tyrode's or similar solutions. Also, it is unusual to continue them for long periods because of the large volumes of medium that would be required. An example of a typical non-recirculating liver perfusion system is given by Thurman *et al.* (2).

A further disadvantage of the recirculating system is that, during a long perfusion, toxic compounds like urea can build up in the perfusate. Graf *et al.* (57) have described a method by which such compounds can be eliminated by continuous dialysis, but such a process is again expensive on reagents.

When investigations are being conducted into the metabolism of compounds which are known to be metabolised slowly, the closed system has advantages. In this system the substrate can be left recirculating through the liver for hours, and perfusate and bile samples collected regularly for analysis. Also, it is possible to use much higher concentrations of substrate than would be possible *in vivo*, when death might follow damage to tissues and organs other than the liver. However, it is also possible to generate metabolites which would not be seen *in vivo*, because compounds normally excreted in the urine remain in the circulation and may be susceptible to further change. Compounds excreted in the urine *in vivo* might also find their way into the bile in a recirculating system.

Although it is possible to generate much information from a single perfusion experiment, recirculating studies are time consuming and each experiment requires its own control. In contrast, non-recirculating perfusions can be performed rapidly and they can very often act as their own control. Normal function can be established in the early stages and the medium or another parameter can then be changed, and the effect on organ function monitored.

2.7 Drug Elimination Kinetics in Isolated Perfused Liver Systems

2.7.1 *Recirculating Systems*

The following two-compartment model can often be used to depict drug elimination from an isolated perfused liver preparation in which k_{12}, k_{21} and k_{23} are the rate constants for primary uptake, reflux and biliary excretion, respectively.

1	k_{12}	2	k_{23}	3
RESERVOIR	$\rightleftharpoons$	LIVER	$\longrightarrow$	BILE
	k_{21}			

The rate of disappearance of a drug from the perfusate of such a system is described by the bioexponential Equation 1 (58):

$$[D]_t = Ae^{-\alpha t} + Be^{-\beta t} \tag{1}$$

where, $[D]_t$ is the drug concentration in the perfusate at time t; A is the zero-time intercept associated with the rapid elimination phase; B is the zero-time intercept associated with the slow elimination phase; α is the rapid elimination rate constant; and, β is the slow elimination rate constant.

A graph of ln $[D]_t$ against time (t) will give a bi-exponential curve (*Figure 5*) in which β is the gradient $(0.693/t\frac{1}{2})$ of the terminal exponential phase, and B is the zero-time intercept of this linear plot extrapolated to time $= 0$. Application of the method of residuals (59) will yield a second straight line with a slope α and a zero-time intercept A.

Once values for A, B, α and β have been determined, they can be used to calculate k_{12}, k_{21} and k_{23} using Equations 2, 3 and 4.

$$k_{12} = \frac{A\alpha + B\beta}{A + B} \tag{2}$$

$$k_{23} = \frac{\alpha\beta}{k_{12}} \tag{3}$$

$$k_{21} = (\alpha + \beta) - (k_{12} + k_{23}) \tag{4}$$

The apparent volume of distribution (V_d) is given by Equation 5:

$$V_d = \frac{\text{dose}}{A + B} \tag{5}$$

The clearance of the drug (CL) can be obtained (60) from:

$$CL = \frac{\text{dose}}{A/\alpha + B/\beta} \tag{6}$$

When $B/\beta >> A/\alpha$, Equation 6 will approximate to:

$$CL \sim \frac{\text{dose}.\beta}{B} \tag{7}$$

Also, if $B >> A$, from Equation 5 it can be seen that:

$$V_d \sim \frac{\text{dose}}{B} \tag{8}$$

Thus, Equation 7 becomes:

$$CL \sim V_d.\beta \tag{9}$$

Such a situation will arise when the proportion of drug in the liver is small, or if it distributes rapidly relative to elimination (60). This monoexponential pattern is

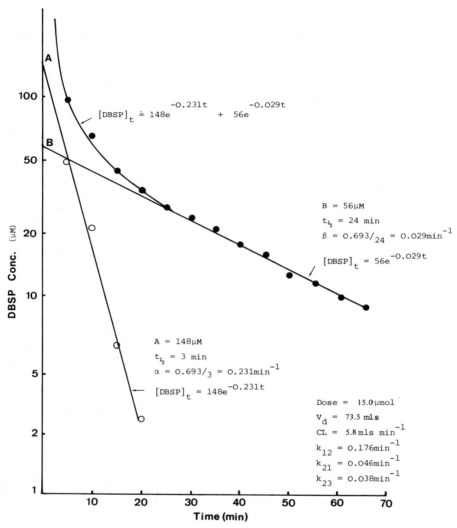

Figure 5. Two-compartment model for the elimination of dibromosulphophthalein (DBSP) from the isolated perfused rat liver. Concentrations of DBSP in a perfusate containing bovine serum albumin (0.5% w/v) and Ficol 70 (2.0% w/v) to avoid extensive protein binding. Experimental values (●), residual values (○). When the perfusion was repeated in the presence of the Na$^+$/K$^+$ ATPase inhibitor, ouabain (0.5 mM), the pharmacokinetic parameters were: V_d, 103 ml; CL, 0.5 ml/min; k_{12}, 0.046/min; k_{21}, 0.033/min; k_{23}, 0.004/min.

exemplified by cyclohexanecarboxylate (Section 4.1), Asulam (61) and d-tubocurarine (62). The elimination rate constant (k_e) is given by the slope of the line ($0.693/t\frac{1}{2}$). Unfortunately, the pharmacokinetic analysis of the plasma disappearance of some drugs, such as ouabain (62) is virtually impossible because of the multiexponential pattern in time, though various computer programs are available which will strip curves described by up to five exponentials (63,64).

The volume of perfusate has a large effect on k_{12}, but the perfusion rate has an effect only with compounds that are rapidly removed from the circulation by the liver (65). Keiding and Steiness (66) have investigated the effect of varying the perfusate flow-rate on the elimination of propranolol from isolated rat liver preparations in order to test the two current models of enzymatic drug elimination from intact livers [the sinusoidal perfusion (parallel tube) model (67); and the venous equilibrium (well-stirred) model (68)]. By infusing the drug into the reservoir at a constant rate and varying the perfusate flow at intervals throughout each experiment they obtained results which were not inconsistent with the sinusoidal model. In contrast, Jones *et al.* (69) produced data which supported the venous equilibrium model, by infusing the same drug directly into the hepatic portal vein. They also varied the protein content of the perfusate in order to alter its protein-binding properties.

Although emphasis has been placed on plasma disappearance data, biliary excretion data should not be ignored. For instance, Blom *et al.* (57) demonstrated monoexponential biliary excretion kinetics for dibromosulphophthalein, with a $t\frac{1}{2}$ similar to that for the second phase of the biexponential plasma disappearance, perhaps indicating that after preliminary distribution, uptake of this dye from the perfusate is dependent upon biliary excretion.

2.7.2 Non-recirculating System

In single-pass perfusions, the concentration of drug in the influent perfusate will be constant, whilst the effluent concentration will vary until it reaches a steady-state. A semilogarithmic plot of the difference between the effluent concentration at steady-state and at times before the steady-state is attained, against time will yield a straight line if the uptake of drug into the liver is a first order process (70).

Expressions for the 'half-time to approach the effluent drug concentration at steady-state' value ($t\frac{1}{2}$) and clearance have been derived (70):

$$t\frac{1}{2} = \frac{0.693 \ V_L.\text{fu}.K\text{pu}}{Q} \tag{9}$$

where, V_L is the volume of the liver; fu is the fraction of drug in the perfusate unbound; Kpu is the ratio of concentration of drug in the liver to the unbound drug concentration in the effluent perfusate; and, Q is the perfusate flow.

$$CL = QE \tag{10}$$

where E is the extraction ratio of the drug.

A single-pass system has been used recently to demonstrate that the route of hepatic administration greatly influences the availability (F) of drugs (71):

$$F = \frac{\text{Rate of appearance of drug in effluent}}{\text{Rate of presentation of drug}} = \frac{Q_{out}.C_{out}}{Q_{in}.C_{in}} \tag{11}$$

where, C_{in} and C_{out} are the concentrations of drug in the influent and effluent respectively; and, Q_{in} and Q_{out} are the respective influent and effluent flow-rates. At a constant flow of 10 ml/min, the availability of the drug lidocaine was 18 times greater when infused through the hepatic artery, than when infused *via* the portal vein.

3. VIABILITY AND FUNCTION TESTS

Once the pefusion is underway keep a close eye on the general appearance of the liver. It should remain a uniform pink colour and not develop any dark red or black areas. There should be no signs of swelling and the edges should be thin. Once the perfusate flow has been adjusted to 1 ml/g liver/min under a constant pressure head (e.g., 30 cm) it should remain constant. A decrease in flow is an indication of increased vascular resistance brought about by the presence of vasoconstrictors or blockage of vessels by blood clots or cellular debris.

Unless the perfusate is continually supplemented with a source of bile salts it is unlikely that the bile flow will remain constant. For a 10 g liver, the bile flow-rate will be about $0.5 - 0.75$ ml/h initially, but it is not unusual for this to fall throughout the perfusion period (*Figure 6.6*).

An indication that the liver is taking up oxygen is given by observing the differences in colour of the perfusate entering and leaving the organ. If an erythrocyte-containing perfusate is being oxygenated efficiently, the blood entering the liver will be characteristically arterial, with a bright red appearance. On leaving the organ it should be noticeably darker. If there are no red blood cells present in the medium, or if a more definitive measurement is required, oxygen electrodes can be inserted to measure the pO_2 of the entering and effluent perfusate. Alternatively, samples can be withdrawn and pO_2 and pCO_2 measured with a gas analyser. Typically, δpO_2 values are 250 mm Hg whereas δpCO_2 values are much less at about 10 mm Hg (44).

Determination of lactate and pyruvate concentrations in the perfusate will also give an indication of the respiratory state of the isolated organ, with the lactate/pyruvate ratio (usually ~ 10) increasing in anoxia. The lactate/pyruvate ratio can be used to calculate cytosolic NAD^+ redox states (72) whereas the mitochondrial NAD^+ redox state can be derived from β-hydroxybutyrate/acetoacetate values (73).

The steady disappearance of glucose (*Figure 6.1*) from the perfusate shows that normal transport and metabolism mechanisms for the sugar are in operation. Transport and biliary excretion mechanisms can also be demonstrated using a dye such as bromo-sulphophthalein (BSP), which will be absorbed and then biliary excreted. However, such an addition to the perfusate is unwise if other drugs are being studied, because of possible competition for transport and metabolism. The addition of a radiolabelled amino acid such as [³H]lysine can be used to show that plasma protein synthesis is occurring satisfactorily (74). A steady increase in radioactivity in the acid-precipitable fraction of the cell-free perfusate will be observed (see *Figure 6.5*). To demonstrate the continued manufacture of glycoproteins, a radiolabelled form of fucose can be added to the perfusion medium, and acid-precipitable radioactivity again determined (75).

Signs of tissue damage can be obtained by monitoring either perfusate potassium or enzyme concentrations. Aspartate (GOT) and alanine (GPT) transaminases are most commonly used to assess liver damage (*Figure 6.3* and *6.4*), but have the disadvantage of also being released from haemolysed red blood cells. As the red blood cell volume is large compared with the blood system of the intact animal, GOT and GPT released in this way will not be insignificant. The utilization of sorbitol dehydrogenase (SDH), which is not found in erythrocytes, as a diagnostic enzyme can overcome this difficulty. Other enzymes which have been used to assess tissue damage in perfused liver systems

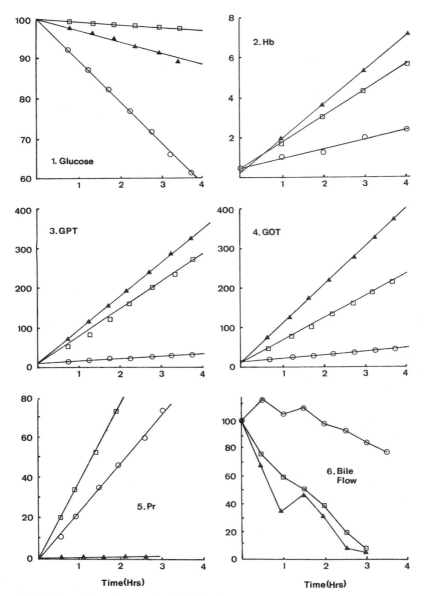

Figure 6. Viability and function tests. Typical values for some viability and function parameters determined for control livers (○) and livers perfused with 0.5 mM ouabain (▲) and 0.5 mM p-hydroxymercuribenzoate (□). **1**, Disappearance of glucose from the perfusate, expressed as a percentage of the initial concentrations (control, 10.4 mM; p-hydroxymercuribenzoate, 11.8 mM; and ouabain, 12.5 mM). **2**, Percent haemolysis of red blood cells. **3**, Perfusate GOT activity (U/l). **4**, Perfusate GPT activity (U/l). **5**, Acid-precipitable radioactivity in the perfusate, expressed as c.p.m. per 25 μl sample. 10 μCi L-4,5-[³H]lysine is added to the perfusate at zero time. Cell-free perfusate samples (25 μl) are added to 0.9% (w/v) NaCl (0.5 ml), followed by 10% (v/v) trichloroacetic acid (0.525 ml). Samples are filtered under suction through glass fibre discs, and washed twice with 5% (v/v) trichloroacetic acid (2 × 20 ml). The discs are then washed in absolute alcohol, dried and added to scintillation fluid (10 ml) for counting. **6**, Bile flow-rates, expressed as a percentage of initial rates (control, 11.8 μl/min; p-hydroxymercuribenzoate, 7.5 μl/min; and ouabain, 7.6 μl/min).

include glutamate dehydrogenase (GLDH), pyruvate dehydrogenase (PDH), and lysosomal enzymes such as acid phosphatases, β-glucuronidase, triglyceride lipase and N-acetyl β-glucosaminidase (76).

When using a peristaltic pump to circulate a perfusate containing erythrocytes, haemolysis is inevitable, but it is wise to follow it carefully to ensure that it is not excessive. This can be done by determining the absorbance of the cell-free perfusate at 420 nm and comparing it with a completely haemolysed sample (*Figure 6.2*).

A number of spectrophotometric and spectrofluorometric techniques are available to measure intracellular pigments from the surface of the perfused liver, though non-recirculatory systems utilising haemoglobin-free perfusates are required for this purpose. Flavoproteins and pyridine nucleotides can be estimated by fluorescence, the flavoprotein content being an indication of the cytosolic redox state, and the pyridine nucleotides reflecting the cytosolic and mitochondrial redox state (77). Type I and Type II binding spectra of substrates to cytochrome P-450 can be measured and the redox states of cytochromes a, c, b_5 and P-450 monitored (78,79).

If necessary, after ligation of the appropriate vessels, complete lobes may be removed from the liver during the course of experiments. This will allow direct determination of experimental parameters at a particular time point by techniques requiring the disruption of the tissue. Also, at the end of the experiment, the tissue can be kept for histochemical examination or electron microscopy, to ascertain the amount of structural damage which may have occurred.

4. SOME EXAMPLES OF APPLICATIONS OF THE ISOLATED PERFUSED LIVER PREPARATIONS

4.1 Cyclohexanecarboxylate Metabolism

This compound is a major intermediate in the metabolism of shikimate, a ubiquitous plant acid which is genotoxic, mutagenic and carcinogenic (80,81). Following ingestion of shikimate and its conversion to cyclohexanecarboxylate by the gut microflora (82), the latter is absorbed, oxidised and conjugated with glycine and glucuronic acid before being excreted predominantly in the urine (83). As it was suspected that the liver might be responsible for the metabolic processes involved, it was decided to study cyclohexanecarboxylate metabolism using isolated perfused rat liver preparations (38).

When [^{14}C]cyclohexanecarboxylate (25 mg) was added to the perfusate of the recycling isolated liver perfusion system, it was eliminated from the medium following first-order kinetics ($t\frac{1}{2}$ = 1.54 h, elimination constant = 0.377/h, CL = 0.063 ml/min/g liver) and after 6 h only 10% of the original compound remained in the perfusate unchanged (*Figure 7*). At this time, there was also hippurate, tetrahydrohippurate, hexahydrohippurate, benzoate and cyclohexylcarbonyl glucuronide present in the perfusate. The latter metabolite was also present in the bile, together with trace amounts of hippurate, tetrahydrohippurate and hexahydrohippurate (*Table 2*). These experiments therefore clearly confirmed that the liver has a major role in the metabolism of cyclohexanecarboxylate.

When the same compound (100 mg/kg body weight) was administered intraduodenally to rats with urethral and bile duct cannulae, 90% of the dose was eliminated in the urine and 8% in the bile after 7 h. Analysis of the urine revealed that the metabolites

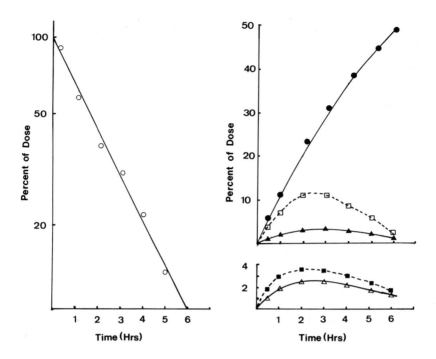

Figure 7. Metabolism of cyclohexanecarboxylate in the isolated perfused rat liver. Disappearance of 25 mg cyclohexanecarboxylate (○) from the perfusate, and appearance of hippurate (●), tetrahydrohippurate (■), hexahydrohippurate (▲), benzoate (△) and cyclohexylcarboxyl-glucuronide (□).

Table 2. Metabolism of Cyclohexanecarboxylate (CCA) in the Whole Rat, the Isolated Perfused Rat Liver and Isolated Rat Kidney Tubules.

	Dose of CCA			
	Whole animal		Perfused Liver	Kidney Cells
Metabolite	(200 mg/kg)	(100 mg/kg)	(25 mg)	(115 μM)
Hippurate	48.2 ± 2.0	57.2	65.8 ± 5.3	48.8 ± 1.7
Tetrahydrohippurate	12.4 ± 0.8	10.2	2.6 ± 1.3	13.1 ± 1.5
Hexahydrohippurate	10.5 ± 0.3	10.3	2.6 ± 1.0	30.1 ± 6.9
Benzoate	0	0	2.6 ± 2.4	2.6 ± 2.1
Benzoylglucaronide	} 18.0 ± 0.6	} 14.0	0	} 1.3 ± 0.5
CCA glucuronide			5.3 ± 1.0	
Bile glucuronides	8.0 ± 0.3	7.7	21.1 ± 1.2	–
% dose metabolized	97.1 ± 4.0	97.3	90.0 ± 3.0	50.4 ± 6.9

For the metabolites, the values quoted (± SEM, n = 3) represent the amount of each metabolite produced expressed as a percentage of the total amount of substrate metabolised. In the isolated perfused liver, 8.0 ± 0.8% of the dose was associated with the erythrocytes, and was not identified. Kidney tubule fragments were prepared by the method of Dawson (84) and incubated for 60 min in medium 199 containing foetal calf serum (10%, v/v) and CCA (115 μM).

were hippurate, tetrahydrohippurate, hexahydrohippurate and the glucuronide conjugates of both cyclohexanecarboxylate and benzoate. These glucuronides also accounted for all the metabolites found in the bile (*Table 2*).

Although the results of the *in vivo* and liver perfusion studies were similar, two differences were observed. Firstly, the isolated liver did not appear to produce benzoyl-glucuronide, although free benzoate was identified, and secondly, far more of the tetra- and hexahydrohippurates were produced *in vivo* than *in vitro*. These observations indicated that, in addition to the liver, other tissues might be involved in the metabolism of cyclohexanecarboxylate and, for this reason, experiments in which cyclohexane-carboxylate was incubated with isolated rat kidney tubules were undertaken. Results of these studies (*Table 2*) indicated that the kidney produces relatively large amounts of tetra- and hexahydrohippurate as compared with the liver. Also, incubation of kidney tubules with benzoate (110 μM) produced hippurate (96% of total metabolites) and benzo-glucuronide (4% of total metabolites). Thus, it seems likely that the kidney is responsible for the relatively high amounts of unsaturated metabolites observed *in vivo*, and may well also be the source of at least some of the benzoyl glucuronide.

4.2 Hepatic IgA Transport

Polymeric IgA present in the blood is rapidly removed by the liver and excreted in the bile. Uptake is dependent upon the IgA combining with secretory component, a protein present on the sinusoidal surface of the hepatocyte and which is continuously being transported as a component of the endocytic vesicles to the bile canalicular suface, where it is transferred to the bile. The secretory component at the plasma membrane is replaced by freshly manufactured protein from the Golgi apparatus. The observation that secretory component continuously followed this route (Golgi to plasma membrane to bile canalicular surface) with or without the addition of IgA, and the role of the cytoskeleton in the process, was established using isolated perfused rat liver preparations (74).

As IgA is a normal constituent of rat blood, it would be impossible to design *in vivo* experiments in which the supply of IgA to the organ was stopped. Not only did the isolated perfused liver preparation permit this, but also bile flow was uninterrupted and it could be collected for analysis. In an experiment in which a rat liver was simply perfused in a recirculating system for 4 h with an artificial medium completely devoid of IgA, analysis of sequentially-collected bile samples revealed increasing amounts of free secretory component, thus confirming that combination with IgA is unnecessary for its transport to the bile canalicular surface and subsequent biliary excretion.

To investigate the role of the cytoskeleton in the process, perfusion experiments were performed in which ^{125}I-labelled monoclonal polymeric IgA was added to the perfusion medium. After a slight delay of about 30 min, IgA appeared in increasing amounts in the bile. Also, addition of a second dose of IgA after 3 h resulted in a further sharp rise in biliary IgA excretion, indicating that the first dose had not saturated the transport system. When the experiment was repeated using a perfusate containing colchicine, which is a known inhibitor of microtubule formation, the amounts of IgA appearing in the bile were greatly reduced, and the second addition of IgA failed to produce a significant effect. Thus the cytoskeleton was shown to have an important function in IgA excretion.

Further analysis of the bile samples showed that in the absence of colchicine, free secretory component concentrations steadily increased with time, reflecting the gradual depletion of endogenous IgA previously bound to secretory component at the sinusoidal surface, and the continued synthesis, transport and excretion of secretory component. In the presence of colchicine, the levels of biliary secretory component remained very low, indicating an inhibition of its transport to and/or from the sinusoid surface. If only transport from the Golgi to the sinusoidal surface were affected, secretory component already present in the plasma membrane would continue to carry IgA until the supply was exhausted, but this did not occur. The rate of appearance of IgA in the bile of colchicine-treated livers suggested that the transport of the secretory component from the sinusoid surface was being inhibited. In these livers although the rate of appearance of IgA in the bile was clearly subnormal, it did rise initially. However, when the plasma membrane became depleted in secretory component, amounts of IgA in the bile fell and the second addition of IgA produced no effect. This observation clearly indicates that the supply of secretory component to the sinusoidal surface was being inhibited. Colchicine did not affect the release of IgA from the endocytic vesicles into the bile; the rate at which this occurred was similar in normal and colchicine-treated livers.

Thus, these perfusion studies showed that the cytoskeleton was involved in transporting free secretory component to the sinusoidal surface, and then carrying the endocytic vesicles to the bile canalicular surface.

4.3 **Release of Cadmium-thionein from Rat Liver**

Administration of cadmium to rats results in uptake by the liver and sequestration as a cadmium-metallothionein complex in the cytosol. If exposure ceases, the cadmium-thionein is eliminated from the hepatocytes at a rate which parallels accumulation in the kidneys. As hepatic release is slow and renal uptake rapid, it is virtually impossible to demonstate this transfer *in vivo*. However, hepatic release of cadmium-thionein has been demonstrated using isolated rat liver preparations (85).

Male rats were fed for 18 months on a diet containing 300 p.p.m. cadmium, and their livers isolated and perfused. Perfusate samples were taken at regular intervals and assayed for cadmium by atomic absorption and metallothionein by a radioimmunoassay. Comparison of these results showed that cadmium was released from the liver as the cadmium-thionein complex, and the rate of release suggested that an active process was involved. Also, the fact that the cadmium-thionein release appeared to cease after a few hours indicates that an equilibrium between release and re-uptake may exist.

4.4 **Some Other Applications**

Isolated rat liver preparations have been employed in numerous toxicological investigations, not only to investigate the hepatic metabolism of foreign compounds (*Table 3*) but also to study the effects of foreign compounds on a variety of liver functions (*Table 4*). Lafranconi and Huxtable (130) used an isolated liver preparation to generate a perfusate containing toxic metabolites of monocrotaline, which was then used to maintain an isolated lung preparation. In this way they were able to assess the pulmonary toxicity of the monocrotaline metabolites. Guaitani *et al.* (131) studied the uptake of

Table 3. Some Xenobiotics whose Metabolic Pathways have been Investigated using Isolated Perfused Liver Preparations.

Xenobiotic	(ref)	Xenobiotic	(ref)
2-Acetamidofluorene	(86)	Methadone	(36)
2-Allyl-2-isopropylacetamide	(87)	Methocarbamol	(97)
Amidopyrene	(88)	Mitoxantrone	(98)
Androstenedione	(89)	Nitrazepam	(92)
Aniline	(35)	p-Nitroanisole	(99)
Asulam	(61)	Nitropyrene	(100)
Benzothiazine derivatives	(90)	Nitrosobenzene	(35)
Bromosulphonephthalein	(91)	Noracetylmethadol	(36)
Carbamazepine	(92)	Norbenzphetamine	(36)
Cyclohexanecarboxylate	(38)	Norethindrone	(101)
Chlormethiazole	(93)	Nortriptyline	(102)
Deptrone methiodide	(94)	Phenol	(103)
Desmethylimipramine	(95)	Phenylhydroxylamine	(35)
Harmol	(40)	Shikimate	(83)
Hexobarbital	(88)	Styrene oxide	(104)
N-Hydroxyamphetamine	(36)	Δ'-Tetrahydrocannabinol	(105)
Imipramine	(95)	Warfarin	(106)
Mestranol	(96)		

cancer cells by perfused livers as a model for cancer-cell dissemination. Thelan and Wendel (132) have used the preparation to measure hepatic ethane production during drug-induced lipid peroxidation.

Techniques are also available to study the heterogeneous response of the isolated perfused liver to hepatotoxins. Metabolic events such as mixed function oxidation (133), sulphation (134) and oxygen uptake (135) can be followed in the periportal and pericentral regions of the liver lobules using microlight guides and miniature oxygen electrodes. Furthermore, zone-specific hepatotoxicity can be assessed using trypan blue. With this technique, Belinsky *et al.* (136) showed that following perfusion with allyl alcohol, infusion of trypan blue (0.2 mM) resulted in stained nuclei in the periportal region of lobules.

5. CONCLUDING COMMENTS

The isolated perfused liver preparation is a technique which offers the opportunity to study hepatic metabolism in a system in which:

(i)   the metabolic and distributional influences of all extrahepatic tissues are removed;

(ii)   the organ remains morphologically intact, and normal uptake, transport, metabolic, excretive and secretive mechanisms remain functional;

(iii)   oxygen, nutrients and substrates are delivered to the cells by the normal physiological route;

(iv)   the composition of the perfusing medium can be carefully defined, so as to allow control of such parameters as

(a)   pH and electrolyte content,

(b)   oxygen-carrying capacity,

Table 4. Some Examples of the Use of Isolated Perfused Rat Liver Preparations to Study the Effects of Xenobiotics on Liver Function.

Process	Xenobiotic	Reference
Glycolysis	Halothane	(107)
	Aminooxyacetate	(108)
	Manganese	(109)
Gluconeogenesis	Aminooxyacetate	(108)
	Aminopyrene	(110)
	Phenformin	(111)
	Salicylate	(112)
	Manganese	(109)
Glycogenolysis	Sulfonylurea	(113)
	Biguanide	(113)
	Verapimal	(114)
Lactate metabolism	Phenformin	(115)
Lipogenesis	Aminopyrene	(116)
	Mestranol	(117)
Fatty acid metabolism	Ethanol	(118)
Plasma protein synthesis	Aflatoxin B_1	(119)
	Phenobarbitone	(120)
	Bromobenzene	(121)
	Carbon tetrachloride	(121)
	Alcohol	(122)
	Actinomycin	(123)
	Puromycin	(123)
	Colchicine	(74)
Mixed function oxidation	Potassium cyanide	(124)
Glucuronidation	Carbon tetrachloride	(125)
	Sorbitol	(126)
	Ethanol	(126)
	3-Methylcholanthrene	(127)
	Phenobarbitone	(127)
Sulphation	3-Methylcholanthrene	(128)
	Phenobarbitone	(128)
Oxygen uptake	Aminopyrene	(47)
Organic anion transport	Nafenopin	(122)
	Phenobarbital	(129)
	Chlorpromazine	(128)
	Phenothiazine	(128)
	Promazine	(128)
	Triflupromazine	(128)

    (c)   protein-binding properties,

    (d)   nutrient and substrate concentrations;

(v)   the rate of delivery of the perfusing medium can be regulated and changed at will;

(vi)   frequent sampling of the perfusate and bile is possible, allowing accurate pharmacokinetic analysis of uptake and elimination processes; and, amounts of substrate in the liver at any point in time can be estimated by difference;

51

(vii)   the functional state of the organ can be monitored continuously. This allows normal function to be established before further investigations are initiated, so that one preparation can act as its own control;

(viii)   relatively large quantities of substrates can be added to the perfusate, without fear of overdosing an experimental animal;

(ix)   the ultimate fate of the substrate can be determined by leaving it to recirculate through the organ for prolonged periods;

(x)   first-pass metabolism can be studied with ease;

(xi)   donor animals may be subjected to pre-treatment, thus allowing metabolic studies in damaged as well as healthy tissue;

(xii)   specific metabolic inhibitors can be added to the perfusate, to permit the study of particular processes.

Thus, it can be seen that the technique has advantages over the *in vivo* and *in vitro* systems which are available. Unfortunately, however, there have been very few direct comparative studies undertaken, though qualitatively the isolated perfused liver has been shown to reflect accurately *in vivo* hepatic metabolism (137). Also, Blom *et al.* (62) have compared hepatic drug transport mechanisms in isolated hepatocytes, the isolated perfused liver and the liver *in vivo*, using dibromosulphophthalein (an organic anion), α-tubocurarine (an organic cation) and ouabain (an uncharged compound). Results, which were compared on the basis that the liver contains 114×10^6 hepatocytes/g wet wt. liver (138), revealed that overall, the perfused liver was the better model for these uptake and excretion mechanisms, compared with isolated hepatocytes which showed lower transport capacities than the liver *in vivo*.

6. ACKNOWLEDGEMENTS

I wish to thank Drs. D. Brewster, A. Hugget and S. Legg for their contributions to the cyclohexanecarboxylate study; Drs. B.M. Mullock, R.H. Hinton and J. Peppard for their collaboration on the IgA work and Dr. M. Dobrota, B. Carter and P. Lloyd for inviting me to help with their metallo-thionein experiments. My thanks also go to Mr. M. Singh for his help in preparing this chapter, and to Mrs. J. Cole for typing the manuscript.

7. REFERENCES

1. Miller,L.L. (1973) in *Isolated Liver Perfusion and its Applications,* Bartosek,I., Guaitani,A. and Miller,L.L. (eds.), Raven Press, New York, p. 11.
2. Thurman,R.G., Reinke,L.A. and Kauffman,F.C. (1979) in *Reviews in Biochemical Toxicology,* Hodgson,E., Bend,J.R. and Philpot,R.M. (eds.), Elsevier/North Holland Inc., New York, p. 249.
3. Franke,H., Huland,H. and Weiss,C.R. (1971) *Z. Ges. Exp. Med.,* **156**, 268.
4. Bowman,R.H. and Maack,T. (1972) *Am. J. Physiol.,* **222**, 1499.
5. Emslie,K.R., Calder,I.C., Hart,S.J. and Tange,J.D. (1982) *Xenobiotica,* **12**, 77.
6. Nicoloyson,L. (1971) *Acta Physiol. Scand.,* **83**, 563.
7. Roth,J.A. (1979) in *Reviews in Biochemical Toxicology,* Hodgson,E., Bend,J.R. and Philpot,R.M. (eds.), Elsevier/North Holland Inc., New York, p. 287.
8 Rosenblom,P.M. and Bass,A.D. (1970) *J. Appl. Physiol.,* **29**, 138.
9. Broadley,K.J. (1972) *Eur. J. Pharmacol.,* **20**, 291.
10. Feuvray,D. and DeLeiris,J. (1973) *J. Mol Cell. Cardiol.,* **5**, 63.
11. Kopp,S.J., Baker,J.D., D'Agrosa,L.S. and Hawley,P.L. (1978) *Toxicol. Appl. Pharmacol.,* **46**, 475.
12. Windmueller,H.G. and Spaeth,E.A. (1972) *J. Lipid Res.,* **13**, 92.

13. Nichols,T.J. and Leese,H.J. (1984) *Biochem. Pharmacol.*, **33**, 771.
14. Grodsky,G.M., Batts,A.A., Bennett,L.L., Vcella,G., McWilliams,N.B. and Smith,P.F. (1963) *Am. J. Physiol.*, **205**, 638.
15. Gerrich,J.E., Frankel,B.J., Fanska,R., West,L., Forsham,A. and Grodsky,G.M. (1974) *Endocrinology*, **94**, 1381.
16. Pascal,J.P. (1972) *Arch. Fr. Maladies L'Appereil Digestif*, **61**, 797.
17. White,R.J. (1971) in *Karolinska Symposia on Research Methods in Reproductive Endocrinology: Perfusion Techniques*, Diczfalusy,E. and Diczfalusy,A. (eds.), Karolinska Institutet, Stockholm, p. 200.
18. Höller,M. and Penin,H. (1984) *Biochem. Pharmacol.*, **33**, 1753.
19. Koide,Y., Kimura,S., Tada,R., Kugai,N. and Yamashita,K. (1983) *Biochem. Pharmacol.*, **32**, 517.
20. Folkman,J. and Gimbrone,M.A. (1971) in *Karolinska Symposia on Research Methods in Reproductive Endocrinology: Perfusion Techniques*, Diczfalusy,E. and Diczfalusy,A. (eds.), Karolinska Institutet, Stockholm, p. 237.
21. Davis,S.R. and Mepham,T.B. (1972) *J. Physiol.*, **222**, 13.
22. Cedard,L. (1971) in *Karolinska Symposia on Research Methods in Reproductive Endocrinology: Perfusion Techniques*, Diczfalusy,E. and Diczfalusy,A. (eds.), Karolinska Institutet, Stockholm, p. 331.
23. Ahren,K., Janson,P.O. and Selstam,G. (1971) in *Karolinska Symposia on Research Methods in Reproductive Endocrinology: Perfusion Techniques*, Diczfalusy,E. and Diczfalusy,A. (eds.), Karolinska Institutet, Stockholm, p. 285.
24. Eik-Nes,K.B., (1971) in *Karolinska Symposia on Research Methods in Reproductive Endocrinology: Perfusion Techniques*, Diczfalusy,E. and Diczfalusy,A. (eds.), Karolinska Institutet, Stockholm, p. 270.
25. Lee,I.P., Mukhtar,H., Suzuki,K. and Bend,J.R. (1983) *Biochem. Pharmacol.*, **32**, 159.
26. Porter,J.E., Mical,R.S., Ondo,J.G. and Kamberi,I.A. (1971) in *Karolinska Symposia on Research Methods in Reproductive Endocrinology: Perfusion Techniques*, Diczfalusy,E. and Diczfalusy,A. (eds.), Karolinska Institutet, Stockholm, p. 249.
27. Varro,V. (1972) *Arch. Fr. Maladies L'Appereil Digesif*, **61**, 799.
28. Ritchie,H.D. (1972) *Ren. Gastroenterol.*, **4**, 159.
29. Bernstein,E.F. (1971) in *Karolinska Symposia on Research Methods in Reproductive Endocrinology: Perfusion Techniques*, Diczfalusy,E. and Diczfalusy,A. (eds.), Karolinska Institutet, Stockholm, p. 44.
30. Hamilton,R.L., Berry,M.N., Williams,M.C. and Severinghaus,E.M. (1974) *J. Lipid Res.*, **15**, 182.
31. Zonta,A. and Dionigi,R. (1973) in *Isolated Liver Perfusion and its Applications*, Bartosek,I., Guaitani,A. and Miller,L.L. (eds.), Raven Press, New York, p. 101.
32. Smith,B.R., Born,J.L. and Garcia,D.J. (1983) *Biochem. Pharmacol.*, **32**, 1609.
33. Curtis,C.G., Powell,G.M. and Stone,S.L. (1971) *J. Physiol.*, **213**, 14P.
34. Guaitani,A., Villa,P. and Bartosek,I. (1983) *Xenobiotica*, **13**, 39.
35. Eyer,P., Kampffmeyer,H., Maister,H. and Rösch-Oehme,E. (1980) *Xenobiotica*, **10**, 499.
36. Gumbrecht,J.R. and Franklin,M.R. (1979) *Xenobiotica*, **9**, 547.
37. Novakova,V., Birke,G., Plantin,L.-O. and Wretlind,A. (1975) *Fed. Proc.*, **34**, 1488.
38. Brewster,D., Jones,R.S. and Parke,D.V. (1977) *Xenobiotica*, **7**, 601.
39. Folkow,B., Hallbäck,M. and Lundgren,Y. (1970) *Acta Physiol. Scand*, **80**, 93.
40. Koster,H., Halsema,I., Scholtens,E., Pang,K.S. and Mulder,G.J. (1982) *Biochem. Pharmacol.*, **31**, 3023.
41. Morrison,J.L., Bloom,W.L. and Richardson,A.P. (1951) *J. Pharmacol. Exp. Therap.*, **101**, 27.
42. Monks,A., Ayers,O. and Cysyk,R. (1983) *Biochem. Pharmacol.*, **32**, 2003.
43. Toprek,H. (1971) *Cancer Res.*, **31**, 1962.
44. Michelakakis,H. and Danpure,C.J. (1984) *Biochem. Pharmacol.*, **33**, 2047.
45. Kaplan,H.R. and Murthy,V.S. (1975) *Fed. Proc.*, **34**, 1461.
46. Bruser,B., Bücher,T., Sies,H. and Versmold,H. (1972) in *Molecular Basis of Biological Activity*, Gaede,K., Horecker,B.L. and Whelan,W.J. (eds.), Academic Press, New York, p. 197.
47. Thurman,R.G. and Scholz,R. (1969) *Eur. J. Biochem.*, **10**, 459.
48. Sies,H. (1978) in *Methods in Enzymology*, Colwick,S.P. and Hoffe,P.A. (eds.) Academic Press Inc., New York and London, Vol. **51**, p. 48.
49. Krone,W., Huttner,W.B., Kampf,S.C., Rittich,B., Seitz,H.J. and Tarnowsky,W. (1974) *Biochim. Biophys. Acta.*, **372**, 55.
50. Clark,L.C.,Jr., Becattini,F., Kaplan,S., Obrock,V., Cohen,D. and Becker,C. (1973) *Science (Wash.)*, **181**, 680.
51. Brown,H. and Hardison,W.G.M. (1975) *Fed. Proc.*, **34**, 1516.
52. Pellegrin,P.L. and Lesne,M. (1984) *Drug Met. Disp.*, **12**, 235.
53. Höller,M. and Breuer,H. (1975) *Z. Klin. Chem. Klin. Biochem.*, **7**, 319.
54. Mayes,P.A. and Felts,J.M. (1966) *Proc. Eur. Soc. Study Drugs*, **7**, 16.
55. Ross,B.D. (1972) *Perfusion Techniques in Biochemistry: A Laboratory Manual in the use of Isolated Perfusion in Biochemical Experimentation*, published by Clarendon Press, Oxford.

56. Bernard,C. (1855) *C.R. Hebd. Séances, Acad. Sci., Paris,* **41**, 461.
57. Graf,J., Kaschnitz,R. and Peterlik,M. (1972) *Res. Exp. Med.,* **157**, 12.
58. Richards,T.G., Tindall,V.R. and Young,A. (1959) *Clin. Sci. (Lond.),* **18**, 499.
59. Gibaldi,M. and Perrier,D. (1975) *Drugs and the Pharmaceutical Sciences: Pharmacokinetics,* published by Marcel Decker Inc., New York.
60. Rowland,M. (1972) *Eur. J. Pharmacol.,* **17**, 352.
61. Heijbroek,W.M.H., Muggleton,D.F. and Parke,D.V. (1984) *Xenobiotica,* **14**, 235.
62. Blom,A., Scaf,A.H.J. and Meijer,D.K.F. (1982) *Biochem. Pharmacol.,* **31**, 1553.
63. Brown,R.D. and Manno,J.E. (1978) *J. Pharm. Sci.,* **67**, 1687.
64. Ruifrok,P.G. (1982) Biopharm. Drug. Disp., **3**, 243.
65. Nagashami,R. and Lavy,G. (1968) *J. Pharm. Sci.,* **57**, 1991.
66. Keiding,S. and Steiness,E. (1984) *Pharmacol. Exp. Ther.,* **230**, 474.
67. Bass,L., Keiding,S., Winkler,K. and Tygstrup,N. (1976) *J. Theor. Biol.,* **61**, 393.
68. Rowland,M., Benet,L.Z. and Graham,G.G. (1973) *J. Pharmacokin. Biopharm.,* **1**, 123.
69. Jones,D.B., Morgan,D.J., Mihaly,G.W., Webster,L.K. and Smallwood,R.A. (1984) *J. Pharmacol., Exp. Ther.,* **229**, 522.
70. Schary,W.L. and Rowland,M. (1983) *J. Pharmacokin. Biopharm.,* **11**, 225.
71 Ahmad,A.B., Bennett,P.N. and Rowland,M. (1984) *J. Pharmacol., Exp. Ther.,* **230**, 718.
72. Bücher,T. (1970) in *Pyridine Nucleotide Dependent Dehydrogenases,* Sund,H. (ed.), Springer, New York, p. 439.
73. Veech,R.L., Raijman,L. and Krebs,H.A. (1970) *Biochem. J.,* **117**, 499.
74. Mullock,B.M., Jones,R.S., Peppard,J. and Hinton,R.H. (1980) *FEBS Lett.,* **120**, 278.
75. Mullock,B.M., Jones,R.S. and Hinton,R.H. (1980) *FEBS Lett.,* **113**, 201.
76. Serrou,B.C., Solassol,C., Meiss,L., Gelis,C. and Romieu,C. (1973) in *Isolated Liver Perfusion and its Applications,* Bartosek,I., Guaitani,A. and Miller,L.L. (eds.), Raven Press, New York, p. 127.
77. Scholz,R., Thurman,R.G., Williamson,J.R., Chance,B. and Bücher,T. (1969) *J. Biol. Chem.,* **244**, 2317.
78. Sies,H. and Brauser,B. (1970) *Eur. J. Biochem.,* **15**, 531.
79. Lubbers,D.L., Kessler,M., Scholz,R. and Bücher,T. (1965) *Biochem. Z.,* **341**, 346.
80. Evans,I.A. and Osman,M.A. (1974) *Nature,* **250**, 348.
81. Jones,R.S., Ali,M., Ioannides,C., Styles,J.A., Ashby,J., Sulej,J. and Parke,D.V. (1983) *Toxicol. Lett.,* **19**, 43.
82. Brewster,D., Jones,R.S. and Parke,D.V. (1976) *Biochem. Soc. Trans.,* **4**, 518.
83. Brewster,D., Jones,R.S. and Parke,D.V. (1978) *Biochem. J.,* **170**, 257.
84. Dawson,A.G. (1972) *Biochem. J.,* **130**, 525.
85. Dobrota,M., Carter,B.A., Lloyd,P. and Jones,R.S. (1984) *J. Human Toxicol.,* in press.
86. Miyata,K., Noguchi,Y. and Enomoto,M. (1972) *Jap. J. Exp. Med.,* **42**, 483.
87. Smith,A. (1976) *Biochem. Pharmacol.,* **25**, 2429.
88. Kvetina,J., Simkova,M., Citta,M. and Deml,F. (1973) in *Isolated Liver Perfusion and its Applications,* Bartosek,I., Guaitani,A. and Miller,L.L. (eds.), Raven Press, New York, p. 235.
89. Eriksson,H., Gustafsson,J.-A. and Ponsette,A. (1972) *Eur. J. Biochem.,* **27**, 327.
90. Lan,S.J., Dean,A.V., Walker,B.D. and Schreiber,E.C. (1976) *Xenobiotica,* **6**, 171.
91. Bel,C. (1969) *C.R. Scand. Soc. Biol.,* **162**, 1949.
92. Garattini,S., Guaitani,A. and Bartosek,I. (1973) in *Isolated Liver Perfusion and its Applications,* Bartosek,I., Guaitani,A. and Miller,L.L. (eds.), Raven Press, New York, p. 225.
93. Herbertz,G., Metz,T., Reinauer,H. and Staib,W. (1973) *Biochem. Pharmacol.,* **22**, 1541.
94. Lavy,U.I., Hespe,W. and Meijer,D.K.F. (1972) *Naunyn-Schmiederbergs Arch. Pharmacol.,* **275**, 183.
95. Bickel,M.H. and Minder.R. (1970) *Biochem. Pharmacol.,* **19**, 2425.
96. Bolt.H.M. and Remmer,H. (1973) *Horm. Metab. Res.,* **5**, 101.
97. Gerber,N. and Seibert,R.A. (1975) *Xenobiotica,* **5**, 145.
98. Ehninger,G., Proksch,B., Hartmann,F., Gärtner,H.V. and Wilms,K. (1984) *Cancer Chemother. Pharmacol.,* **12**, 50.
99. Thurman,R.G., Marazzo,D.P., Jones,L.S. and Kauffman,F.C. (1977) *J. Pharmacol. Exp. Ther.,* **201**, 498.
100. Bond,J.A., Medinsky,M.A. and Dutcher,J.S. (1984) *Toxicol. Appl. Pharmacol.,* **75**, 531.
101. Back.D.J., Macnee,C.M., Orme,M.L'E., Rowe,P.H. and Smith,E. (1984) *Biochem. Pharmacol.,* **33**, 1595.
102. Bahr,C.Von., Borga,O., Fellenius,E. and Rowland,M. (1973) *Pharmacology,* **9**, 177.
103. Powell,G.M., Mitler,J.J., Olaveson,A.H. and Curtis,C.J. (1974) *Nature,* **252**, 234.
104. Van Anda,J., Bend,J.R. and Fouts,J.R. (1977) *The Pharmacologist,* **19**, 191.
105. Halldin,M.M., Isaac,H., Widman,M., Nilsson,E. and Ryrfeldt,A. (1984) *Xenobiotica,* **14**, 277.

106. Losito,R.J. and Owen,C.A. (1972) *Mayo Clin. Proc.*, **47**, 731.
107. Biebuyck,J.F., Lund,P. and Krebs,H.A. (1972) *Biochem. J.*, **128**, 711.
108 Walli,A.K. and Schimassek,H. (1973) in *Isolated Liver Perfusion and its Applications*, Bartosek,I., Guaitani,A. and Miller,L.L. (eds.), Raven Press, New York, p. 187.
109. Wimhurst,J.M. and Manchester,K.L. (1973) *FEBS Lett.*, **29**, 201.
110. Scholz,R., Hansen,W. and Thurman,R.G. (1973) *Eur. J. Biochem.*, **38**, 64.
111. Lloyd,M.H., Iles,R.A., Walton,B., Hamilton,C.A. and Cohen,R.D. (1975) *Diabetes*, **24**, 618.
112. Woods,H.F., Stubbs,W.A., Johnson,G. and Alberti,K.G.M.M. (1974) Clin. Exp. Pharmacol. Physiol., **1**, 535.
113. Büber,V., Felber,J.-P. and Magnenat,P. (1973) in *Isolated Liver Perfusion and its Applications*, Bartosek,I., Guaitani,A. and Miller,L.L. (eds.), Raven Press, New York, p. 199.
114. Koide,Y., Kimura,S., Tada,R., Kugai,N. and Yamashita,K. (1973) *Biochem. Pharmacol.*, **32**, 517.
115. Haeckel,R. (1973) in *Isolated Liver Perfusion and its Applications*, Bartosek,I., Guaitani,A. and Miller,L.L. (eds.), Raven Press, New York, p. 199.
116. Thurman,R.G. and Scholz,P. (1973) in *Isolated Liver Perfusion and its Applications*, Bartosek,I., Guaitani,A. and Miller,L.L. (eds.), Raven Press, New York, p. 187.
117. Weinstein,I. (1972) *Proc. Soc. Exp. Biol. Med.*, **140**, 319.
118. Scholz,R., Kalstein,U. Schwabe,U. and Thurman,R.G. (1974) in *Alcohol and Aldehyde Metabolizing Systems*, Thurman,R.G., Yonetani,T., Williamson,J.R. and Chance,B. (eds.), Academic Press, New York, p. 315.
119. John,D.W. and Miller,L.L. (1969) *Biochem. Pharmacol.*, **18**, 1135.
120. Matern,S., Fröhling,W. and Bock,K.W. (1972) *Naunyn-Schmiederbergs Arch. Pharmacol.*, **273**, 242.
121. Davis,D.C., Hashimoto,M. and Gillette,J.R. (1973) *Biochem. Pharmacol.*, **22**, 1989.
122. Kirsch,R.E., Firth,L.O'C., Stead,R.H. and Saunders,S.J. (1973) *Am. J. Clin. Nutr.*, **26**, 1911.
123. John,D.W. and Miller,L.L. (1966) *J. Biol. Chem.*, **241**, 4817.
124. Meijer,A.J., Woerkom,G.M.van, Williamson,J.R. and Tager,J.M. (1975) *Biochem. J.* **150**, 205.
125. Otani,G., Abon-El-Makarem,M.M. and Bock,K.W. (1976) *Biochem. Pharmacol.*, **25**, 1293.
126. Reinke,L.A., Belinsky,S.A., Kauffman,F.C., Evans,R.K. and Thurman,R.G. (1982) *Biochem. Pharmacol.*, **31**, 1621.
127. Homada,N. and Gessner,T. (1974) *The Pharmacologist*, **16**, 282.
128. Toth,I., Kendler,J., Nagpaul,C. and Zimmerman,H.J. (1972) *Proc. Soc. Exp. Biol. Med.*, **140**, 1467.
129. Sorrel,M.F., Tuma,D.J. and Barak,A.J. (1974) *Pharmacology*, **11**, 365.
130. Lafranconi,W.M. and Huxtable,R.J. (1984) *Biochem. Pharmacol.*, **33**, 2479.
131. Guaitani,A., Kvetina,J. and Garattini,S. (1972) *Eur. J. Cancer*, **8**, 79.
132. Thelan,M. and Whelan,A. (1983) *Biochem. Pharmacol.*, **32**, 1701.
133. Ji,S., Lemasters,J.J. and Thurman,R.G. (1981) *Mol. Pharmacol.*, **19**, 513.
134. Conway,J.G., Kauffman,F.C., Ji,S. and Thurman,R.G. (1982) *Mol. Pharmacol.*, **22**, 509.
135. Matsumura,T. and Thurman,R.G. (1983) *Am. J. Physiol.*, **244**, 656.
136. Belinsky,S.A., Popp,J.A., Kauffman,F.C. and Thurman,R.G. (1984) *J. Pharmacol. Exp. Ther.*, **230**, 755.
137. Garattini,S., Guaitani,A. and Bartosek,I. (1973) in *Isolated Liver Perfusion and its Applications*, Bartosek,I., Guaitani,A. and Miller,L.L. (eds.), Raven Press, New York, p. 225.
138. Seglen,P.O. (1973) *Exp. Cell. Res.*, **82**, 391.
139. Sloviter,H.A. (1975) *Fed. Proc.*, **34**, 1484.

CHAPTER 3

Preparation and Culture of Mammalian Cells

DIANE J.BENFORD and SUSAN A.HUBBARD

1. INTRODUCTION

Mammalian cells are increasingly being used as alternatives to animals in toxicology. There are several reasons for this trend: increasing pressure from the public for reducing the numbers of animals used in experiments; the high financial cost of conventional animal testing which may be significantly reduced by preliminary non-animal studies; dissatisfaction with poor correlations between results in laboratory animals and humans; biochemical mechanisms of toxicity are often more effectively investigated in cell cultures where the environment can be more easily controlled than with the whole animal.

There are basically two approaches to *in vitro* studies, either using an established, rapidly growing cell line, which will give an indication both of general cytotoxic effects on all cells and of effects on dividing tissues, or using highly differentiated cells, whether primary cultures or specialised cell lines, for studying specific effects on different cell types or functions. It is obviously beyond the scope of this chapter to cover in detail the vast range of techniques which may be useful for these subjects. Instead we will describe in detail the techniques used for two of the cell types with which we routinely work in our laboratory, with occasional reference to other cells. The examples presented here are Chinese Hamster ovary (CHO-K1) cells, which are an established cell line, and adult rat hepatocytes, which in suspension or maintenance culture retain much of their normal differentiated function. Cell lines such as CHO have little capacity for drug metabolism, particularly cytochrome P450 dependent activity, and therefore may be insensitive to toxic chemicals requiring metabolic activation. This problem is partially circumvented by incubation of the cells either with a liver subcellular fraction (e.g., S9) or with hepatocytes. These techniques and their limitations are also discussed in this chapter. No attempt has been made to cover techniques used in genotoxicity assays as these are covered in another volume of this series (1).

The methods outlined in this chapter have been shown to be practicable in our laboratory. Variations may, of course, occur from laboratory to laboratory. More detailed descriptions may be found in other reviews, most notably by Freshney (2).

2. BASIC TECHNIQUES

2.1 Equipment and Apparatus

Some of the specific equipment required for the maintenance of cell cultures is listed in *Tables 1 − 6*. A further description of the essential equipment will be given here.

2.1.1 *Incubation Facilities*

In general, many cell lines can be maintained in an atmosphere of 5% CO_2:95% air at 99% relative humidity. The concentration of the CO_2 is kept in equilibrium with the sodium bicarbonate in the medium. Different media have differing buffering capacity and this is discussed further below. If a CO_2 controlled incubator is not available, or cultures must be kept sealed in flasks (i.e., after treatment with some volatile substances), then cells may be maintained in flasks sealed after gassing with 5% CO_2/95% air, or vessels kept in boxes gassed and then sealed with pressure sensitive tape. In the case of boxes, the humidity must be maintained with a dish of water.

Various media may be used so that a controlled CO_2 atmosphere is not required and in this case a CO_2 incubator is not necessary. Hepatocytes in primary culture are often maintained in Leibovitz L-15 medium which does not require a CO_2 atmosphere, however, flasks must not be sealed (as the hepatocytes require a high O_2 tension which is reduced with time in sealed ungassed vessels). Most cell lines are maintained at 36.5°C, although some cultures, such as skin cultures may require lower temperatures. Cultured cells can generally survive lower temperatures, but rarely survive temperatures greater than 2°C above normal, and therefore the incubator should be set to cut out at approximately 38.5°C to prevent cell death. Incubators are designed to regulate an even temperature and this is more important than accuracy, i.e., temperature should be ±0.5°C. Most incubators have areas of differing temperature, therefore fan assisted incubators are preferable to help maintain even temperature distribution.

2.1.2 *Fridges and Freezers (−20°C)*

Both items are very important for storage of media (liquid media) at 4°C and for enzymes (e.g., trypsin) and some media components (e.g., glutamine and serum) at −20°C.

2.1.3 *Microscopes*

A simple inverted microscope is essential so that cultures can be examined in flasks and dishes. It is vital to be able to recognise morphological changes in cultures since these may be the first indication of deterioration of a culture. A very simple light microscope with ×100 magnification will suffice for routine cell counts in a haemocytometer, although a microscope of much better quality will be required for chromosome analysis or autoradiography work.

2.1.4 *Sterile Work Area*

A laminar flow hood offers the best sterile protection available. If a hazardous chemical is to be handled a Class II Biohazard Cabinet which has a vertical laminar flow should be used. However, for primary cultures and also if no laminar flow hood or sterile room is available, an area for sterile work should be set aside, where there is no throughfare. If aseptic techniques are adhered to and the area kept clean and tidy, sterility can be easily maintained. Details of checks for culture contamination are given in later sections.

2.1.5 *Washing Up and Sterilising Facilities*

A wide range of plastic tissue culture ware is available reducing the amount of washing up necessary. However, glassware such as pipettes should be soaked in a suitable detergent (e.g., DECON), then passed through a stringent washing procedure with thorough soaking in distilled water prior to drying and sterilising. Pipettes are often plugged with non-absorbent cotton wool before putting into containers for sterilising.

Glassware, such as pipettes, conical flasks, beakers (covered with foil) are sterilised in a dry heat oven at 160°C for one hour. All other equipment, such as automatic pipettor tips and bottles (lids loosely attached) are autoclaved at 121°C. Indicator tubes (e.g., Brownes' tubes, sterile test strips) are necessary for each sterilising run to ensure that the machine is operating effectively. Autoclave bags are available for loose items. Aluminium foil also makes good packaging material. A wide range of autoclaves are available from laboratory ware suppliers.

2.1.6 *Tissue Culture Ware*

A variety of tissue culture plastic ware is available, the most common being specially treated polystyrene. Although all tissue culture plastic ware should support cell growth adequately, it is essential when using a new supplier or type of dish to ensure that cultures grow happily in it. The tests to ensure this, such as growth curves and time of reaching a confluent monolayer, are similar to those used to ensure that serum batches are satisfactory.

Cells can be maintained in Petri dishes or flasks (25 cm² or 75 cm²) which have the added advantage that the flasks can be gassed and then sealed so that a CO_2 incubator need not be used; this is particularly useful if incubators fail. Tissue culture ware is chosen to match the procedure.

Sometimes it may be necessary to condition a surface by pretreating it with 'spent' medium which has been used with another culture. The choice of vessel depends on several factors: whether the culture is in suspension or grows as a monolayer (only monolayer cultures are dealt with in this chapter); the cell yield; whether it needs CO_2 or not; and what form of sampling is to take place. Cost can also be a limiting factor. Cell yield is proportional to available surface areas. It is important to ensure that an even layer can grow, especially in the currently popular multiwell dishes (24 and 96 wells).

For attaching cells such that histological stains may be used, Thermanox cover slips (L.I.P., Shipley, UK) fit into multiwell dishes and can be removed and treated with the various organic solvents required in staining (e.g., xylene). Normal tissue culture ware is not resistant to organic solvents.

2.1.7 *Liquid N_2*

Invariably for established and finite cell lines, samples of cultures will need to be frozen down for storage. It is important to maintain continuity in cells to prevent genetic drift and to guard against loss of the cell line through contamination and other disasters. The procedure for freezing cells is described in Section 2.5.2 and is general for all cells in culture. They should be frozen in exponential phase of growth with a suitable preservative, usually dimethylsulphoxide (DMSO). The cells are frozen slowly at

1°C/min to −50°C and then kept either at −196°C immersed in liquid N_2 (in glass sealed ampoules) or above the liquid surface in the gas phase (screw top ampoules). Deterioration of frozen cells has been observed at −70°C (3), therefore −196°C (liquid N_2) seems to be necessary.

To achieve slow freezing rates a programmable freezer can be obtained, or an adjustable neck plug or freezing tray (as marketed by Union Carbide) for use in a narrow-necked liquid nitrogen freezer. Alternatively, ampoules may be frozen in a polystyrene box with 1″ thick walls. This will insulate the ampoules to slow the freezing process to 1°C/min in a −70°C freezer.

2.1.8 *Water Still or Reverse Osmosis Apparatus*

A double distilled or Reverse Osmosis water supply is essential for preparation of media, and rinsing glassware. The pH of the double distilled water should be regularly checked as in some cases this can vary. Variations in the quality of the water used may account for variation in results, therefore water from one source should be used. Water is sterilised by autoclaving at 121°C for 20 min. The distilled water must be glass distilled and stored in glass if it is to be used for preparation of media. Storage in plastic may result in leaching of toxic substances from the plastic into the water.

2.1.9 *Filter Sterilisation*

Media that cannot be autoclaved must be filter sterilised through a 0.22 μm pore size membrane filter. These are obtainable in various designs to allow a wide range of volumes to be filtered (e.g., Millipore, Gelman). They can be purchased as sterile, disposable filters, or they may be sterilised by autoclaving in suitable filter holders. Culture media, enzymes, cofactors and bicarbonate buffers are examples of non-autoclavable substances.

2.1.10 *Facilities for Counting Cells*

It is possible to monitor cell growth by eye (looking for confluency), however, more accurate cell counts are required for most experimental purposes.

A haemocytometer (Improved Neubaur) is the simplest, cheapest method of counting. A coverslip slightly wet at the edges is placed over the counting grid such that interference patterns ('Newtons rings') formed when glass attaches to glass indicate that it is securely in place. Introduce the medium containing the cells under the coverslip from a Pasteur pipette by capillary action. Fluid will run into the grooves of the chamber to take away the excess. The cells in the central grid area can then be counted. The haemocytometer is designed such that when properly set up an area of 1 mm² and 0.1 mm deep is filled − this is a central area of 25 smaller squares bounded by triple parallel lines. Each of the 25 squares is further subdivided into 16 to aid counting. For routine subculture about 100 − 300 cells should be counted, but 500 − 1000 for more precise counts. For the Improved Neubaur chamber, the cell concentration, calculated from the number of cells (n) in the central 25 square area (1 mm², depth 0.1 mm), is $n \times 10^4$ cells/ml. Viability determinations (described in Section 3.2) can be easily carried out after diluting cells with dye. This counting method is open to error in a number of ways. Errors may be introduced by incorrect preparation of the chamber, incorrect sampling of cells, and aggregation of cells. However, it has the advantage of allowing visual inspection of the morphology of the cells.

An electronic particle counting method using a Coulter Counter can also be used. In this method the cells are drawn through a small orifice which changes the current flow through the orifice producing a series of pulses which are sorted and counted. The counter will need calibrating for the size of cell being used.

Other lengthier methods such as plating or cloning efficiency are described in Section 3.3.

2.1.11 *General Small Items of Equipment*

A number of small items of equipment are useful for performing tissue culture. A vacuum pump is helpful to enable medium to be quickly aspirated from cultures. The pump should be adequately protected from back movement of liquid.

In addition automatic pipettes, such as Gilsons, Oxfords and Finnpipettes, are suitable for treating cultures with chemicals as they utilise disposable, autoclavable, plastic tips. For larger volumes using glass or plastic pipettes, some form of suction aid such as 'Pipet-aid' or 'pi-pump' is necessary.

2.2 **Media**

The choice of culture medium is dependent on the requirements of the cell line. The media details for the cell cultures dealt with in this chapter are given in Sections 2.5 and 2.6. Only a short summary of the basic media requirements is given here. Actual levels of additions are given only for the specific cultures discussed. Supplier's catalogues give more detailed information.

2.2.1 *Basic Media*

A wide variety of culture media is available. The most basic media are balanced salt solutions (BSS), e.g., phosphate-buffered saline (PBSA), which may be used for washing cells and for short incubations in suspension. More complex defined media are used for growth and maintenance. Defined media can also vary in complexity, by the addition of a number of constituents, e.g., from Eagle's minimum essential medium (MEM) which contains essential amino acids, vitamins and salts, to McCoy's media, which contains a larger number of different amino acids, vitamins, minerals and other extra metabolites (such as nucleosides).

2.2.2 *Buffering Capacity*

A number of supplements to the basic media are necessary to enable them to be used for culturing cells. Cell cultures have an optimum pH for growth, generally between pH 7.4 – 7.7. The type of buffering that is used for the media depends on the growth conditions. When cells are incubated in a CO_2 atmosphere an equilibrium is maintained between the medium and the gas phase. A bicarbonate -CO_2 buffering system is most often used due to its low toxicity towards the cells. Hepes, a much stronger buffer, may also be used, however, in this case much greater concentrations of Hepes than bicarbonate are required when used in a CO_2 atmosphere.

Each type of media has a recommended (by the supplier) bicarbonate concentration and CO_2 tension to achieve the correct pH and osmolarity. Nevertheless this may vary slightly between laboratories, therefore a sample of media should be left under the normal incubating conditions and monitored overnight, the buffering can then be adjusted accordingly.

Hepes buffer should normally be used in conjunction with bicarbonate for which a relationship between the Hepes and bicarbonate exists for differing CO_2 levels, although, Hepes alone can maintain pH in the absence of exogenous CO_2. The addition of 5 mM pyruvate to the medium increases the endogenous cellular production of CO_2 and limits the need for a CO_2 atmosphere. Some defined media have been devised for this purpose, e.g., Leibovitz L-15 medium used for hepatocytes. Cells which produce large amounts of endogenous CO_2 under certain incubation conditions may require Hepes to buffer this CO_2 product.

The density of the culture may affect the CO_2 requirement, however, in general phenol red in the medium will indicate the state of the pH at any given time.

2.2.3 Glutamine and Amino Acids

In addition to buffering the medium, there are other growth requirements including amino acids, the requirement for which may vary with cell culture type. Commonly the necessary amino acids include cysteine and tyrosine, but some non-essential amino acids may be needed. Glutamine is also required by most cell lines and it has been suggested that cultured cells use glutamine as an energy and carbon source in preference to glucose, although glucose is present in most defined media. Glutamine is usually added at a final concentration of 2 mM, however, once added to the medium the glutamine is only stable for about 3 weeks at 4°C.

2.2.4 Serum

Although there is much research aimed at attempting to reduce the requirements of cells for serum, by alternative supplementation of the media, it is apparent that most cell lines still require serum for adequate growth. Various sources of serum may be used such as calf, foetal calf, and horse. Many continuous cultures utilise calf serum, but often foetal calf serum provides the best growing conditions. The level of serum used depends on the particular cell line and should be determined empirically.

Batches of serum may vary considerably in their ability to support cellular growth. It is therefore important to test batches of serum and have sufficient quantities stored at $-20°C$ of a batch that shows suitable growth supporting characteristics. To check these properties, cloning efficiency and growth characteristics (morphology, growth patterns) should be carried out over a range of 2% to 20% serum as this wide range will indicate if any alterations in the concentration of serum is possible to give optimal growth characteristics for a particular cell line. Details of these techniques are given in Section 3.3.2.

2.2.5 Antibiotics and Antimycotics

Unless good sterile conditions can be maintained (e.g., using laminar flow cabinets) it is necessary to incorporate antibiotics and antimycotics into the media. A wide range of suitable preparations are available from relatively specific antibiotics, e.g., penicillin/streptomycin solutions, to broader spectrum antibacterial/antimycoplasmal agents such as kanamycin. The antibiotics chosen should clearly not be toxic to the cells in culture and may depend on the type of contamination experienced in the individual laboratory.

2.2.6 *Supply of Media and Preparation*

The choice of culture medium used will depend on the cell line and the incubation conditions. However, it is best to start with the medium recommended by the original supplier of the cells. To change the medium necessitates investigating factors such as growth curves and cloning efficiency as described below. To effect a change in culture conditions from one medium to another it is advisable to 'condition' the cells, by increasing the ratio of new to old medium with successive passages.

Culture media have a limited storage life and the recommendations indicated by the supplier should be followed. Liquid defined media may have a storage life at +4°C of up to one year, while glutamine lasts only 3 weeks at +4°C. Serum lasts for about one year at −20°C.

Culture media can be supplied in powder form which requires dissolving and filter sterilising, as a 10x concentrate liquid which requires dilution prior to use, or as a 1x liquid media. Media preparation requires high quality water as described in Section 2.1.8.

Bottles of media should be supplemented in small batches, for instance two weeks supply at a time. This will ensure that constituents such as glutamine do not have time to deteriorate and also that if contamination should occur it is confined to a few bottles.

2.3 Biological Starting Materials

2.3.1 *Cell Lines*

Most cell lines in common use are commercially available from companies such as Flow Laboratories (Irvine, UK) and Gibco Ltd. (Paisley, UK). Full details of recommended media, growth conditions and split ratios for subculturing are supplied with the cells. Less common cell types may be obtained from the American Type Culture Collection (Rockville, MD, USA) and the National Collection of Animal Cell Culture (NCACC, Porton Down, UK).

2.3.2 *Primary Cultures*

Fetal or neonatal tissues are the most easily dissociated and are more likely to produce dividing cultures but they tend to be less differentiated and adult tissues may be more relevant for specific studies. Tissue dissociation is generally achieved by digestion with enzymes such as trypsin, collagenase or other proteases. Maximal exposure to the enzymes may be achieved either by perfusing an entire organ or by dissecting the tissue into thin slices or small fragments which are then incubated with the enzymes. Tissues should be as fresh as possible. Accident victims, suitable for donating organs for transplant purposes are probably the best source of human tissue.

2.4 Sterile Techniques

It is essential that all procedures are carried out using aseptic or sterile techniques.

As described in Section 2.1.4, laminar flow facilities or sterile rooms provide a suitable environment, but even then sterile techniques should be employed. If a non-sterile environment is used, then the reliance on sterile techniques is very high. Basic aseptic technique is to ensure that the work area is clear, swabbed down regularly with 70% ethanol and that all the equipment used has been sterilised.

Clean laboratory coats are also essential. All manipulations of media may require flaming the pipettes and bottle necks if working on the open bench. It is not possible to describe in detail sterile techniques, these are better learnt by watching an experienced worker.

2.5 Culture of Chinese Hamster Ovary Cells

The techniques described here apply to most established cell lines which grow in monolayer culture, except that the preferred culture media will vary. Reagents and equipment required for the routine maintenance of CHO cells are shown in *Tables 1* and *2*.

2.5.1 Subculturing

When monolayer cultures have grown to confluency (occupy all the available substratum) they must be subcultured. This is usually achieved by trypsinisation as follows:

(i)   Remove the medium from the cells by careful aspiration, wash with PBSA and then add sufficient trypsin/EDTA solution to cover the cells.

(ii)   Leave for 15 s then remove by pipette and incubate the cells at 37°C for 5 to 15 min, until they have a rounded appearance and slide freely.

(iii) Add fresh culture medium and gently resuspend by titurating with a pipette.

Table 1. Equipment for Culture of CHO Cells.

1.	CO_2 incubator at 37°C with 5% CO_2:95% air, and 99% relative humidity.
2.	Phase contrast inverted microscope.
3.	Light microscope and haemocytometer or a Coulter counter.
4.	Liquid nitrogen freezer.
5.	Programmable freezer[a].

[a]Optional.

Table 2. Reagents for Culture of CHO Cells.

1.	Phosphate buffered saline (PBS A)
	137 mM NaCl
	2.68 mM KCl
	1.47 mM KH_2PO_4
	8.10 mM Na_2HPO_4
	or prepare from tablets obtained from Flow Labs[a].
2.	McCoys 5A culture medium[a].
	Supplemented with 10% fetal calf serum[a], 100 μg/ml Kanamycin sulphate[a], glutamine and sodium bicarbonate as instructed by the supplier.
3.	Trypsin/EDTA (0.05%:0.02%)[a].
4.	Dimethylsulphoxide.

[a]Supplied by manufacturers of tissue culture media such as Flow Labs. Ltd. (Irvine, Scotland) and Gibco (Paisley, Scotland).

(iv) Count the cells (see Section 2.1.10), dilute to 5×10^4 cells/ml and plate out in new culture vessels at 10^4 cells/cm^2. CHO cells plated at this density should grow to confluency in 3 to 4 days although other cell types may require different plating densities.

Cells that do not detach by this method may respond to either (a) use of 0.25% crude trypsin instead of the usual trypsin/EDTA solution (*Table 2*) or (b) incubation in excess trypsin/EDTA or trypsin alone followed by addition of an equal volume of complete medium (serum contains a trypsin inhibitor). The suspension must then be centrifuged to pellet the cells, the medium discarded and the cells resuspended in fresh medium before plating out.

2.5.2 *Storage*

(i) Cells should be grown until not quite confluent (i.e., still in the exponential phase of growth) and checked for the absence of contamination (Section 3.1).

(ii) Trypsinise the cells as above (Section 2.5.1) and resuspend in fresh medium at 2.5×10^6 cells/ml.

(iii) Add DMSO to fresh medium in equal volumes to give 50% DMSO and then mix one part of this with 4 volumes of the cell suspension. This gives a final concentration of 10% DMSO and 2×10^6 cells/ml.

(iv) Dispense into liquid nitrogen ampoules using heat-sealed ampoules for liquid phase storage or screw-top ampoules for vapour phase storage.

(v) Cool the cells at approximately 1°C per min down to -70°C as described in Section 2.1.7. Cells are then stored at -196°C.

(vi) To thaw the cells after storage, immerse the ampoules in water at 37°C.

(vii) Transfer the cells to a clean vessel and slowly dilute by the addition of 19 volumes of fresh culture medium. This reduces the DMSO concentration to 0.5% which is nontoxic to most cell types.

(viii) Plate the cells out into culture dishes or flasks and then after a 24 h attachment/recovery period, replenish the medium.

The cells can then be grown to confluency and subcultured as normal. It is advisable to passage cells once after thawing to ensure that they are growing normally before use in toxicity studies.

2.6 Isolation of Rat Hepatocytes

There are many methods of hepatocyte isolation, mostly variations based on the collagenase perfusion method of Berry and Friend (4). The method described here requires the minimum of specialized equipment and was adapted in our laboratory from the method of Rao *et al.* (5). The equipment and reagents required for the isolation of adult hepatocytes are listed in *Tables 3, 4* and *5*. An alternative procedure for fetal rat hepatocytes, where whole liver perfusion is impracticable, is described in Appendix I. This procedure is unsuitable for adult hepatocytes, giving a low cell yield with poor viability. Hepatocytes may be used either freshly isolated in suspension for periods of up to 5 h, or in maintenance cultures for longer (up to one week). If freshly isolated cells are being used the preparation need not be carried out under aseptic conditions.

Table 3. Equipment for Hepatocyte Isolation

1.	Peristaltic pump[a] (with flow rates 10−50 ml/ml), e.g., Watson-Marlowe 502S.
2.	37°C reciprocal shaking water bath.
3.	5% CO_2:95% air.
4.	Dissection equipment.
5.	18G and 20G cannulae.
6.	Suture cotton.
7.	Nylon mesh (such as bolting cloth) 125 μm pore size ('Nybolt', John Staniar and Co., Manchester).
8.	Bench centrifuge.

[a]Sterilise perfusion tubing by flushing with 70% ethanol followed by sterile bicarbonate buffer.

Table 4. Reagents for Hepatocyte Isolation.

1. 70% ethanol.
2. Ca^{2+} free bicarbonate buffer
    - 142 mM NaCl
    - 4.37 mM KCl
    - 1.24 mM KH_2PO_4
    - 24.0 mM $NaHCO_3$
    - 0.62 mM $MgSO_4$
    - 0.62 mM $MgCl_2$
    + phenol red until just pink in double distilled or reverse osmosis water.
3. Collagenase solution.
    75 mg collagenase[a] in 150 ml of bicarbonate buffer plus 5.0 mM $CaCl_2$.
4. Phosphate buffered saline (PBS A).
    - 137 mM NaCl
    - 2.68 mM KCl
    - 1.47 mM KH_2PO_4
    - 8.10 mM Na_2HPO_4
    or prepare from tablets obtainable from Flow Labs.
5. Trypan blue dye.
    0.5% in 0.85% saline.

[a]*N.B.* Sources and batches of collagenase vary considerably and may result in poor yields and viabilities. Boehringer (Lewes, Sussex) and Worthington (Harrow, Middlesex) are generally considered to be the reliable suppliers.

In addition, freshly isolated hepatocytes may be stored for 24 h in the presence of substrates and albumin at 4°C with no apparent loss of functional viability (6).

2.6.1 Perfusion

See Chapter 2 for a more complete description of perfusion technique.

(i)   The bicarbonate buffers should be maintained at 37°C and bubbled with 5% CO_2:95% oxygen to ensure full oxygenation and correct pH throughout the perfusion.

(ii) Administer a lethal dose of Sagatal (May and Baker, Dagenham, UK) or other suitable anaesthetic to the rat (virtually any size of animal can be used with this method).

(iii) When under deep anaesthesia swab the abdomen with 70% ethanol and open the peritoneal cavity by a mid-lateral incision.

Table 5. Media for Hepatocyte Incubation.

Medium	Constituents	Recommended Use
1. Krebs-Henseleit	118 mM NaCl 4.8 mM KCl 3.2 mM CaCl$_2$ 1.2 mM MgSO$_4$ 1.0 mM KH$_2$PO$_4$ 24 mM NaHCO$_3$ 2.5 mM Hepes[a]	Basic balanced salt solution for short incubation of isolated hepatocytes.
2. Supplemented Krebs-Henseleit	as above + 0.1% glucose and 2% bovine serum albumin Essential amino acids may also be required for certain purposes	Maintains viability of isolated cells for longer than basic Krebs-Henseleit. Protein may cause interference with some assays.
3. Leibovitz L-15 medium	As supplied commercially by manufacturers of tissue culture media[b]	Will maintain viability of isolated cells for longer than basic Krebs-Henseleit. Contains phenol red which may interfere with some assays.
4. Complete L-15 medium	Leibovitz L-15[b] 10% fetal calf serum 10% tryptose phosphate broth 100 μg/ml 100 U/ml penicillin/ streptomycin 10^{-6} M insulin[c] 10^{-5} M hydrocortisone[c]	

[a]Hepes may be omitted but if so solution requires gassing with 5% CO_2 before use in order to attain correct pH.
[b]L-15 medium may be supplied with or without glutamine. Without glutamine the medium has a longer shelf-life but requires addition of 2 mM glutamine before use.
[c]Optional.

(iv)   Displace the intestines to the right to expose the hepatic portal vein.

(v)   Place a ligature around the inferior vena cava but do not tighten.

(vi)   Cannulate the hepatic portal vein (20G cannula), allow blood to fill the cannula and then connect it to tubing primed with bicarbonate buffer so that no air bubbles pass into the liver.

(vii)   Perfuse the bicarbonate buffer using a peristaltic pump at 25 – 30 ml/min.

(viii)   Allow the liver to swell briefly then sever the inferior vena cava below the ligature. The liver should blanche rapidly and evenly.

(ix)   Open the thoracic cavity and dissect away the central portion of the rib cage.

(x)   Cannulate the vena cava through the right atrium (18G cannula) then tighten the ligature around the inferior vena cava so that the perfusate voids through the upper cannula and runs to waste.

(xi)   After perfusing with bicarbonate buffer for approximately 20 min, stop the peristaltic pump and transfer both the inlet and outlet tubes to the collagenase solution.

(xii)   Restart the pump at a flow rate of 10 – 15 ml/min and allow the collagenase to recycle for 15 – 20 min until the liver is soft.

2.6.2 *Separation of Hepatocytes*

(i)   Carefully dissect out the liver, avoiding rupture of the outer capsule and transfer to a beaker of PBS A.

(ii)   Comb through the liver with open scissor points and agitate gently to disperse the cells.

(iii)   Filter the cell suspension through nylon mesh and then allow the cells to settle under gravity.

(iv)   Aspirate off the supernatant (containing the majority of Kupffer cells, fibroblasts and non-viable heptocytes) and wash the hepatocytes twice by resuspending in PBSA and centrifuging at 50 g for $1-2$ min.

(v)   Finally resuspend in $50-100$ ml of the required medium (see *Table 5*) and determine the yield and viability by trypan blue exclusion as in Section 3.2.1. A 250 g rat should yield $4-8 \times 10^8$ hepatocytes with viability greater than 90%.

2.7 **Primary Maintenance Cultures of Rat Hepatocytes**

(i)   Isolate the hepatocytes as in Section 2.6 using aseptic techniques throughout.

(ii)   Resuspend hepatocytes in complete L-15 medium (*Table 5*) and dilute to a cell density of 0.5×10^6/ml.

(iii)   Plate out at approximately 8×10^4 cells/cm^2 in tissue culture vessels such as Falcon Primaria 25 cm^2 flasks or 24 well plates.

(iv)   Incubate at 37°C (CO$_2$ is not required) for $2-3$ h during which time $60-70$% of the cells will attach to the plastic surface, then carefully remove the medium and replace with a fresh volume.

(v)   Cultures may be treated with test chemicals at this stage or allowed a further recovery period overnight before treatment.

(vi)   The medium should be changed the following day and thereafter may be renewed daily if required but this is not essential.

In a healthy culture the cells will spread out after 24 h with obvious intercellular spaces. The monolayer should be almost confluent but some spaces will remain. Few rounded dead cells should be present. Hepatocytes from adult rats are generally non-dividing so mitotic figures will not be common. After $4-5$ days in culture small colonies of fibroblasts may become apparent and may start to exclude the hepatocytes. This feature is a major constraint on the useful life-time of primary hepatocyte cultures.

2.8 **The Problem of Cytochrome P450 Maintenance**

The problem of the maintenance of 'normal' levels of cytochrome P450 in hepatocyte cultures is obviously significant if they are to be used as a model system for toxicological studies. Unfortunately we cannot detail a method of avoiding this problem but we believe it is too important not to be mentioned here. Control forms of cytochrome P450 decline rapidly between 8 and 24 h of culture. Depending on the culture conditions, an increase may be observed of highly active multiple molecular forms, similar to those induced by polycyclic aromatic hydrocarbons. Several attempts to maintain the normal population of cytochromes P450 have been made, mainly by manipulation of the culture medium. In particular Paine and colleagues have shown that omission of the cysteine and cystine and supplementation with 5-aminolevulinic acid partially maintains the apparent P450 levels (determined according to the reduced carbon monoxide difference spectrum), but the activity towards different substrates changes with time of culture (7,8). Other methods such as hormonal supplementation or growth on collagen rafts have had similar results (9,10). Co-culture of hepatocytes with a rat liver epithelial cell line appears to be more promising, but a full study of the molecular forms of P450 in the cultures has not yet been made (11).

Thus, although several laboratories are investigating this problem, it is still not possible to maintain the levels of P450 isozymes in the same proportions as found *in vivo*. Cultured hepatocytes are therefore of limited value for studies of P450 mediated metabolism of foreign compounds. However, toxicity of most chemicals requiring metabolic activation may still be observed in this model as cultured hepatocytes will be exposed to a compound for much longer than either freshly isolated cells (which lose viability after a few hours) or the liver in the whole animal (where the exposure time depends on the rates of absorption, excretion etc.).

Probably the ideal solution to cytochrome P450 maintenance requires genetic engineering to produce a cell line which combines rapid division with the normal differentiated function of freshly isolated hepatocytes. This would obviate the requirement for a supply of fresh liver for each experiment, and also provide a suitable line for studies requiring dividing cells and a metabolising system (see below).

2.9 Metabolic Activation

Chemicals may exert their toxic effects either directly or through formation of one or more active metabolites. Established cell lines such as CHO cells tend to have very low levels of some of the enzymes involved in metabolic activation. In particular the population of cytochromes P450 is different from, and much lower than, that found in most tissues *in vivo*. Consequently cell lines may be insensitive to toxic chemicals which are activated by these routes. Liver is the major site of xenobiotic metabolism but other tissues, such as intestine, lung, kidney and skin, may also be important. Within a particular tissue the enzymes may be concentrated in specific cell types (e.g., hepatocytes in the liver, Clara cells and epithelial type II cells in the lung). Within the cells different enzymes may be associated with different subcellular organelles or intracellular compartments. Cytochrome P450-dependent enzymes are located primarily in the endoplasmic reticulum, whereas some conjugating enzymes (phase II metabolism) such as glutathione and sulphur transferases are predominantly cytosolic. It is often considered that cytochrome P450-dependent oxidation is normally an activation route, and conjugation a detoxication route. However, this is by no means inevitable; examples of the converse situation are also known.

Ideally, a metabolising system should produce all the possible metabolites that could be formed in any tissue in man. In practice this cannot be achieved and the compromise of using rodent liver is commonly made. For mutagenicity studies in particular, the potential for metabolite production is often increased by pretreating rats with Aroclor 1254 (a mixture of polychlorinated biphenyls) which induces virtually all major forms of hepatic cytochromes P450. Rat liver preparations which may be used as metabolising supplements to cell culture systems include: whole homogenate; 9000 g superna-

Table 6. Specific Equipment and Media for Preparation of S9 Fraction and S9 Mix.

1. Aroclor 1254 (200 mg/ml in corn oil).
2. 1.15% KCl in 0.1 M sodium phosphate buffer.
3. Potter-Elvehjem homogeniser fitted to a motor with a speed of 2400 r.p.m.
4. Refrigerated centrifuge capable of 9000 g average (e.g., MSE High Speed 18).
5. *S9 Mix*

	Volume	Final Concentration
Serum-free culture medium	28.7 ml	
S9 (30 mg/ml)	1.0 ml	1 mg/ml
NADP (80 mM)	0.15 ml	0.4 mM
G6P (100 mM)	0.15 ml	0.5 mM
Total	30 ml	

tant (S9, post mitochondrial supernatant); 15 000 g supernatant (S15); 104 000 g pellet (microsomal fraction); or intact cells (usually hepatocytes). Each of these preparations has some disadvantages. Subcellular fractions are easy to prepare in large batches and can be stored frozen until required, but require the further addition of cofactors. It is generally only possible to optimise the conditions and cofactors for one type of metabolic route (e.g., oxidation), which may be inappropriate if an unknown chemical is being investigated. Intact cells maintain the metabolising enzymes in their physiological concentrations and spacial relationships, and remain viable for longer than subcellular fractions, but are technically more difficult to prepare. They must be prepared immediately before use, and highly reactive metabolites may bind to components within the metabolising cell itself and therefore will be less available to the response cell system.

Only a detailed description of the preparation of aroclor-induced rat liver S9 fraction will be given here. Descriptions of the preparation and use of microsomal fractions preparation are given in Chapter 8.

2.9.1 *Preparation of Rat Liver S9 Fraction*

(i) The equipment and solutions required to obtain aroclor-induced S9 fractions are shown in *Table 6*.

(ii) Give rats a single i.p. injection of 500 mg/kg aroclor 1254 in corn oil (Mazola) dissolved at 200 mg/ml, five days before sacrifice.

(iii) Remove the livers aseptically and place into preweighed sterile ice-cold 1.15% KCl (in 0.1 M phosphate buffer pH 7.4) and weigh.

(iv) All subsequent procedures are carried out at 4°C and under sterile conditions.

(v) Wash the liver with two changes of buffer, and finally add 3 ml of buffer/g wet weight of liver.

(vi) Finely chop the liver with sterile scissors and then homogenise using no more than 8 strokes of a Potter-Elvehjem homogeniser. During the homogenisation, surround the vessel by ice water to keep the temperature of the homogenate at 4°C.

(vii) Then centrifuge the homogenate at 9000 g for 10 min in a high-speed centrifuge at 4°C and decant off the supernatant (S9). The S9 fractions from a minimum of three animals should be pooled to eliminate individual animal variation.

2.9.2 *Characterisation of S9 Fraction*

(i) Determine the protein in the S9 fraction by the method of Lowry (12).
(ii) Normally the fraction is diluted to give a protein concentration of 30 mg/ml for storage, although it may be stored at the initial protein concentration without loss of activity.
(iii) Determine the sterility of the preparation, either by:
  (a) Spread 0.5 ml of the fraction onto a nutrient agar plate (Difco, East Moseley, UK), incubate it for 48 h at 37°C and then visually inspect looking for both bacterial and fungal growth; or
  (b) Add 0.5 ml of the S9 fraction can be added to 5 ml of the culture medium that is to be used in assays, and incubate in a suitable culture vessel (e.g., Petri dish) at 37°C for 48 h to assess contamination.

To ensure that the preparation is metabolically competent, it is necessary to characterise it prior to use. In addition, this serves as a means to ensure that each new preparation is of similar quality. The minimum requirement for assessing metabolic capability is to measure the cytochrome P450 binding spectrum. This is achieved in an S9 fraction by measuring the difference in absorbance at 450 nm minus 510 nm of the difference spectrum of the CO-complex of reduced cytochrome P450 minus the non-reduced CO-complex using an extinction coefficient of $100 \text{ mM}^{-1} \text{ cm}^{-1}$. In this way interference by haemoglobin present in the fraction is prevented (13).

In addition to a cytochrome P450 binding spectrum, enzyme activities may be determined to characterise the preparations. This is discussed in Chapter 9 with respect to microsomal fractions and can equally be applied to S9 fractions.

2.9.3 *Storage of S9 Fraction*

The S9 fraction may be stored either in a $-70°C$ freezer, or a $-196°C$ (liquid N_2) freezer. It is usually stored in 2 ml aliquots in sterile screw-top ampoules and then rapidly frozen directly in the appropriate freezer. Storage of S9 fractions (or microsomes) for periods of up to 7 months is possible (14) although loss of some metabolic capability may occur. S9 fractions should be thawed at room temperature and then kept left on ice until required.

2.9.4 *Use of S9 with Cultures*

The S9 fraction requires NADPH for metabolic activity. This is provided by adding an NADPH regenerating system consisting of glucose-6-phosphate and NADP, thus utilising the glucose-6-phosphate dehydrogenase present in the soluble fraction to produce NADPH. While a number of variations in cofactor concentrations may be used, the following recipe is used for co-incubation with mammalian cells in our laboratory. NADP (80 mM) and glucose-6-phosphate (100 mM) are prepared in distilled water and filter sterilised through a 0.22 μm filter. Excess solution can be stored at $-20°C$ for one week, but should not be refrozen after thawing. The proportions used with S9 fraction to give the S9 mix are given in *Table 6*. The S9 mix is prepared with serum-free culture medium.

To add to the cells in culture, the culture medium is removed from the cells to be treated and replaced with S9 mix and incubated for 2 h. After 2 h, the S9 mix is removed and the cells washed twice with complete (serum added) medium and then the complete medium is replaced.

Incubation for longer than 2 h results in toxicity of the S9 mix towards the cells. In some instances, if excessive toxicity is observed, the use of S15 (with less lysosomal enzymes present) will reduce the toxicity of the metabolising fractions.

2.9.5 *Co-cultures of Hepatocytes with Other Cells*

Primary hepatocytes may be co-cultivated with a variety of other cells including V79 Chinese hamster lung fibroblasts (15) and human fibroblasts (16). Co-cultivation techniques will vary slightly with cell type, but in general the following technique will be suitable for fibroblast co-cultivation.

(i)   To co-cultivate with hepatocytes, incubate the cells in culture in 25 cm² flasks at 5×10^5 cells/flask for 18 h, after which time the medium is replaced by 2×10^6 viable hepatocytes in L-15 medium containing 25 mM Hepes buffer (15).

(ii)   After the incubation period (maximum of 18 h), dissociate the cell layer with trypsin/EDTA (Section 2.5.1).

(iii)   The hepatocytes may be distinguished from the cultured cells by the smaller size of the cultured cells.

(iv)   The cultured cells may then be reseeded at the appropriate concentration for use.

(v)   The hepatocytes are unlikely to survive the enzyme dissociation and reseeding step, however, a slow centrifugation (50 g for $1-2$ min) will ensure removal of the majority of the hepatocytes prior to reseeding.

This method has been shown to be suitable for V79 cells, and is likely to be appropriate for CHO cells as well. Minor variations may be necessary for use with other cell lines.

3. CHARACTERISATION OF CULTURES

The characterisation of a cell line serves two purposes. It ensures that any transformation or other changes in the culture may be noticed, the cultures terminated and a fresh culture established from frozen stocks. Also, changes in culture conditions (e.g., the media), that may affect the cell growth can be observed. For media components such as serum, that can vary widely from source to source it is important to maintain uniformity in culture conditions. The presence of unseen contaminants may be indicated by a change in culture characteristics.

3.1 **Contamination**

Cultures can be contaminated by fungus or bacteria, which most often become visible to the naked eye or under a low-power microscope, or by mycoplasma, which is visible only if specially stained with fluorescent dyes such as Hoechst 33258 and observed under a microscope, or can be identified in scanning electron micrographs of cultures. Contamination may alter the growth characteristics drastically, causing pH changes and growth retardation. It is often detected as a pH change, scum or film on surfaces of medium, cloudiness in the medium, or spots on the growth surface that dissipate when the flask is moved. Unseen contamination may affect cellular metabolism if left undetected.

3.1.1 *Identification of Mycoplasma*

Mycoplasma contamination cannot be seen by the naked eye, and therefore has to be checked regularly (3 monthly). A number of methods are available, but the easiest and most reliable method is to stain the DNA by a fluorescent stain, Hoechst 33258, which reveals mycoplasma as fine particles or filamentous staining over the cell cytoplasm at ×500 magnification. Other microbial contamination can also be detected in this way as they will of course also be stained. The nuclei of the cultured cells will be also stained. An advantage of this method is that kits are available with control slides for comparison from the standard media suppliers (e.g., Flow or Gibco).

3.1.2 *Handling of Contamination*

Basic good sterile techniques do not always prevent contamination. If contamination occurs, affected cultures should be disposed of into 2.5% hyperchlorite solution. Media bottles known, or suspected, to be contaminated should be disposed of also. Some basic preventative measures may be taken: checking sterilising procedures (autoclave and oven procedures); checking sterility of laminar flow hoods; regular checks on cultures; each worker having his own set of media; and disposal of contaminated cultures rather than attempting to decontaminate them.

3.2 **Viability Indices**

When cells are freshly isolated from a tissue the proportion of living, or viable, cells should be determined before they are used. This is most often determined by assessment of membrane permeability, under the assumption that a cell with a permeable membrane has suffered severe, irreversible damage.

3.2.1 *Trypan Blue Exclusion*

This is a rapid test for gross damage which is conveniently combined with determining cell number.

(i)   Mix a small aliquot of cell suspension with an equal volume of 0.5% trypan blue solution.

(ii)   Then, within 1 − 5 min, introduce the suspension into a haemocytometer chamber as described in Section 2.1.10.

(iii)   Non-viable cells appear blue, with the nucleus staining particularly darkly.

(iv)   Count the viable (unstained) cells and the total number of cells as described in Section 2.1.10.

(v)   Express the number of viable cells as a percentage of the total. Calculate the cell number as described in Section 2.1.10 after multiplying by 2 to allow for the dilution with dye.

Naphthalene black can be used as an alternative to trypan blue.

Table 7. Reagents and Equipment for Determination of LDH Leakage.

1.	Solubilizing Agent
	0.9% NaCl
	0.1% Triton X-100
	0.1% Bovine serum albumin
2.	Substrate
	3.5 g K_2HPO_4
	0.45 g KH_2PO_4
	31.0 mg sodium pyruvate
	in 450 ml distilled water (may be stored at $-20°C$)
3.	NADH
	42 mg in 4.5 ml 1% $NaHCO_3$, freshly prepared.
4.	Recording spectrophotometer with cell thermostat equilibrated to 37°C.

3.2.2 *Enzyme Leakage*

Leakage of cytosolic enzymes such as lactate dehydrogenase (LDH) provides a test that is similar in sensitivity to dye exclusion and can be more accurately quantitated but takes rather longer to perform. The reagents and equipment required for this test are shown in *Table 7*.

(i)   Centrifuge an aliquot of cell suspension at a speed suitable for sedimenting all the cells without causing cell damage (e.g., 50 *g* for 2 min for hepatocytes).

(ii)   Transfer the supernatant medium to a clean tube and keep on ice.

(iii)   Add an equal volume of solubilising agent to the cell pellet and vortex mix. Samples should be kept on ice and the assays performed on the same day.

(iv)   To a cuvette, add 3 ml of substrate at 37°C, 50 μl of NADH and $25-200$ μl of sample (depending on the cell type).

(v)   Mix rapidly and follow the increase in absorbance at 340 nm at 37°C.

(vi)   Calculate the initial change in absorbance per min for the medium and for the solubilised cells and express the activity of the medium as a percentage of the total (i.e., to give percent leakage).

3.2.3 *Other Indices*

Other indices that may be measured include:

(i) *Latency of cytosolic enzymes.* LDH activity is measured in whole cell suspensions. Addition of exogenous NADH stimulates LDH activity only in damaged cells, as intact membranes are impermeable to reduced cofactors. Comparison of activity in intact cells and in detergent solubilised cells permits determination of percentage viability (17).

(ii) *Chromium release.* $^{51}Cr^{3+}$ is taken up by viable cells, reduced to $^{51}Cr^{2+}$ and then retained. Non-viable cells release the $^{51}Cr^{2+}$, which can be detected by gamma counting of the cell free supernatant (18).

(iii) *Plating efficiency.* The viability parameters described so far are all based on membrane integrity and are therefore indications of gross damage. Some cells, apparently viable by these methods may not be completely healthy. This will be demonstrated by an inability to attach in culture to the surface of the vessel. Cells should be plated out at normal density and allowed to attach for 24 h before trypsinizing off and counting. Hepatocytes of $90-95\%$ viability according to trypan blue exclusion normally have a plating efficiency of $60-70\%$.

(iv) *ATP content.* The most convenient method is the luminescent assay using luciferin-luciferase (19). Preparations of freshly isolated cells with an ATP content less than 2 μmol/g wet weight should be discarded.

3.3 Assessment of Normality

During the continuous passage of cell lines, changes may take place, the cultures becoming increasingly less like the original cells. It is important for each cell line that series of parameters of 'normality' are recorded so that deviations from this normal behaviour can then be observed. Deviations from this normal behaviour can then be dealt with by going back to frozen stocks of cells. When using the cultures for a specific purpose, if changes in characteristics occur, reproducibility cannot be expected. To monitor for normality a number of parameters may be measured, from simple morphology and growth characteristics to more difficult chromosome characterisation techniques (karyotyping). This latter technique is too specialised to be considered here (see ref. 2).

3.3.1 *Morphology*

This is the simplest method for assessment of cell normality and may be regularly checked using an inverted microscope at least at every passage time. Factors such as growth surface, media and temperature may affect morphology and it is advisable to have photographs of the cell line as a record of normal morphology. It is impossible to describe what a line should look like, but the terms 'fibroblast' and 'epithelial' are loosely used to describe morphology. A fibroblast-like cell, when in monolayer, has a length which is more than twice its width, while a cell that is polygonal with more regular dimensions is regarded as epithelial-like. Cultures which deviate from normal morphology, particularly when they reach monolayer stage, should be discarded.

3.3.2 *Growth Characteristics*

Changes in growth characteristics can easily be detected. The time between medium changes and passaging cells varies between cell lines depending on the growth rate, but should remain constant for each cell line. Alterations in time to reach confluency (i.e., monolayers) indicate changes in the growth characteristics of the cells. A growth cycle pattern should be established for each cell line. This is dependent on the number of cells plated. The cells in a culture are all at different stages of the growth cycle, a clonal growth analysis is required to look at individual cellular growth rates. The cloning efficiency (if each colony grows from a single cell) or plating efficiency can be determined by seeding a low number of cells ($2-50$ cells/cm^2). These will then grow to form small discrete colonies and can be washed, fixed with methanol, then stained with Giemsa stain (2 ml/25 cm^2), washed with tap water and the colonies counted.

The plating efficiency:

$$\frac{\text{Number of colonies}}{\text{Number of cells seeded}} \times 100 = \text{plating efficiency}$$

A colony is only counted over a certain size, i.e., a colony of less than 30 cells is not normally counted.

The plating efficiency can be used to monitor normal cellular growth, recovery from freezing, effects of serum batches.

It is probably not necessary to determine routinely in great detail generation times provided the cells are growing normally, with normal morphology, growth pattern and plating efficiency. Other special methods are available (2).

3.3.3 *Other Parameters*

Other parameters to characterise the cell cultures include chromosome analysis (number, state), DNA, RNA and protein content. Chromosome number is more stable in normal cells than in transformed cells and can be a good indicator that a cell has altered (i.e., transformed). Particular care is required using cell lines such as CHO in that the chromosome number can vary around the normal number of 22. Therefore only large variations may be observed. Details of methods are beyond the scope of this chapter as they are described elsewhere in this book.

4. USE OF MAMMALIAN CELLS IN TOXICOLOGY STUDIES

The diverse applications of mammalian cells in toxicity studies are covered in two reviews (20,21). They are used in a number of different areas of toxicology and can be designed to serve some very specific purposes. Some general points in using mammalian cells are described here.

4.1 **Choice of Cell Type**

The choice of cell type will depend on the purpose of the study. If general toxicity screening for comparison of a series of samples is required then a rapidly growing, easily managed cell line is normally used. The choice may be influenced by a requirement for cells originating from human tissues or from another particular species or from tumour or non-tumour tissue. Fibroblastic or epithelial-like cells may be preferred. Cell lines such as CHO, V79, HeLa, BHK and L929 have frequently been used (20). It should be remembered that established lines such as these may be genetically unstable and a diploid finite line may give more consistent results over the long term in different laboratories [e.g., BCL-Dl human embryo lung cells are used in the FRAME cytotoxicity programme (22) for this reason].

If mechanistic studies on a particular tissue or function are being performed it may not be possible to use cell lines. Many normal cell functions are progressively lost with time in culture. Thus, mechanistic studies are often performed on short term maintenance (non-dividing) cultures in order to retain normal functions as close to the *in vivo* situation as possible.

4.2 Application of the Test Substance

Many test chemicals of low water solubility will need to be dissolved in an organic solvent before addition to the culture medium. The solvent may adversely affect the cells and should be limited to a total concentration of 1% or less. Appropriate solvent controls and medium controls should be included in the test. If S9 mix is required, the enzyme activity of the fraction may be inhibited by many solvents and it is advisable to limit the solvent concentration to 0.1% if possible. Insoluble samples such as particulates and oils present special problems. They may be administered either as suspensions in the medium or dissolved in solvent before addition. These alternatives may result in very different toxicity data. Oily substances such as safrole and clofibrate may, if used at high concentrations, dissolve in the plastic of the culture vessels and the observed toxicity may be partly due to leaching of toxic substances from the plastic.

4.3 Positive Controls

The reproducibility of the response of the test system should be checked by use of positive controls chosen for a particular effect being investigated and the reliability of the effect. Cyclophosphamide is often used as a positive control for metabolic activation but its effect may lack consistency due to its instability. Cycloheximide or dinitrophenol can be used for cytotoxicity screens. For mechanistic studies the controls will depend on the required effect (e.g., clofibric acid or monoethylhexylphthalate for peroxisomal proliferation).

4.4 Parameters of Toxicity

Any of the characterisation methods described in Sections 3.2 and 3.3 may also be used to assess toxicity. The sensitivity of routine toxicity screens may often be increased by allowing cells to continue growing for a few days after treatment. In this situation many parameters dependent on the total number of cells present (such as protein or DNA content) will give similar results. In our laboratory, protein content is mainly used as a convenient criterion of toxicity as large numbers of assays can be performed simultaneously and protein determinations need not be performed immediately. An example of results obtained in our laboratory with CHO as part of the FRAME cytotoxicity programme (22) is shown in *Table 8* .

Toxicity indices for mechanistic studies depend on the cell type in use and are too numerous to cover in detail here (see 20,21). They can be divided into several main categories and some examples of the effects commonly measured in cultured hepatocytes are shown in *Table 9*. The interrelationship of the different parameters has not been fully investigated and the precise sequence of events leading to cell death is not yet known.

5. SIGNIFICANCE AND EVALUATION OF IN VITRO RESULTS

Toxicity in man or in the laboratory animal is subject to a myriad of variables that can not be reproduced in the cell culture system. These variable factors include absorption, distribution, interactions of metabolism and toxicity in different tissues, excretion, and variations due to endogenous (e.g., genetic or hormonal) and exogenous (e.g., diet or environment) factors. Thus, it will never be possible to predict accurately

Table 8. Cytotoxicity Testing in CHO Cells.

Test Chemical	ID50[a] ($\mu g/ml$)	LD50[b] (mg/kg)
Vincristine	0.03	1.3
Cycloheximide	0.018	2
Dinitrophenol	25	30
p-Chloromercuribenzoic acid	32	25

[a]ID50 represents the dose causing a decrease in growth rate resulting in the total protein content being 50% of control after 3 days.
[b]LD50 in rat (mouse for PCMB) by oral route (25).

Table 9. Toxicity Indices in Cultured Hepatocytes.

Type of Effect	Example
1. Membrane damage	Trypan blue uptake. Cytosolic enzyme leakage. Cr^{2+} release. Changes in 'Calcium pump' mechanism.
2. Alterations in synthesis or degradation of macromolecules	[^{14}C]Leucine incorporation into protein. [^{3}H]Uridine incorporation into RNA.
3. Modifications in metabolism capacity	ATP levels. NADP/NADPH ratios. Glutathione content. O_2 consumption. Rate of oxidative metabolism from [^{14}C]glucose to $^{14}CO_2$. Drug metabolising enzyme activities. Lipid peroxidation activity.
4. Morphology	Light and electron microscopy.

non-toxic levels of chemicals from their *in vitro* effects. However, a qualitative extrapolation is more feasible. The ranked order of toxicity of a series of related compounds will often be similar *in vivo* to that in *in vitro* systems, permitting elimination of the most toxic of a series following a preliminary *in vitro* screen.

Another obvious value of using mammalian cells is that many cell lines are derived from human cells and the culturing techniques may be applied (with minimal modifications) to human tissues thereby eliminating the problem of extrapolating from laboratory animal data to man. An effect that occurs in both human and animal cell cultures that occurs in the animal *in vivo* is very likely also to occur in man *in vivo*.

It is tempting to compare *in vitro* data with laboratory animal studies but as the latter do not always agree with human toxicity it may be more valuable to correlate *in vitro* and human data directly, whenever possible. Mechanistic studies performed *in vitro* may be valuable in explaining observed toxicity *in vivo* thereby permitting prevention or development of antidotes to the toxic effect.

Despite the more obvious drawbacks of using mammalian cells in culture for toxicity studies, *in vitro* assays are morally and economically more acceptable than *in vivo* systems (23). Well designed specific assays, may in fact, be more sensitive than *in vivo* systems and variables which may affect toxicity can be more readily examined (24).

Studies are progressing to establish the usefulness and limitations of cytotoxicity tests

78

(22). The area of toxicity in cell cultures is expanding fast, however, caution is necessary until it can be shown that *in vitro* tests can accurately predict *in vivo* responses.

6. REFERENCES

1. Venitt,S. and Parry,J.M. (1984) *Mutagenicity Testing: A Practical Approach*, published by IRL Press, Oxford and Washington DC.
2. Freshney,R.I. (1984) *Cultures of Animal Cells: A Manual of Basic Techniques*, published by Alan R. Liss, Inc., New York.
3. Green,A.E., Athreya,B., Lehr,H.B. and Coriell,L.L. (1967) *Proc. Soc. Exp. Biol. Med.*, **124**, 1302.
4. Berry,M.N. and Friend,D.S. (1969) *J. Cell. Biol.*, **43**, 506.
5. Rao,M.L., Rao,G.S., Holler,M., Breuer,H., Schattenberg,P.J. and Stein,W.D. (1976) *Hoppe-Seyler's Z. Physiol. Chem.*, **357**, 573.
6. Farrell,R. and Lund,P. (1983) *Biol. Sci. Rep.*, **3**, 539.
7. Paine,A.J., Hockin,L.J. and Allen,C.M. (1982) *Biochem. Pharmacol.*, **31**, 1175.
8. Lake,B.G. and Paine,A.J. (1982) *Biochem. Pharmacol.*, **31**, 2141.
9. Dickins,M. and Peterson,R.E. (1980) *Biochem. Pharmacol.*, **29**, 1231.
10. Michalopoulos,G., Sattler,G.L. and Pitot,M.C. (1976) *Life Sci*, **18**, 1139.
11. Begue,J.M., Guguen-Guillouzo,C., Pasdeloup,N. and Guillouzo,A. (1984) *Hepatology*, **4**, 839.
12. Lowry,P.H., Roseborough,N.J., Farr,A.L. and Randall,N.J. (1951) *J. Biol. Chem.*, **193**, 265.
13. Estabrook,R.W., Peterson,J., Baron,J. and Hildebrandt,A. (1973) *Methods Pharmacol.*, **2**, 303.
14. Hubbard,S.A., Brooks,T.M., Gonzalez,L. and Bridges,J.W. (1985) in Arlett,C.F. and Parry,J.M. (eds.), *Comparative Mutagenicity Testing: The 2nd UKEMS Collaborative Study*, MacMillan Scientific Publications Ltd., London, p. 413.
15. Jones,C.A. and Huberman,E. (1980) *Cancer Res.*, **40**, 406.
16. Michalopoulous,G., Biles,C. and Russel,F. (1979) *In Vitro*, **15**, 796.
17. Moldeus,P., Hogberg,J. and Orrenius,S. (1978) in *Methods in Enzymology*, Fleischer,S. and Packer,L. (eds.), Academic Press, Inc., London and New York, Vol. **52**, p. 60.
18. Zawydinski,R. and Duncan,G.R. (1978) *In Vitro*, **14**, 707.
19. Stanley,P.E. and Williams,S.G. (1969) *Anal. Biochem.*, **29**, 381.
20. Stammati,A.P., Silano,V. and Zucco,F. (1981) *Toxicology*, **20**, 91.
21. Roberfroid,M. and Krack,G. (1984) *Feasibility of In Vitro Toxicity Testing*, Report to the Commission of the European Communities.
22. Balls,M. and Bridges,J.W. (1984) in *Alternative Methods in Toxicology*, Mary Anne Leibert Inc., New York, Vol. **2**, p. 63.
23. Bridges,J.W. (1980) in *Towards Better Safety of Drugs and Pharmaceutical Products*, Breimer,D.D. (ed.), Elsevier/North-Holland, Amsterdam, New York and Oxford, p. 283.
24. Bridges,J.W., Benford,D.J. and Hubbard,S.A. (1983) in *The Role of Animals in Scientific Research: An Effective Substitute for Man*, Turner,P. (ed.), The MacMillan Press Ltd., Basingstoke, p. 47.
25. Registry of Toxic Effects of Chemical Substances, DHHS (NIOSH) Publication No. 81-116.

APPENDIX

Isolation of Fetal and Neonatal Rat Hepatocytes

Keith Snell and Carole A.Evans

1. INTRODUCTION

Hepatocytes from fetal rats have been prepared by perfusion techniques using either lysozyme (1) or collagenase (2) for enzymatic digestion of the liver tissue. Because perfusion of fetal rat livers is technically demanding, more recent procedures have used chopped or minced fetal liver incubated *in vitro* with collagenase (3), lysozyme (4) or trypsin (5). Since fetal liver *in vivo* contains a high percentage of haematopoietic cells and only about 34% of hepatocytes, by cell number (6), special considerations are involved in obtaining a relatively pure hepatocyte preparation. If the ultimate aim is to culture the hepatocytes, the problem is a minor one since haematopoietic cells do not attach to the culture vessel surface and are therefore removed when the hepatocyte culture medium is replaced after the initial attachment period (see Section 2.7). If the hepatocytes are to be used in the freshly isolated state, then repeated washings and low speed sedimentation are required to enrich the preparation with the higher density and larger cell volume parenchymal hepatocytes. By this means the proportion of hepatocytes can be increased to about 95% by number.

Another problem experienced with fetal hepatocytes is the result of the high DNA content of the fetal cells. DNA liberated from damaged cells as a result of the chopping or mincing of the tissue causes a marked tendency for cells to clump together, creating anoxic cell aggregates and a consequent loss of cell viability and yield. This problem can be overcome by including deoxyribonuclease I in the enzymatic digestion medium (7).

1.1 Isolation Procedure

The equipment and reagents required for hepatocyte isolation are detailed in *Tables 1* and *2*.

Table 1. Equipment for Fetal Hepatocyte Isolation.

1. McIlwain Automatic Tissue Chopper	(Mickle Laboratory Engineering Co. Ltd., Gomshall, Surrey, UK)
2. Dissection Equipment	
3. Reciprocal Shaking Water-bath (set at 37°C)	
4. 5% CO_2-95% O_2 supply	
5. Nylon mesh, 125 μm mesh size	('Nybolt bolting cloth', John Staniar & Co., Manchester, UK)
6. Plastic 50 ml beakers	
7. Bench centrifuge	
8. Pasteur pipettes	

Table 2. Reagents for Fetal Hepatocyte Isolation.

1. Phosphate-buffered saline solution (PBSA)
     137   mM NaCl
     2.68 mM KCl
     1.47 mM KH_2PO_4
     8.10 mM Na_2HPO_4
    for preservation of cellular glycogen levels the medium should be supplemented with 5.5 mM glucose.
2. Solution 1 supplemented with 0.5 mM-EGTA.
3. Hanks balanced salt solution
     137   mM NaCl
     5.4   mM KCl
     0.44 mM KH_2PO_4
     0.34 mM Na_2HPO_4
     5.5   mM glucose
    Solutions $1-3$ should be well gassed with $5\% CO_2 - 95\% O_2$ immediately before use.
4. Collagenase-deoxyribonuclease solution[a].
    Dissolve 5 mg of collagenase and 1 mg of deoxyribonuclease I in 10 ml of solution 3, add 0.2 ml
    of 250 mM $CaCl_2$, a drop of phenol red as indicator and adjust to pH 7.5 by the dropwise addition
    of 2.8% $NaHCO_3$ (N.B. The amount of collagenase used is dependent on the age of the animals: fetal
    day 17 livers, 2 mg; fetal day 18 or 19 livers, 3.3 mg; fetal day $20-22$ and neonatal livers, 5 mg).
5. Trypan blue dye solution
     0.5% in 0.85% saline

[a]Collagenase preparations from Boehringer (Lewes, Sussex, UK) or Sigma (Poole, Dorset, UK) and deoxyribonuclease I from Boehringer have been found to be satisfactory.

(i)   Fetal animals are obtained from pregnant female rats of the appropriate gesta-
        tional age ($15-22$ days) by caesarian section and laparotomy.

(ii)   Sacrifice the fetal or neonatal ($0-2$ days *post partum*) rats by decapitation, dissect
        out the livers and pool in ice-cold PBS solution.

(iii)   Take batches of the pooled livers ($1-2$ g) and cut into 0.5 mm³ cubes using
        an automated tissue chopper, or by slicing with a hand-held microtome blade,
        in two directions at 90° to one another.

(iv)   Incubate the liver pieces in 10 ml of phosphate buffered saline solution (solution
        1) in a 250 ml glass erlenmeyer flask in a shaking water bath (40 strokes/min)
        at 37°C for 10 min.

(v)   Decant off the saline solution and discard, replace with 10 ml of fresh PBSA
        and incubate for a further 10 min as above.

(vi)   Continue this washing procedure for a further two times with 10 ml portions
        of PBSA-EGTA (solution 2).

(vii)   At this time the collagenase-deoxyribonuclease solution (solution 4) should be
        freshly prepared.

(viii) Incubate this enzyme solution with the liver pieces at 37°C for 45 min in the
        shaking water bath (90 strokes/min).

(ix)   At the end of this time, filter the cell digest through nylon mesh (125 μm mesh
        size) fastened over a plastic beaker with an elastic band.

(x)   Agitate the residual liver fragments gently with a Pasteur pipette to loosen any
        detached cells and wash the nylon mesh with about 20 ml of PBSA solution (solu-
        tion 1).

(xi)   Discard any tissue fragments which are not easily dispersed in this way.

(xii)   Centrifuge the cell suspension in plastic 50 ml conical tubes in a bench centrifuge at room temperature at 20 *g* for 2 min.

(xiii) Remove the supernatant using a Pasteur pipette and discard.

(xiv)   Resuspend the cell pellet in about 15 − 20 ml of PBSA (solution 1) by gently swirling the centrifuge tube.

(xv)   If this is not fully effective, dispersion can be aided by gently sucking the suspension in and out of a Pasteur pipette.

(xvi)   Repeat the centrifugation and resuspension steps a further three times, the final wash using the medium in which the cells are to be incubated (e.g., Krebs-Ringer bicarbonate buffer containing 2% defatted albumin).

(xvii) Filter the final resuspension through nylon mesh to remove any large cell aggregates and then dilute with incubation medium to give a cell density of about 10^6 cells/ml, or as appropriate.

(xviii) A portion of the final resuspension is used to determine the cell number and viability by trypan blue exclusion (see Section 3.2.1).

The yield of hepatocytes by this method is about $10 − 20 \times 10^6$ cells/g wet weight of liver. For livers from animals younger than day 17 of gestation, it is recommended that the enzymic digestion step above is replaced by a step involving dispersion of the cells in Hanks balanced salt solution (solution 3) using suction in a Pasteur pipette (8).

2. REFERENCES

1. Hommes,F.A., Oudman-Richters,A.R. and Molenaar,I. (1971) *Biochim. Biophys. Acta.*, **244**, 191.
2. Ferre,P., Satabin,P., El Manoubi,L., Callihan,S. and Girard,J. (1981) *Biochem. J.*, **200**, 429.
3. Yeoh,G.C.T., Bennet,F.A. and Oliver,I.T. (1979) *Biochem. J.*, **180**, 153.
4. Van Lelyveld,P.H. and Hommes,F.A. (1978) *Biochem. J.*, **174**, 527.
5. Schulze,H.P., Huhn,W., Franke,H. and Dargel,R. (1984) *Biomed. Biochim. Acta*, **43**, 1227.
6. Greengard,O., Federman,M. and Knox,W.E. (1972) *J. Cell. Biol.*, **52**, 261.
7. Devirgitiss,L.C., Dini,L., Di Perro,A., Leoni,S., Spagnuolo,S. and Stafanini,S. (1981) *Cell. Mol. Biol.*, **27**, 687.
8. Oliver,I.T., Martin,R.L., Fisher,C.J. and Yeoh,G.C.T. (1983) *Differentiation*, **24**, 234.

CHAPTER 4

Post-implantation Embryo Culture for Studies of Teratogenesis

STUART J.FREEMAN, MARY E.COAKLEY and NIGEL A.BROWN

1. INTRODUCTION

It is axiomatic in teratology to regard those structural abnormalities of foetuses arising from the actions of exogenous agents as being due to a teratogenic insult that has occurred during the early organogenesis period of embryonic development. It is, therefore, of obvious value to developmental toxicologists to study the embryo at this sensitive stage of development and the devising of techniques to culture early organogenesis-stage embryos offers considerable benefits. Although techniques for culturing post-implantation embryos of a number of mammalian species have been developed (1), the most successful methods utilise rat and mouse embryos and this chapter deals exclusively with these.

In rodent embryos, the organogenesis period extends throughout the second trimester of pregnancy (day 7–15 in the rat; day 6–12 in the mouse) beginning immediately after the relatively undifferentiated embryo has implanted in the uterine wall. In addition to describing culture techniques for organogenesis embryos, we have indicated the various applications and limitations of embryo culture in teratological research. For further details of certain applications of embryo culture, the reader is referred to several reviews (1–3).

As used in this chapter the word 'embryo' is meant to include not only those tissues destined to form the foetus, but also the following extra-embryonic structures: visceral yolk sac, amnion, chorion, allantois, placenta (ectoplacental cone), all of which are preserved in cultured embryos. It does not include additional extra-embryonic tissues: parietal yolk sac, Reichert's membrane and trophoblast, which are removed before culture of embryos.

2. BASIC TECHNIQUES

Embryos may be cultured in static liquid medium (watchglass cultures) (1) but superior development is obtained when the culture medium is agitated or circulated. The circulation system developed by New (1) requires rather complex apparatus in which the explanted embryo is fixed in position and the medium circulated to flow over it. Whilst the method is useful for the continuous observation of embryonic development, a simpler and more versatile technique uses a system of rotating bottles for the culture of embryos. Continuous rotation of the bottles permits efficient oxygenation of the liquid medium and further assists embryonic respiration by gently swirling the embryos about in the medium. It is this latter procedure that has been widely employed in teratological studies and is described below.

The details given in this chapter are based upon cultures of rat conceptuses aged 9.5 days at explantation and maintained for 48 h. This is generally the most convenient species and embryonic age for routine or initial studies. As we shall discuss, organogenesis embryos from the rat appear to be easier to maintain *in vitro* than those from the mouse. In addition, the period of 9.5 − 11.5 days embryonic age is that during which the major organ primordia are established and also during which development *in vitro* most closely parallels that *in utero*. Essentially, similar techniques are used for embryos 1 or 2 days younger or 1 day older than this routine explantation stage, and for the mouse at equivalent stages. Important differences are noted where appropriate.

2.1 Equipment

2.1.1 *Rollers and Rotators*

The apparatus used in our laboratory is featured in *Figure 1*. It consists of a set of motor-driven haematological rollers (30 − 60 r.p.m.), housed in a standard tissue culture incubator (4), on which sealed culture bottles are rotated. The incubator is of the anhydric type which we have found more suitable for embryo culture than the water-jacketed incubator. The insulating properties of the latter are so efficient that maintenance of a temperature at 37 − 38°C is made difficult by the extra heat generated by the roller motor. With two sets of rollers in operation, even the anhydric incubator is unable to regulate the temperature and we have found it necessary in these circumstances to leave the incubator door open (internal glass door closed). This problem does not arise when the roller motor is located outside the incubator, an arrangement characteristic of the commercially available (BTC Engineering Ltd., Cambridge) embryo culture apparatus. The latter unit comprises a single set of rollers encased within a Perspex incubator with thermostatically-controlled temperature regulation. Its use may be preferred when only a few cultures are to be maintained at any one time or where there is a restriction in available laboratory space − the unit takes up little room on a laboratory bench.

An alternative arrangement favoured in a number of laboratories uses a rotating rather than rolling apparatus. In its simplest form, the motor-driven rotating plate is designed to accommodate sealed culture bottles, and is contained within a temperature-regulated incubator (2). In a more sophisticated system the culture bottles are unsealed and continuously exposed to fresh gas mixture. The bottles are inserted into a hollow rotating drum through which the gas is circulated. The rotating shaft of the drum projects through the side of the incubator, most often constructed from Perspex, and is driven by an external motor (5). The system was devised because it was anticipated that continuous gassing of culture bottles would provide better conditions for embryo development than would sealed bottles gassed at intervals. However continuous gassing has not been shown to have any clear advantage (5). It may be considered for experiments in which the pCO_2, pO_2 or pH is critical although the osmolarity of the medium has been shown to be more unstable. The rotator apparatus is not commercially available.

Figure 1. Apparatus for embryo culture used in our laboratory. Standard anhydric incubator containing haematological rollers, adapted as described in the text.

2.1.2 *Culture Bottles*

In selecting suitable bottles for embryo culture the following considerations are important. The bottle should have a smooth outer surface to permit even rolling and a smooth inner surface to prevent possible damage to embryos during culture. A bottle with a short neck and deep shoulder is desirable; this allows a reasonable volume of culture medium to be added to the bottle without the danger of droplets of medium, and therefore embryos, becoming lodged in the neck. It also provides for a large volume of gas to be contained within the bottle. The best results are obtained with sterile culture ware and therefore autoclavable bottles and caps are preferable. We have experimented with the use of sterile plastic universal containers as culture vessels, which we have adapted

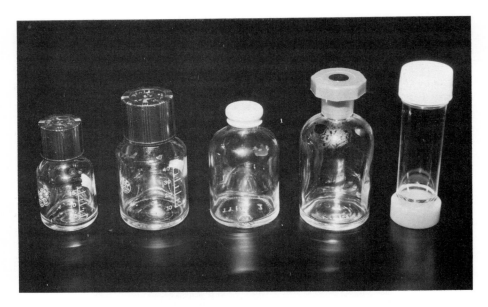

Figure 2. A variety of bottles that can be used as culture vessels. The Sovirel borosilicate bottles with screw-cap tops (50 ml and 25 ml), preferred in this laboratory, are featured at left. At right is the adapted sterile plastic Universal container which has proved useful in certain types of experiments (see text).

for use by adding a plastic cap to the bottom of the tube to permit even rolling, and lining the screw-cap top with a Teflon disc to improve sealing (see *Figure 2*). Such containers are less reliable than the glass bottles but might be preferred for use when radioactive compounds or toxins are included in the culture medium, since they are then easily disposed of. The range of culture bottles that can be used is featured in *Figure 2*. We have obtained consistently good results with the screwcap Sovirel bottles and these are recommended for use. A recent introduction has been the New Brunswick swivel cap (6) which can replace the conventional screw-cap top. The former has five inlets which allow for the continuous infusion and removal of gases and the sampling or changing of the medium without otherwise disturbing the culture.

2.1.3 *Microscopes*

For the fine dissection of the early embryo from decidual tissues a binocular dissecting microscope is essential. The microscope should have a good range of magnification (6 × to 50 ×) to suit explants of various sizes and should provide good access for the dissector to the preparations. Illumination of the microscope stage is best provided from a fibre optic source. To observe embryos in bottles during the culture period an inverted binocular microscope can be used.

2.1.4 *Sterile Area*

The use of sterile conditions for most of the manipulations of embryos and culture serum is advisable, particularly when antibiotics are not added to the serum. In the preparation of serum (see Section 2.4), it is advisable to carry out the entire process, with the exception of withdrawal of blood from the animal, in a sterile area. For explantation of the embryos (Section 2.5), once the individual implantation sites have been isolated, removal of embryos should be performed in sterile conditions, wherein standard tissue culture practice applies.

Laminar flow cabinets provide a sufficient level of containment for all manipulations and are available in various sizes to suit the needs of the laboratory. Alternatively, the custom-built Bass modular downflow hood may be preferred where several operators are working at once. The hood is a free-standing structure enclosing a sterile area within flexible plastic walls and can be constructed to a size suitable for any laboratory.

2.1.5 *Instruments*

Standard surgical instruments are used for most of the explantation procedure. Separation of embryos from the surrounding decidua, removal of Reichert's membrane from embryos and post-culture manipulation of embryos require the use of fine watchmaker's forceps. The forceps must be in pristine condition: damage to the tips drastically reduces their usefulness. For younger embryos (7.5 or 8.5-day rat; 6.5-day mouse) tungsten needles may be preferred.

2.1.6 *Serum Preparation*

Optimum development of embryos in culture is only achieved with serum prepared from blood that has been immediately centrifuged on withdrawal (Section 2.4). It is essential that centrifugation is carried out before the blood has begun to clot. Therefore it is important that plastic ware be used for the preparation since this retards the clotting process. Bleeding of ether-anaesthetised animals (for rat and mouse embryo culture serum, most commonly rats) is performed under a fume hood with a high suction rate, and the blood is withdrawn (Section 2.4) into a suitably-sized plastic syringe (20 ml for male rats; 10 ml female rats; 5 ml mice). A long large bore needle (19 gauge $\times$ 2 inches) permits rapid collection of blood. Fresh blood is immediately transferred to plastic conical centrifuge tubes (12 ml) and centrifuged. Once prepared, the serum is stored in sterile plastic universal containers. Standard bench top centrifuges with swing-out buckets are quite satisfactory for serum preparation. The only additional equipment is a water bath, set at 56°C, for heat-inactivation of serum before use in culture.

2.2 Media and Buffers

2.2.1 *Culture Media*

In the original description of the roller culture method for rat embryos the culture medium comprised 100% homologous rat serum. It was subsequently discovered (1) that the method of preparing the serum was critical to the success of a culture: serum prepared from blood centrifuged immediately after collection gave results superior to conventionally prepared delayed-centrifugation serum. It makes no difference to the success of culture whether serum is prepared from male animals or females, pregnant or non-pregnant (*Table 1*). No adequate replacement for homologous serum has yet been found, although many investigators, including ourselves, have to some extent been able to dilute rat serum without any major adverse effects on cultures. Indeed, some dilution of the serum improves development in culture of 12.5- and 13.5-day-old rat embryos (7).

Table 1. Comparison of Ability of Homologous Sera to Support Rat Embryonic Development in Culture.

Parameter		Pregnancy day on which sera was obtained				Non-pregnant female serum	Male serum	F-value
		10	12	15	17			
Embryo	Protein (μg)	368.1 ± 29.6	332.8 ± 19.9	348.5 ± 16.8	312.5 ± 23.7	316.9 ± 15.4	356.9 ± 18.9	1.15
	DNA (μg)	33.5 ± 2.4	30.9 ± 2.0	35.0 ± 2.3	36.5 ± 2.9	35.3 ± 1.7	40.5 ± 2.3	1.75
	Somite No.	29.8 ± 0.5	29.5 ± 0.4	30.0 ± 0.4	29.5 ± 0.3	29.3 ± 0.3	29.5 ± 0.5	0.43
	Crown-rump length (mm)	5.5 ± 0.1	5.1 ± 0.2	5.2 ± 0.1	5.1 ± 0.2	5.1 ± 0.1	5.5 ± 0.1	1.94
	Head length (mm)	2.6 ± 0.1	2.4 ± 0.1	2.5 ± 0.1	2.4 ± 0.1	2.4 ± 0.1	2.6 ± 0.1	0.91
	Diameter (mm)	6.0 ± 0.1	5.9 ± 0.1	6.0 ± 0.1	6.0 ± 0.2	5.9 ± 0.2	6.1 ± 0.2	0.15
Yolk sac	Protein (μg)	196 ± 17.7	184.6 ± 21.8	193.8 ± 8.1	217.7 ± 17.1	198.7 ± 11.9	231.1 ± 15.9	1.19
	DNA (μg)	2.5 ± 0.5	7.7 ± 0.5	7.7 ± 0.7	6.9 ± 0.6	6.1 ± 0.5	6.9 ± 0.7	1.04

All values are mean ± S.E. Analysis of variance showed no significant differences between values ($p > 0.1$).

In our laboratory, rat serum is routinely diluted to 75% with minimal essential medium (MEM) (Eagles), which we have found to be superior to CMRL-1066, MCDB 104, MDME or DMEM as diluent. Many workers dilute serum to 90% with water and it is possible that serum becomes concentrated through water evaporation during heat-inactivation (see Section 2.4) so that some dilution is necessary to restore the osmolarity of the serum. We have found that reduction of the serum content of the culture medium to less than 75% results in a significant reduction in embryo growth proportional to the extent of dilution, although for short-term cultures further dilution might be permissible. Sheep serum, prepared in the same way as rat serum (Section 2.4) and commercially available horse, bovine, foetal and newborn calf serum have been tested in our laboratory for their suitability as culture media. Partial or total substitution of any of these sera for rat serum in our standard culture medium yields inferior results.

The only heterologous serum so far found to approach the efficacy of rat serum is human serum. Glucose supplementation of human serum to 3 mg/ml is necessary to support embryonic development in culture (8). Even then, human serum is not a complete substitute for rat serum and is apt to give variable results. A mixture of 90% human, 10% rat serum has been found to be a satisfactory culture medium (9); though no more defined than rat serum, its attraction is the reduced requirement for rats, and hence reduced cost. A culture medium of 25% human serum, 25% rat serum, 50% Weymouth's medium has been used successfully in the 24-h culture of 9.5-day rat embryos (10). Mouse embryos have been cultured in various media but the best results are obtained with mouse or rat sera (11). Rat serum is most widely used because of the obvious difficulties in obtaining sufficient volumes of mouse serum.

Little progress has been made in the development of a defined medium for embryo culture. The observations of Cockroft (12), using dialysed rat serum as a culture medium, clearly demonstrated the need for glucose and certain vitamins for normal embryonic development in culture. Our own findings, that glucose- and vitamin-supplemented dialysed rat serum is not a complete substitute for our standard culture medium, suggest that additional micromolecular components of serum, so far unidentified, are essential for normal development.

2.2.2 *Explanation Media*

Embryos are explanted from the uterine decidua in a defined medium containing glucose. In our, as in most, laboratories, Hank's balanced salt solution (HBSS) is used, but MEM is equally good for this purpose.

2.2.3 *Gases*

The oxygen tension of the gas phase in culture bottles is critical to the normal development of the embryo. The requirement for oxygen increases with embryonic age as the embryo develops oxidative metabolic pathways (1); the change in oxygen requirement is outlined in *Table 2*. The gas mixtures used are:

$$5\% \; O_2; \; 5\% \; CO_2; \; 90\% \; N_2$$
$$20\% \; O_2; \; 5\% \; CO_2; \; 75\% \; N_2 \; (\text{or } 95\% \text{ air}; \; 5\% \; CO_2)$$
$$40\% \; O_2; \; 5\% \; CO_2; \; 55\% \; N_2$$
$$95\% \; O_2; \; 5\% \; CO_2$$

Gas mixtures may be obtained as special gases from British Oxygen.

Table 2. Gassing Schedule for Post-implantation Embryo Culture.

Gestational age: Rat	8.5	9.5		10.5		11.5		12.5
Mouse		8		9		10		
Somite number		0 − 4		12 − 15		26 − 29		
pO$_2$ of gas mixture[a]	5% ⟶		20%[b] ⟶		40% ⟶		95% ⟶	

[a]All gas mixtures have pCO$_2$ of 5%; the balance is N$_2$.
[b]This gas mixture may be replaced with 5% CO$_2$:95% air.

2.3 Biological Starting Material

2.3.1 *Timed-pregnant Females*

In our laboratory, the rat colony (Wistar-Porton strain) is maintained in a 10 h/14 h light/dark cycle, the dark cycle beginning at 5 pm. For mating, each male is caged with two females in the late afternoon and the following morning mating is confirmed by checking females for the presence of a vaginal plug. (In the absence of a plug, a vaginal smear will readily detect sperm if present.) Mating is assumed to have occurred at the mid-point of the dark cycle (in our case at midnight) and fertilisation 2 h subsequently (2 am). Regarding the morning of plug detection as the first day of pregnancy, explantation of 9.5-day (headfold) embryos takes place between 2 and 3 pm on the 10th day of pregnancy.

The time scale of mouse embryonic development varies considerably between strains. Embryos of the MF1-F1 strain used in our laboratory reach headfold stage late on the 8th day of pregnancy, embryonic age approximately 8.0 days. For convenience, therefore, mice are maintained under a reversed light/dark cycle, the dark cycle ending in the late afternoon. Regarding the following day as the first day of pregnancy, embryos reach headfold stage at noon of the 8th day. In our hands, mouse embryos develop poorly in culture if explanted before the 4 − 6 somite stage, which is reached by the middle of the same afternoon.

2.3.2 *Animals for Serum Preparation*

Embryos grow equally well in serum prepared from male and pregnant and non-pregnant female rats (*Table 1*). Obviously the larger the animal, the more blood can be obtained; a 500-g rat will usually yield 16 − 20 ml blood, a 250-g rat about 10 ml. Generally speaking, we use the largest males available, preferably retired breeding stock.

2.3.3 *Adult Rats for S9 Preparation*

Young male rats weighing between 150 and 250 g are used in the preparation of a liver S9 mix or purer microsomal fraction (see also Chapter 8).

2.4 Preparation of Serum

(i)   Under a fume hood, anaesthetise the animal by placing it in a gas jar containing ether-soaked cotton wool. Once the righting reflex is lost (1 − 2 min), remove the animal and place it on its back, maintaining anaesthesia by placing a beaker containing a pad of ether-soaked cotton wool over its head.

(ii)   Open the ventral body wall by making an incision in the centre of the lower abdomen and cut in a V-shape up towards the fore limbs. Pull the flap of skin over the chest to reveal the intestines which should be displaced gently to one side. Using a piece of gauze or cotton wool gently rub away underlying fat and mesentery to reveal the dorsal aorta. (Do not confuse the aorta with the vena cava which lies immediately below it and is darker.)

(iii)   Locate the region in which the aorta bifurcates into the iliac vessels and insert the needle of the syringe into the aorta at the point to a distance of about 1 cm. Gently withdraw the plunger to facilitate uptake of blood into the syringe. Gentle pressure need only be applied to the plunger as the pumping action of the heart will force the blood into the syringe. Indeed, too much pressure will collapse the vessel. In such an event insert the needle further into the aorta, beyond the collapsed area; likewise when the initial vigour of blood flow has subsided. If the animal stops ventilating during blood withdrawal, apply gentle pressure to the chest wall to facilitate collection of remaining blood. Of course, if a large volume of blood (10 − 12 ml in male rat) has already been collected, the animal will stop ventilating. It is important not to linger at this stage. The time taken to collect a few more drops of blood may be just long enough for the blood already collected to have begun clotting.

(iv)   When no further blood can be withdrawn, remove the needle swiftly from the aorta and pull back on the plunger to draw blood from the needle into the syringe. Remove the needle from the syringe and quickly dispense the blood into ice-cold sterile plastic screw-cap centrifuge tubes. If more than 8 ml of blood is collected, divide the volume equally between two tubes. Centrifuge the tubes at 2000 *g* for 3 min.

    The procedure is more efficiently carried out by two people: one person concentrates on bleeding the animals while the other anaesthetises the animals and manages the centrifuge. Since the whole procedure from anaesthesia to placing the tubes in the centrifuge takes only 2 − 3 min, it is useful to have a second centrifuge available so that the whole operation is continuous. The bleeding procedure is shown in *Figure 3*.

(v)   After centrifugation the tubes are allowed to stand, at room temperature, for 30 − 60 min or until the serum clot begins to retract from the sides of the tube. Under sterile conditions (see Section 2.1.4), squeeze the clot with forceps to release serum and discard the residual fibrinous material. Re-centrifuge the tubes at 2000 *g* for 5 min.

(vi)   After centrifugation, decant and pool the serum from all the animals used. Aliquot the serum into sterile plastic containers and freeze at −20°C. The serum can be stored for 1 − 2 weeks. For longer storage periods, freeze the serum at −70°C. Antibiotics (penicillin 5000 i.u./ml; streptomycin 5000 μg/ml) may be added to serum before freezing although it is possible that these may influence any teratogenic action under investigation.

(vii) For use in culture, thaw the serum at room temperature or 37°C, and heat-inactivate for 30 min in a 56°C water bath. Loosen the cap on the serum container during heat-inactivation to allow the ether to evolve.

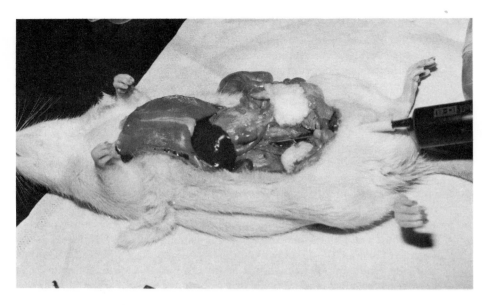

Figure 3. Bleeding procedure in preparation of serum for culture medium. The tip of the needle can be seen just within the aorta.

(viii) Centrifuge the serum for 5 min at 2000 g to sediment the precipitate that forms during freezing. Alternatively the serum may be filtered through a 0.45 μm filter. The serum is now ready for use in culture.

2.5 Explantation and Culture of Embryos

2.5.1 *Explantation*

The various steps in explantation and dissection of embryos are illustrated in *Figure 4*.

(i) Sacrifice the pregnant animal by cervical dislocation or gassing with carbon dioxide and open the ventral body wall as described above [Section 2.4 (ii)].

(ii) Displace the intestines to reveal the uterus containing swellings, and cut across the cervix. Lifting the cervix with forceps, tear the uterus-associated fat and mesentery away with a second pair of forceps, and explant the uterus by cutting just below each oviduct.

(iii) Lay the uterus on a clean sheet of tissue paper and using fine scissors cut through the uterine wall along the antimesometrial edge the full length of the uterus. Care should be taken not to damage embryos by keeping the points of the scissors raised so as not to pierce the implantation sites. Using two pairs of fine forceps, gently pull the uterine wall free from each pear-shaped decidual swelling, and excise each swelling by cutting off that portion of the uterus to which it is attached.

(iv) Transfer the swellings to a dish of sterile HBSS, warmed to 37°C, for dissection of embryos from decidua. All subsequent operations require the use of a binocular microscope and watchmaker's forceps, and should be performed under sterile conditions (Section 2.1).

(v) Tease off the small piece of attached uterus to free the pear-shaped decidual swell-

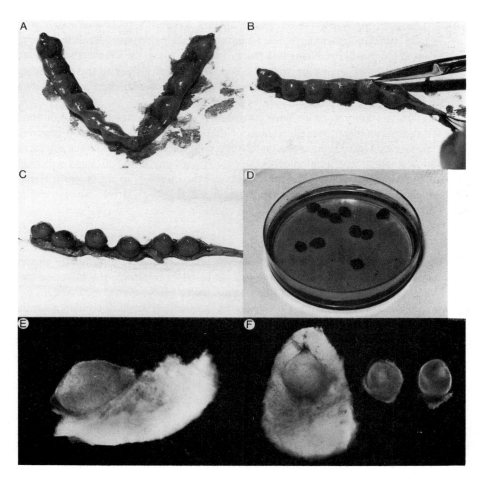

Figure 4. Sequence illustrating explantation of headfold embryos. These embryos, possessing about seven somites, are at a slightly more advanced stage than the 9.5-day embryos usually explanted for culture (see *Figure 5*). (**A**) Freshly explanted uterus containing decidual swellings; (**B**) cutting through wall of the uterus; (**C**) displacement of uterine wall to reveal pear-shaped decidual swellings; (**D**) dish of freshly removed decidual swellings; (**E**) side view of dissected decidual swelling showing position of embryos; (**F**) embryo at various stages of dissection from decidua and outer membranes. Note the Reichert's membrane of the central specimen standing proud of the embryonic endoderm at the embryonic pole. The specimen on the right has been divested of Reichert's membrane and much of the ectoplacental cone and is ready for culture.

ing. To dissect the implant, pierce the bottom of the decidua with forceps in the region where the red colour of the ectoplacental cone is just visible through the decidual tissue. Pull the decidua gently apart, separating the two halves of the 'pear' along the midline; one half will contain the implanted embryo. Extreme care should be taken with this operation so as not to damage the embryonic membranes, particularly when using somite-stage embryos.

(vi)   Separate into two halves that piece of decidua containing the embryo by gripping the decidua with forceps in the region below the ectoplacental cone and gently pulling apart.

(vii)   At the pole opposite to the ectoplacental cone, Reichert's membrane stands proud of the embryonic endoderm. This membrane must be removed to allow embryos to develop in culture. Using two pairs of forceps, grasp the Reichert's membrane at this point of separation from the endoderm, peel it back to the point where it is attached to the ectoplacental cone. The membrane can then be teased away from the remaining decidua, releasing the embryo.

(viii) In our own laboratory we have consistently obtained better results when much of the trophoblast (reddish region) of the ectoplacental cone is also removed, taking care not to rupture the underlying yolk sac membrane. To the unpractised, the whole explantation procedure is difficult and initial attempts will likely prove frustrating. Once dissected, transfer the embryos to fresh warm Hank's medium. Before culture, all embryos should be checked for damage and for developmental stage assessment.

2.5.2 *Culture of Headfold-stage Embryos*

(i)   Typically, within a single litter, embryos will vary with regard to their precise stage of development at explantation. In order to minimize the effect of such variation on the experimental parameters under study, divide the embryos between control and treatment groups such that each group will contain a similar range of embryonic stages. Unless specifically required, early headfold-stage embryos should be rejected since these develop less well in culture; mid- or late-headfold-stages are to be preferred. This variation of embryonic age is illustrated in *Figure 5*.

(ii)   Prepare the culture medium by adding 1 ml MEM (Eagles) to 3 ml rat serum in a 50 ml Sovirel bottle. Gas the bottle for 5 min with a mixture of 5% O_2, 5% CO_2, 90% N_2. (The gas is humidified and filtered by passage through distilled water and a 0.22 μm filter before entering the culture bottle *via* a sterile plastic or Pasteur pipette.) The flow of the gas should be such that the surface of the medium is slightly disturbed. In each bottle place two embryos (1 embryo/2 ml culture medium; this is routine practice but a maximum of three embryos per bottle may be used). Replace the screw-cap on the culture bottle and slowly roll the bottle by hand to coat the inner surface with culture medium and ensure that the embryos are freely suspended. Place the bottle on the roller apparatus in an incubator maintained at $37-38°C$.

(iii)   After 18 h of incubation re-gas the culture bottles with a mixture of 95% air, 5% CO_2 for 2 min. Further re-gas after $24-25$ h of culture with a mixture of 40% O_2, 5% CO_2, 55% N_2 for 2 min. The gassing schedule is outlined in *Figure 6*. For embryos explanted at stages later than headfold, a different gassing schedule will be required, which can be determined by reference to *Table 2*.

   In order to observe embryos during development in culture, bottles can be removed from the incubator for short periods and viewed through a binocular inverted microscope. Re-gassing of culture bottles is a convenient time to do this.

(iv)   After 48 h of incubation (or alternative time point), harvest the conceptuses according to the procedure described in Section 3.4.

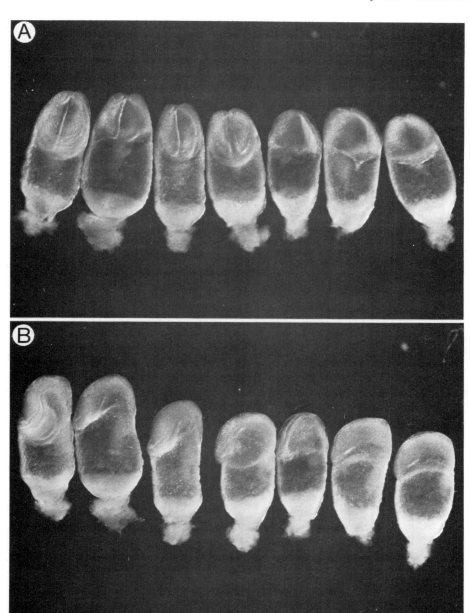

Figure 5. An illustration of the variation in developmental stage reached at 9.5 days gestational age. In **A**, embryos are oriented to show the neural groove, those on the right being more advanced in development. In **B**, the side view of the embryos demonstrates the progressive development, from left to right, of the headfold. In selection of embryos for culture, the more developed embryos fare better and are preferred. Of the featured embryos, the four on the right are sufficiently developed to be included in the culture.

	0	1	2	3	4	5	SCORE
A YOLK SAC CIRCULATORY SYSTEM	no visible, or scattered, blood islands	Corona of blood islands w or w/o anastamoses	Vitelline vessels with few yolk sac vessels	Full yolk sac plexus of vessels	Yolk stalk obliterated. vitelline artery & vein well separated		
B ALLANTOIS	Allantois free in exocoelome	Allantois fused with chorion	Umbilical vessels	Separate aortic origins of umbilical and vitelline vessels			
C FLEXION	Ventrally convex	Turning	Dorsally convex	Dorsally convex with spinal torsion			
D HEART	Endocardial rudiment not visible, or visible but not beating	Beating 's' shaped cardiac tube	Convoluted cardiac tube	Bulbus cordis, atrium commune and ventriculus communis	Dividing atrium commune		
E CAUDAL NEURAL TUBE	Neural plate or neural folds	Closing, but unfused neural folds (groove)	Neural folds fused at level of somites 4/5	Posterior neuropore formed, but open	Posterior neuropore closed		
F HIND BRAIN	Neural plate	Rhombomeres A and B	Anterior neuropore formed but open	Anterior neuropore closed. rhombencephalon formed	Pronounced pontine flexure with transparent roof of 4th ventricle		
G MID BRAIN	Neural plate	Mesencephalic brain folds	Closing or fusing mesencephalic folds	Completely fused mesencephalon	Visible division between mesencephalon & diencephalon		
H FORE BRAIN	Neural plate or no visible prosencephalon	Prosencephalic brain folds	Completely fused prosencephalon	Visible telencephalic evaginations	Well elevated telencephalic hemispheres		
J OTIC SYSTEM	No sign of otic development	Flattened or indented otic primordium	Otic pit	Otocyst	Otocyst with dorsal recess	Otocyst with endolymphatic duct	
K OPTIC SYSTEM	No sign of optic development	Sulcus opticus	Elongated optic primordium	Primary optic vesicle with open optic stalk	Indented lens plate	Lens pocket or lens vesicle	
L OLFACTORY SYSTEM	No sign of olfactory development	Olfactory plate	Olfactory plate with rim	Distinct olfactory ridges	Lateral nasal process and medial rim		
M BRANCHIAL BARS	None visible	I visible	I and II visible	I, II and III visible	II overgrowing and obscuring III		
N MAXILLARY PROCESS	No sign of maxillary development	Maxillary process demarcated, visible cleft anterior to bar I.	Maxillary process fused with nasal process				
P MANDIBULAR PROCESS	No sign of mandibular development from bar I	First brancial bars fused and forming mandibular process					
Q FORE LIMB	No sign of fore limb development	Distinct evagination of wolffian crest at level of somites 9-13	Fore limb bud	Paddle shaped fore limb bud	Distinct Apical ridge on fore limb bud		
R HIND LIMB	No sign of hind limb development	Distinct evagination of wolffian crest at level of somites 26-30	Hind limb bud	Paddle shaped hind limb bud			
S SOMITES	0 - 6	7 - 13	14 - 20	21 - 27	28 - 34	35 - 41	

Figure 6. The morphological scoring system. To use the system, all of the developmental parameters listed in the left hand column are examined and allocated an appropriate score (top row). Addition of individual scores gives a total morphological score, an objective and quantitative indicator of embryonic development. Further details are provided in the text (reproduced with permission: Brown and Fabro, 1981).

2.6 Preparation of Drug-metabolising System

The formation of teratogens through metabolism of otherwise harmless agents is now recognised as an important aspect of teratological research. Since the embryo itself has apparently little drug-metabolising capacity (see Section 4.2), the addition of a metabolising system to embryo cultures may be necessary. Crude homogenate, post-mitochondrial supernatant (S9) or a pure microsomal fraction is freshly prepared from the livers of adult rats pre-treated with Aroclor-1254 (a mixture of polychlorinated biphenyls with mixed monooxygenase-inducing properties). Details of these procedures are described in Chapter 8 of this volume. Culture of embryos and hepatocytes has been suggested as an alternative method of studying the role of drug metabolism in teratogenesis (13,14).

3. ASSESSMENT OF EMBRYONIC DEVELOPMENT

3.1 Morphological Development

Generally speaking, the viability of early organogenesis-stage embryos in culture increases with the age of embryos at explantation. Thus, only about 50% of rat embryos explanted at egg-cylinder stages (7.5 or 8.5 days) develop normally to the 30 − 40 somite stage (11.5 − 12 days), whereas almost 100% of embryos explanted at headfold stage (9.5 days) develop 30 somites, growing and developing at a rate indistinguishable from

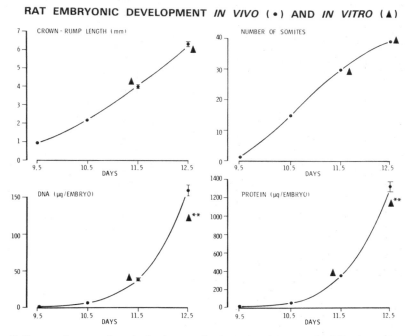

RAT EMBRYONIC DEVELOPMENT *IN VIVO* (•) AND *IN VITRO* (▲)

Figure 7. Comparative rat embryonic development *in vivo* (●) and *in vitro* (▲) following explantation at 9.5 days. Between gestational ages 9.5 and 12.5 days, the increase in crown-rump length and somite number is similar *in vivo* and *in vitro*. However, between 11.5 and 12.5 days, the increase in DNA and protein contents of cultured embryos is significantly less (**) than those that have developed *in vivo*, as judged by Student's t-test.

that of *in vivo* littermates (1). Mouse embryos perform similarly in culture, except at the earlier stages (egg-cylinder to headfold) where development is inferior to that obtained with rat (11). Beyond the 30-somite stage, embryos do continue to develop for a short time but the growth rate declines (*Figure 7*). Because of the apparent equivalence of *in vitro* and *in vivo* development of embryos between headfold and 30-somite stages, culture of embryos over this period is most routinely employed in toxicological research.

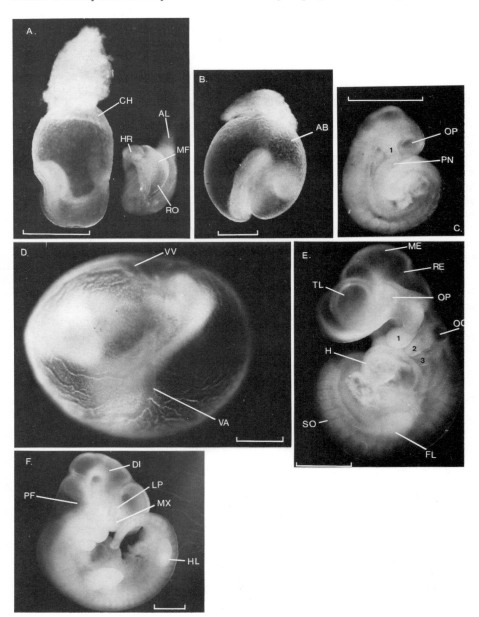

During the period of organogenesis, considerable differentiation of a number of organ primordia occurs and these serve as markers for determining the success of culture, and/or the basis for assessing toxic or teratogenic effects of culture conditions. The extent to which these numerous morphological criteria are applied in the determination of embryonic development *in vivo* varies between laboratories. The most complete system, which encompasses all of the others, is that put forward by Brown and Fabro (15). In this system, 17 developmental features are selected for examination, and each feature further sub-divided into the various stages, up to six in all, of its morphogenetic sequence (*Figure 6*). Quantitation of embryonic development is achieved by assigning to each sub-division a score $(0-5)$; addition of the 17 individual scores from each morphological parameter provides a total score, an objective and quantitative indicator of morphogenesis. We have found the procedure to be valuable and quite practicable in routine use.

3.2 Embryonic Growth

Both morphological and biochemical measurements are taken to estimate embryonic growth. In the former category, measurements of the yolk sac diameter (at the midline, horizontal to the placenta/ectoplacental cone) and crown-rump (the maximum length of the embryo) and head (frontal tip of prosencephalon to most dorsal part of mesencephalon) lengths are readily achieved with the use of a micrometer eyepiece in a dissecting microscope. The number of somites (regarding that somite adjacent to the mid-point of the fore limb bud as number 9) also provides an indication of embryonic growth.

Biochemical indicators of growth that are routinely applied to culture embryos are estimates of total protein and DNA, assayed by established techniques. Either, or both, measurements provide good estimates of the growth of embryos in culture.

Figure 7 presents a comparison of *in vivo* and *in vitro* growth and morphological data of rat embryos aged $9.5-12.5$ days, and *Figure 8* features embryos at various developmental stages between the same ages.

3.3 Evaluating Embryonic Development in Practice

From the foregoing discussion it will be evident that there are numerous and varied parameters that can be used to assess embryonic development in culture. As a guide

---

Figure 8. Rat embryos at various stages of development between gestational ages 9.5 and 12.5 days. **A:** Headfold stage embryos with (left) and without (right) extra-embryonic membranes. **B** and **C:** 10.5-day embryo with and without yolk sac. Rotation of the embryo to the dorsally convex position has occurred. The head shows considerable differentiation; several somites have formed and the heart has by now begun beating. **D** and **E:** 11.5-day embryo with and without yolk sac. The yolk sac is highly vascularized. The embryo is developing the telencephalon and has also developed optic vesicles, otocysts, three branchial arches, a 3-chambered heart, about 28 somites and fore-limb buds. **F:** 12.5-day embryo with pronounced positive flexure of the hind brain and formed hind-limb buds. The embryo has nearly 40 somites at this stage. Scale bars equal 1 mm. Abbreviations: AB-anastomosing blood islands; AL-allantois; CH-chorion; DI-diencephalon; FL-forelimb bud; H-3-chambered heart; HL-hindlimb bud; HR-heart rudiment; LP-lens placode; ME-mesencephalon; MF-mesencephalic brainfold; MX-maxillary process; OC-otocyst; OP-optic primordium (C); optic vesicle (E); PF-pontine flexure; PN-posterior neuropore; RE-rhombencephalon; RO-rhombomere; SO-somites; TL-telencephalon. Numbers 1 (C) and 1, 2, 3 (E) indicate branchial arches.

to the investigator, the routine procedure used in our own laboratory is detailed below.

(i) Immediately after removal of the culture bottles from the incubator, but before harvesting of embryos, check the overall appearance of embryos on an inverted microscope; normal embryos should have a vigorously beating heart and yolk sac circulation. Poor yolk sac vasculature and general appearance is evidence of sub-optimal development (see Section 5).

(ii) Remove the embryos from the culture vessel to a small dish of saline. Care should be taken not to damage the extra-embryonic membranes at this stage since measurements of yolk sac diameter are required as an estimate of growth (see Section 3.2); transfer of conceptuses is readily achieved by the use of a wide-bore Pasteur pipette. Yolk sac diameter is then measured with the micrometer eyepiece.

(iii) Using fine forceps, remove the extra-embryonic tissues and measure the crown-rump and head length of the embryo. At this point the various embryonic structures listed in *Figure 6* are examined. With practice, examining all of these features can be accomplished in 1 or 2 min. As a minimum, the embryo should be checked for normal axial rotation to the dorsally convex position — the flexure of the trunk should be such that the tail passes to the right side of the head; closure of anterior neuropore and fusion of the neural tube — the neural suture line should be checked to ensure that it is not irregular or in any other way abnormal; the presence of optic and otic vesicles; the presence of branchial arches and fore-limb buds; somites should be counted and their overall appearance noted.

(iv) Subsequent steps are determined by experimental requirements. The embryo can be solubilised in NaOH for protein and/or DNA assays. Portions of the total solution need only be used for this, as would be the case, for example, if incorporation of radiolabel into embryonic tissues were being investigated; the remaining solution could then be used for radioactive counting. Protein or DNA assays are routinely carried out in this laboratory. Alternatively, histological examination of embryos can be performed, in which case fixation of embryos by standard techniques would follow. Photography of whole embryos is best achieved using fresh specimens immediately after harvesting. Following fixation, the structural detail of the embryo is less well defined and photography less successful. A major advantage of evaluating the morphological development of embryos in culture along the lines of Brown and Fabro (15), is that subtle dysmorphogenesis of the embryo can be identified. *Figure 9* shows some examples of such defects in cultured embryos (in the absence of gross malformation and severe growth retardation) which are manifest as malformations at term.

4. APPLICATION OF WHOLE EMBRYO CULTURE IN TERATOLOGY

4.1 Screening for Teratogens

The availability of simple and reliable techniques for the culture of rodent embryos during the early organogenesis phase has inevitably aroused the interest of reproductive toxicologists seeking to replace, or reduce in practice, the currently used whole animal screens for potential teratogenicity/embryotoxicity of drugs and chemicals. Several reviews (2, 3, 8) have discussed this application of the culture technique.

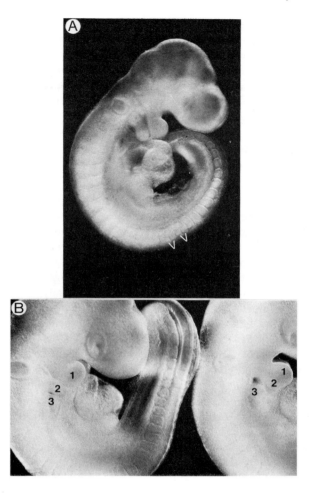

Figure 9. Specific structural defects of 11.5-day rat embryos induced by direct teratogen treatment *in vitro*. **(a)** Treated embryo with two pairs of fused somites (numbers 14 + 15 and 16 + 17). *In vivo*, these lesions lead to rib and vertebral malformations. **(b)** Control embryo (left) illustrating normal morphology of branchial arches. Treated embryo (right) showing fusion of first and second arches. *In vivo*, such a defect produces a spectrum of mandibular, hyoid, ear and cranial malformations. The treated embryo also shows an abnormal otic primordium.

However, we have recently argued (16) that the routine use of embryo culture in teratogenicity testing of large numbers of chemicals is impracticable for reasons of cost, effort and time involved, although the technique can be of great value in the testing of small numbers of chemicals. There have been no reports of large scale validation studies of embryo culture as a suitable technique for teratogenicity screening. Furthermore, the effect of a test agent on an embryo separated from its mother may lead to erroneous conclusions of the embryopathic nature of the agent. Early organogenesis-stage rat and mouse embryos possess negligible xenobiotic-metabolising activity (but see Section 4.2.2) and yet biotransformation, a maternal function, is known to be an

101

essential feature in teratogenesis caused by certain drugs (see Section 4.2.1). Admittedly, there are methods, such as addition of drug-metabolising systems to the culture medium, that help overcome this problem, but including tests of this kind in the overall screening strategy using embryo culture would increase further the burden of effort. Even so, embryo culture can and should perform a valuable role in the in-depth study of teratogenic agents, serving as a supplement to the growing range of *in vitro* teratogenicity tests (16) as well as the established safety evaluation protocols.

4.2 Studies of Mechanisms of Teratogenesis

4.2.1 *Identification of Teratogenic Agent*

The particular value of embryo culture is to be found in its use to study the events of teratogenesis: defining the nature of the teratogenic lesion. The opportunity to study the growing and developing embryo in isolation from its mother simplifies and extends the range of studies that can be conducted to this end. In such studies it is first of all necessary to identify the teratogenic agent since the observation that a substance administered to a pregnant animal results in the production of malformed offspring need not necessarily mean that that substance is itself the teratogen. Maternal metabolism may act to convert a teratogenically inert substance to a potently teratogenic metabolite; conversely an agent directly harmful to the embryo may be rapidly and efficiently detoxified by the mother to the extent that it poses little teratogenic hazard to the embryo. Some knowledge of the metabolism and pharmacokinetics of the drug or chemical under study is therefore clearly desirable in establishing the identity of a teratogenic agent.

The isolation of the embryo in culture, regarded as a disadvantage in screening (Section 4.1), becomes an advantage in the identification of the proximal teratogen. Several approaches can be used to determine whether an administered compound itself, or one of its metabolites, is teratogenic, and the procedures are outlined below.

(i) *In vivo/in vitro comparison of effects.* Dose the dam with test compound at a specified gestational time (e.g., day 8.5 or 9.5 in rat) and permit the embryos to continue to develop *in vivo* for a specified period (e.g., 48 h). Using a separate group of pregnant animals, explant the embryos at the same gestational age as those treated *in vivo* and culture *in vitro* for the same length of time as for *in vivo* treatment in medium to which test compound has been added.

Compare the *in vivo* and *in vitro* embryonic development at harvesting. Similar effects in the two groups would indicate that the test compound itself, and not a maternal metabolite, is the proximal teratogen.

(ii) *Comparison of sera from treated and untreated adults.* Compare the development in culture of embryos in serum from animals treated with the test compound, with embryos in serum from untreated animals to which the test compound has been added directly. Abnormal embryonic development in serum from treated animals only would indicate that biotransformation of test compounds is necessary for teratogenesis.

(iii) *Use of drug-metabolising system.* Compare the development in culture of embryos in medium containing test compound to which a drug-metabolising system ('S-9') with or without cofactors has been added, with embryos cultured in medium containing test compound only. Teratogenesis resulting only from the former experiment would confirm the need for biotransformation of test compound to induce malformations of embryos.

(iv) *Direct addition of metabolites of test compound.* Determine the effects of isolated metabolites of test compound added directly to the culture medium on embryo development. The concentration of added metabolite should be equimolar to that of parent compound required, after biotransformation, to induce teratogenesis *in vitro*. Equivalent teratogenic effects of biotransformed parent compound with one or more metabolites, would identify the latter as the proximal teratogen(s).

Cyclophosphamide (CP) teratogenesis in rats provides a good example of the application of these regimes (3). Although a potent teratogen in rats, this anti-neoplastic agent failed, when added directly to the medium, to induce malformations in embryos cultured between 9.5 and 11.5 days, suggesting that teratogenesis *in vivo* was caused by a metabolite of CP. This conclusion was reinforced by the findings that embryos cultured in serum from rats treated with CP developed abnormally, and that inclusion of S9 and cofactors (NADPH and glucose-6-phosphate) with CP in cultures resulted in the production of abnormal embryos. Direct addition of stable metabolites of CP to embryo cultures demonstrated that phosphoramide mustard exactly mimicked the teratogenic action of bioactivated CP; acrolein had no effect on embryo growth and development; 4-ketocyclophosphamide had no effect on embryo growth but did produce some embryos with malformations, the nature of which differed from those produced by bioactivated CP. On this basis, phosphoramide mustard is suggested as the proximal teratogen.

Two limitations of embryo culture in determining the nature of teratogens are worth noting. Firstly, it has been found that inclusion of a drug-metabolising system in a culture may fail to generate a teratogenic metabolite. For example, the cytostatic drug procarbazine is teratogenic in rats but fails to induce abnormalities in embryos cultured *in vitro* either in the presence or absence of 'S9' and cofactors (17). Secondly, it should be borne in mind when using as a culture medium serum from drug-treated animals that such serum may differ from control serum in addition to the presence of the drug and its metabolites. A toxic effect of the drug upon the treated animals may change the serum in a way which, of itself, causes poor development of embryos in culture.

Methods 2 and 3 above offer the attractive possibility of inclusion of human material in the experimental system, for example, the addition of a human drug-metabolising system to the culture medium. Mention has already been made of the ability of human serum to support growth and development of rat embryos in culture (Section 2.2). By using serum from patients undergoing drug therapy, it would be possible to evaluate the teratogenic effects of human drug metabolites on cultured embryos. Indeed this approach has already been demonstrated as a workable one with the discovery that serum from patients treated with certain cancer chemotherapeutic agents or anti-convulsants exerted significant toxic effects on cultured rat embryos (8).

4.2.2 *Approaches to the Study of Teratogenic Lesions*

The design of experimental studies of teratogenic effects at the cellular or molecular level will be determined by several factors, such as a knowledge of the chemical nature of the teratogen, and of its biochemistry and toxicity in adult systems. Consideration of the nature of the defect(s) induced by a teratogen will focus the attention of the investigator on specific tissues or organ primordia of the developing embryo. Since the extent of such information differs for each teratogen, there is no universally applicable strategy for investigating teratogenic mechanisms.

However, several approaches will commonly be adopted and the embryo culture technique is especially suited to a number of these. For example, the tissue distribution and subcellular localization of teratogens in embryos is readily measured after uptake from the culture medium of small amounts of radio- or fluorescent-labelled teratogen followed by autoradiography, fluorescence microscopy or direct assay of isolated organs or subcellular fractions. Information on the localisation of a teratogen within the embryo or its associated membranes can provide clues to its mechanism of action.

The observations that the teratogens trypan blue, anti-yolk sac antibody and leupeptin localise almost exclusively in the apical vacuoles (lysosomes) of the visceral yolk sac endoderm of rats led to the hypothesis that induced yolk sac dysfunction was the primary mechanism of teratogenic action (18). Embryo culture experiments demonstrated that the yolk sac performed an essential role during early organogenesis (19). Pinocytic uptake and intralysosomal digestion of exogenous protein by yolk sac provides a source of amino acids for embryo growth and development. Inhibition of uptake and/or digestion processes has been implicated in the teratogenesis caused by these agents (18). However, simply demonstrating the presence of a teratogen in a certain tissue or organ need not always imply that this is its site of action. Unpublished experiments from our own laboratory have shown that the teratogenic anti-convulsant, valproic acid at certain concentrations, is fairly uniformly distributed throughout the tissues of the cultured rat embryo, yet is responsible for only a few rather specific defects, principally those resulting from an action on the somites or pre-somitic mesoderm (20).

Another aspect of teratogenic action that will often need to be investigated is the extent to which biochemical modification or interaction, or metabolism of the teratogen in embryonic cells is involved in inducing abnormal morphogenesis. The use of the isolated embryo in culture avoids the confounding influence of maternal modification of the teratogen.

Although our knowledge of the metabolism of the early post-implantation embryo is poor, there are some examples of teratogens that exert their actions only after modification within embryo cells. Using embryo culture, Nebert and colleagues have presented compelling evidence (21) that benzo[a]pyrene teratogenicity in embryos of certain mouse strains is due not to the polycyclic hydrocarbon itself, but to its chemically active intermediates generated by cytochrome P450-mediated aryl hydrocarbon hydroxylase. The activity of the latter was shown to be induced by benzo[a]pyrene, as evidenced by sister chromatid exchange in embryonic cells. Another polycyclic hydrocarbon inducer of P450-mediated monooxygenases, 3-methylcholanthrene, has been demonstrated to dramatically increase the incidence of abnormalities caused by 2-acetylaminofluorene in cultured rat embryos (22), suggesting metabolism of the latter compound to a teratogenically active form. The generation of unstable highly reactive intermediates within embryo cells may be an important factor in the teratogenicity of many xenobiotics.

Such unstable intermediates resulting from maternal metabolism would probably be detoxified before reaching the embryo, but the passage of residual parent compound into embryo cells might well result in the formation of short-lived toxic species at the site of teratogenic action.

Intermediary metabolism has also been implicated as a target for teratogens. It has been shown that replacement of glucose with mannose as a source of metabolic fuel in rat embryo culture medium, induces severe growth retardation and dysmorphogenesis of embryos, an effect that can be remedied by adding glucose or increasing the pO_2 of the medium to promote oxidative metabolism (23). The teratogenic thymidine analogue, 5-bromodeoxyuridine, has been shown to be teratogenic in mice by metabolism to the nucleotide derivative then incorporation into embryonic cellular DNA, with the resultant interference with gene expression and synthesis of cell-specific products (25).

These few examples among many (see ref. 25) demonstrate the value and scope of embryo culture as a tool in teratological research. Further refinements of the technique, such as the finding that rat embryos can be cultured for short periods ($2 - 3$ h) in serum-free medium, permitting brief exposures to compounds whose activity may be influenced by the large amount of serum in the culture medium, or the discovery of suitable organic solvents of low embryotoxicity for the delivery of water-insoluble compounds to the culture medium (26) extend the range of experiments that can be performed.

Very considerable gaps in our knowledge of teratogenic mechanisms still remain, particularly with regard to the metabolic properties of embryos and the varying responses of embryonic tissues to teratogenic insult. There can be little doubt that embryo culture will have a major part to play in the study of these and related phenomena.

5. SIGNIFICANCE OF TERATOLOGICAL STUDIES USING WHOLE EMBRYO CULTURE

The most obvious and important criterion for acceptance of whole embryo culture as a suitable technique for teratological studies is that the embryo retains, as far as possible, its normal morphological and biochemical characteristics. The gross morphological and growth data presented in *Figure 7* indicate that the culture system satisfies this criterion at least at this level of examination. In addition to these data, histological comparisons of *in vivo* and *in vitro* embryos have been made. New *et al.* (27) reported that histological differentiation of rat embryos in culture ($9.5 - 11.5$ days embryonic age) was very similar to that of *in vivo* littermates, although in certain respects (i.e., optic vesicle development and closure of posterior neuropore) development *in vitro* was slightly in advance of that *in vivo*. These findings were largely corroborated by Herken and Anschutz (28) who demonstrated equivalent *in vivo* and *in vitro* development between 10.5 and 12 days embryonic age. On continued culture, *in vitro* histological differentiation, like macroscopic development, begins to lag behind *in vivo* development. Although there are few data comparing *in vivo* with *in vitro* biochemical development of embryos during early-organogenesis, normal development may be inferred on two counts. First, normal morphogenesis in culture can be regarded as a corollary of normal biochemical development. Secondly, the limited data available clearly show a close similarity in the *in vivo* and *in vitro* embryonic response to teratogenic treatment with hydroxyurea (29), and excess vitamin A (30), indicating that the affected biochemical functions of the embryo are preserved in culture. Furthermore, the well-known ameliorative effect of zinc on cadmium embryotoxicity has been reproduced in the embryo culture system (2), again arguing for the biochemical integrity of the cultured embryo. Some direct comparisons of *in vivo* and *in vitro* enzyme development have been made. A histochemical (31) and biochemical (32) study of yolk sac enzymes showed that cultured embryos possessed levels of activity very similar to age-matched embryos that had developed *in vivo*. In contrast, levels of ornithine decarboxylase activity in cultured embryos were significantly lower than those measured in age-equivalent *in vivo* embryos (33). Although this is the only published report of sub-optimal development of cultured embryos, the conclusion drawn from the study, that the culture environment in some way interferes with the expression of enzyme activity, should serve as a caution to workers investigating biochemical mechanisms of teratogenic action.

While recognising the great value of embryo culture in teratology, it is important to recognise also some of the limitations of the technique, a number of which have already been discussed (Section 4). One of the major drawbacks is that the period during which embryonic development *in vitro* is comparable with that *in vivo* represents only a portion of the total organogenesis phase. Consequently, teratogenic insults occurring at a stage of development not encompassed by the culture period cannot be studied with this technique, although other *in vitro* systems such as embryonic organ or cell cultures enable the study of particular structures during later organogenesis. Another consequence of this limited experimental period is that the sequellae of dysmorphogenesis observed in cultured embryos cannot accurately be investigated. Thus it is often difficult to determine how malformations of the cultured embryo would be manifest at later gestational ages and at term. It is anticipated that the severity of some abnormalities observed during early organogenesis would result in the subsequent death or resorption of the embryo. Conversely, minor defects of the embryo might prove to be only transient phenomena and capable of repair so that at term the foetus appears healthy and normal. It is therefore not always possible to equate *in vitro* with *in vivo* teratogenicity, a point worth remembering in the application of embryo culture as a teratogenicity screen.

In interpreting *in vitro* data, the viability of the yolk sac must also be considered. Agents that induce embryopathy in culture also very often cause abnormal development of the yolk sac at equal, or even lower, doses. Since an embryo that develops within an abnormal yolk sac for any length of time becomes malformed, effects on the yolk sac must be taken into account; an indication of the health of the yolk sac (e.g., circulation) in experimental treatments resulting in embryonic malformation should be provided.

Finally, most embryo-culture teratology studies are performed by adding a chemical to the culture medium and leaving it there for the whole culture period. Since the culture system is essentially 'closed', this exposes the developing embryo to a constant concentration of toxin for 48 h. Observation of an apparently embryotoxic action under such conditions does not necessarily predict that the compound will be teratogenic *in vivo*, where embryonic exposure may be quite different. The pharmacokinetic properties of the chemical and species under study, and the exposure conditions, determine embryonic exposure which may, for example, show short-term high exposures separated by long periods of low exposure. If exposure profiles *in vivo* are known these can, of course, be mimicked *in vitro* by changing culture media (e.g., ref. 29). Thus, the observation that a chemical is dysmorphogenic *in vitro*, but not in the same species *in vivo*, is likely to be a reflection of differing exposure and does not in any way invalidate the culture technique.

6. REFERENCES

1. New,D.A.T. (1978) *Biol. Rev.*, **53**, 81.
2. Sadler,T.W. and Warner,C.W. (1984) *Pharmacol. Rev.*, **36**, 1455.
3. Shepard,T.H., Fantel,A.G., Mirkes,P.E., Greenaway,J.C., Faustman-Watts,E., Campbell,M. and Juchau,M.R. (1983) in McLeod,S.M., Okey,A.M. and Spielberg,S.P. (eds.), *Developmental Pharmacology: Progress in Clinical and Biological Research*, Vol. **135**, Alan R. Liss, Inc., New York, p. 147.
4. Priscott,P.K. (1979) *Experientia*, **35**, 1414.
5. New,D.A.T. and Cockroft,D.L. (1979) *Experientia*, **35**, 138.
6. Tarmaltzis,B.C., Sanyal,M.K., Biggers,W.J. and Naftolin,F. (1984) *Biol. Reprod.*, **31**, 415.
7. Priscott,P.K., Yeoh,G.C.T. and Oliver,I.T. (1984) *J. Exp. Zool.*, **230**, 247.
8. Klein,N.W. and Pierro,L.J. (1983) in Johnson,E.M. and Kocchar,D.M. (eds.), *Handbook of Experimental Pharmacology*, Vol. **65**, Springer-Verlag, Berlin, p. 315.
9. Lear,D., Clarke,A., Gulamhusein,A.P., Huxam,M. and Beck,F. (1983) *J. Anat.*, **137**, 279.
10. Mirkes,P.E., Fantel,A.G., Greenaway,J.C. and Shepard,T.H. (1981) *Toxicol. Appl. Pharmacol.*, **58**, 322.
11. Sadler,T.W. (1979) *J. Embryol. Exp. Morphol.*, **49**, 17.
12. Cockroft,D.L. (1979) *J. Reprod. Fertil.*, **57**, 505.
13. Brown,N.A. and Kram,D. (1982) *Teratology*, **25**, 30A.
14. Oglesby,L.A., Ebron,M.T., Carver,B. and Kavlock,R.J. (1984) *Teratology*, **29**, 49A.
15. Brown,N.A. and Fabro,S. (1981) *Teratology*, **24**, 65.
16. Brown,N.A. and Freeman,S.J. (1984) *ATLA*, **12**, 7.
17. Schmid,B.P., Trippmacher,A. and Bianchi,A. (1982) *Toxicology*, **25**, 53.
18. Lloyd,J.B., Freeman,S.J. and Beck,F. (1985) *Biochem. Soc. Trans.*, **13**, 82.
19. Freeman,S.J., Beck,F. and Lloyd,J.B. (1981) *J. Embryol. Exp. Morphol.*, **66**, 223.
20. Brown,N.A. and Colhoun,C.W. (1984) *Teratology*, **29**, 20A.
21. Galloway,S.M., Perry,P.E., Meneses,J., Nebert,D.W. and Pedersen,R.A. (1980) *Proc. Natl. Acad. Sci. USA*, **77**, 3524.
22. Faustman-Watts,E., Giachelli,C., Greenaway,J., Fantel,A., Shepard,T. and Juchau,M. (1984) *Teratology*, **29**, 28A.
23. Freinkel,N., Lewis,N.J., Akazawa,S., Roth,S.I. and Gorman,L. (1984) *N. Eng. J. Med.*, **310**, 223.
24. Skalko,R.G., Packard,D.S., Schwendimann,R.N. and Raggio,J.F. (1971) *Teratology*, **4**, 87.
25. Wilson,J.G. and Fraser,F.C. (eds.) (1977) *Handbook of Teratology*, Vol. **3**, published by Plenum, New York/London.
26. Kitchin,K.T. and Ebron,M.T. (1984) *Toxicology*, **30**, 45.
27. New,D.A.T., Coppola,P.T. and Cockroft,D.L. (1979) *J. Embryol. Exp. Morphol.*, **36**, 133.
28. Herken,R. and Anschutz,M. (1981) in Neubert,D. and Merker,H.-J. (eds.), *Culture Techniques*, Walter de Gruyter, Berlin/New York, p. 19.
29. Warner,C.W., Sadler,T.W., Shockey,J. and Smith,M.K. (1983) *Toxicology*, **28**, 271.
30. Steele,C.E., Trasler,D.G. and New,D.A.T. (1983) *Teratology*, **28**, 209.
31. Miki,A. and Kugler,P. (1984) *Histochemistry*, **81**, 409.
32. Freeman,S.J. (1981) PhD thesis, University of Keele, UK.
33. Huber,B.E. and Brown,N.A. (1982) *In Vitro*, **18**, 599.

CHAPTER 5

The Identification and Assessment of Covalent Binding *In Vitro* and *In Vivo*

CARL N.MARTIN and R.COLIN GARNER

1. INTRODUCTION

Many of the toxicological consequences of human or animal exposure to foreign compounds are as a result of covalent binding of the compound to cellular macromolecules (e.g., 1 – 5). There are a variety of methods to assess such binding including radiotracer-, antibody- and physico-chemical procedures. In this chapter we shall concentrate on the use of radiolabelled compounds to measure covalent binding with cellular macromolecules, and in particular DNA.

The choice of DNA as the target molecule stems from its crucial role in carcinogenesis and mutagenesis and the difficulty many workers have in obtaining highly purified preparations for study.

2. BASIC TECHNIQUES

The study of covalent binding involves a variety of techniques. The various crucial steps are listed below, beginning with a general description of high pressure liquid chromatography as applied to this area of study.

(i)   High pressure liquid chromatography.
(ii)   Preparation of radiolabelled starting material and assessment of purity.
(iii)   Dosing of test system.
(iv)   Extraction and purification of DNA.
(v)   Preparation of DNA for quantitation and for liquid scintillation counting.
(vi)   High pressure liquid chromatographic analysis of hydrolysed DNA.
(vii)   Determination and synthesis of reactive intermediates.
(viii) Reaction of postulated reactive intermediate with DNA *in vitro*.
(ix)   Enzymic hydrolysis of adducted DNA to deoxyribonucleoside/carcinogen adducts.
(x)   Proof of structure of biologically produced adduct.

2.1 High Pressure Liquid Chromatography

2.1.1 *Equipment*

All of the chromatography procedures described below and elsewhere in this chapter have been performed on Pye-Unicam PU 4000 series chromatography equipment (*Figure 1*). We recommend the use of a two-pump gradient elution system for reproducibility

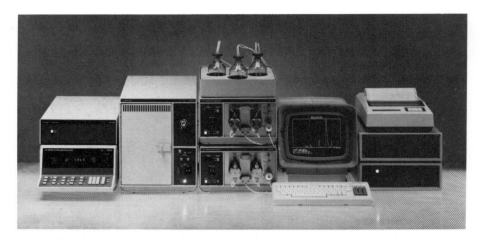

Figure 1. Typical h.p.l.c. system incorporating column oven, automatic injector, two pump high pressure mixing solvent system, computer-controlled solvent gradient, u.v. detection and computer-assisted data analysis.

of chromatography runs.

(i) Injection of sample is *via* an injection loop system using syringes of between 25 and 1000 microlitre capacity.

(ii) Suitable gradient systems are chosen to give adequate resolution of the compound of interest. Chromatography runs should not be so long that only a few samples can be analysed in a day. Runs greater than 120 min in length should be avoided. Samples can be injected using an automatic sample injector in order that the chromatography equipment can be used continuously.

(iii) Monitoring of the eluent is normally by ultraviolet or fluorescence detection, or by determination of radioactivity. If the compound has a characteristic absorption which is of higher wavelength than 280 nm then monitoring can best be performed at this wavelength in order to avoid interference with contaminating compounds. Otherwise monitoring at the lambda max should be carried out.

(iv) Connection of the detector to a video monitor and computer has advantages over the normal 10 mV recorder, in that sample peaks can be integrated and relative concentrations determined. In addition, relevant information apertaining to a chromatography run can be entered. *Figure 2* shows a typical chromatography print-out from a commercial apparatus.

(v) It is often necessary to collect the chromatographic peak of interest and this can be done using a fraction collector. Many modern fraction collectors can collect on the basis of time, number of drops, peaks or a combination of these. The fraction collector can be electrically connected to the h.p.l.c. pumps and the automatic sample injector so that repeated runs can be performed. By use of appropriate racks, many fraction collectors enable collection directly into scintillation vials which aids in speeding up the analysis of chromatography runs.

2.1.2 *Columns*

With the large number of columns and packing materials available, it is often the preju-

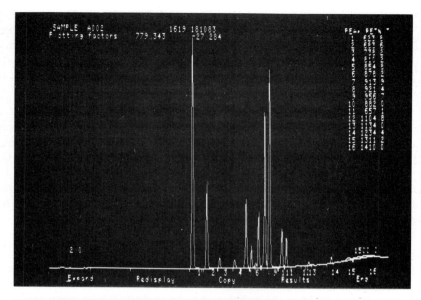

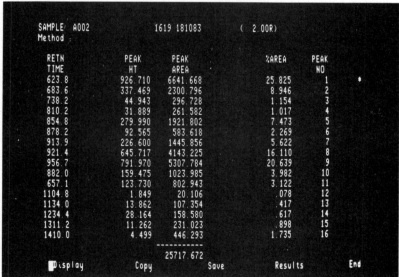

Figure 2. Typical output from an on-line computer analysis of u.v. absorption profile of an h.p.l.c. run.

dices of the researcher which determines the one to be used. We, in our laboratory, routinely use either analytical columns (25 × 0.3 cm) packed with Partisil-10 ODS2 (Whatman), or Techsil-ODS (HPLC Technology), or Shandon columns (25 × 0.6 cm) packed with the same material. We pack our own columns using a Shandon column packer.

111

We have found that reverse-phase chromatography is suitable for almost all separations of interest to us. A whole range of compounds can be separated by reverse-phase chromatography using water/methanol gradients. Ionised molecules which elute in the void volume even when water is used as the eluent, can be retained by addition of an ion-pairing agent.

2.1.3 *Separation Conditions*

These will vary depending on the sample to be analysed. No system can be applied to all compounds. Generally one should attempt to elute the compound of interest in as fast a time as possible compatible with good resolution. Gradients starting at 100% water and running up to 100% methanol quickly are to be discouraged since a long period of time is required to re-equilibrate the column. In order to sharpen up chromatographic peaks and to prevent tailing, a few drops of formic acid can be added to both the water and methanol solvent bottles.

2.2 **Preparation of Radiolabelled Starting Material**

The choice of starting material will, to an extent, depend on some knowledge of elements of the metabolism of the study chemical. In addition, a choice has to be made between using tritium or carbon-14 as the radiotracer to be studied. Whilst many individuals favour the use of C-14 because there is no possibility of exchange taking place, we in our laboratory have tended to use tritium because we can obtain material labelled to a higher specific radioactivity. As a rough 'rule of thumb' *in vivo* studies of covalent binding should use material of at least 100 mCi/mmol specific activity.

There are as many ways of synthesising radioactive compounds are there are organic synthetic routes. Certain carcinogens or other study compounds can be purchased from commercial sources: Amersham International plc, White Lion Road, Amersham, Buckinghamshire HP7 9LC, UK; CEA, Labelled Compounds Division, BP.2, Gil-sur-Yvette, 91190 France; and NEN Chemicals GmbH, Postfach 401240, 6072 Dreieich, FRG. In addition, these companies provide custom labelling and tritium exchange labelling services.

The radiochemical purity of the synthesised compound is clearly crucial to studies on its ability to bind to macromolecules. This can be measured in a variety of ways but one of the best and most convenient is the use of h.p.l.c. Samples are co-injected with authentic standard compound and fractions collected and counted for radioactivity. One is then looking for a single radioactive peak which co-elutes with the standard compound.

2.3 **Dosing of Test System**

2.3.1 *In Vitro Studies*

When studying a chemical suspected of having carcinogenic activity, it may be of interest to determine the structure of proximate or ultimate carcinogenic species and the consequent product of reaction with DNA. This problem can often be approached *in vitro* by incorporating a suitable activating system. In some instances purified DNA such as calf thymus DNA can be utilised as the target molecule whilst activation can be catalysed using a liver microsomal preparation or post-mitochondrial supernatant fraction (see also Chapter 9). In other cases a whole cell system is required such as hepatocytes or liver slices in which activation and DNA binding occur within the intact cell.

(i) *Liver post-mitochondrial supernatant or microsomal activation systems using purified DNA as the target.* Generally either mice or rats are used to prepare suitable liver preparations. In some circumstances one might want to use other species, such as the hamster or even man. It is best to use animals that have been pre-treated with an inducing agent such as phenobarbitone, 3-methylcholanthrene or Aroclor-1254.

Induction by phenobarbitone. Administer a solution of sodium phenobarbitone (1 mg/ml) to young adult animals in the drinking water for at least 7 days before sacrifice. Keep a watch on the animals to ensure that they do not become so sedated that they fail to eat. Rats of less than 100 g weight should not be used. Seven days pre-treatment allows cytochrome P-450 to be maximally induced.

Induction by 3-methylcholanthrene. Inject animals with a single intraperitoneal injection on day 0 (40 mg/kg of 3-methylcholanthrene made up as a 40 mg/ml solution in corn oil). Repeat the dose of 3-methylcholanthrene 48 h after the initial injection was administered and kill the animals 24 h later.

Induction by Aroclor-1254. Inject animals with a single intraperitoneal injection of Aroclor-1254 on day 0 (500 mg/kg of Aroclor 1254 made up as a 200 mg/ml solution in corn oil) and kill 5 days later.

Chemical sources: sodium phenobarbitone − British Drug Houses, Poole, Dorset; 3-methylcholanthrene − Koch-Light Limited, Colnbrook, Bucks; Aroclor-1254 − not commercially available, but enquire from Monsanto Chemical Co.

Preparation of liver homogenate.

(a) Starve animals for 24 h previously. Then anaesthetise, exsanguinate by cutting the carotid artery and kill the animals by cervical dislocation.

(b) Remove the livers, wash in ice-cold 0.9% sodium chloride and mince with scissors.

(c) Add 5 g portions of liver to 15 ml of 150 mM ice-cold potassium chloride and homogenise the whole using either a Potter-Elvejheim homogeniser with a Teflon pestle or a blade homogeniser (Ultra-Turrax, Janke and Kunkel, Darmstadt, FRG) or equivalent. We homogenise the liver with 10 passes of the revolving pestle or by using the blade homogeniser for 8 sec.

(d) Immediately the liver is homogenised, place the resulting homogenate on ice in tubes. The liver should at no time be allowed to warm up in order to prevent enzyme inactivation.

Preparation of liver post-mitochondrial supernatant (PMS or S9).

(a) Pour the homogenate as obtained above into centrifuge tubes and spin at 9000 g for 10 min at 4°C. Carefully decant off the supernatant, ensuring that none of the pellet is carried over and either keep on ice if required for use immediately or pour into screw-capped cryotubes (Nunc, ordered through Gibco Biocult, Paisley, UK) and store at −90°C until needed. We would recommend that each batch is given a number and the date of preparation recorded. Any S-9 unused after 6 months should be discarded.

Preparation of liver microsomes.

(a) Pipette the S9 fraction into ultracentrifuge tubes, cap and spin for 60 min at 100 000 g at 4°C.

(b) Discard the supernatant, wipe off any lipid adhering to the sides of the tube with a tissue, and remove the pellet with ice-cold 150 mM potassium chloride by scraping with a glass rod.

(c) Resuspend the pellet by homogenisation and make up to the original volume, i.e., if a 10-ml aliquot was centrifuged then resuspend the microsomal pellet to 10 ml with KCl solution.

(d) In certain circumstances an extremely 'clean' microsomal fraction is required. Re-centrifuge the resuspended pellet for 30 min at 100 000 g at 4°C and resuspend as above.

(e) Stand the microsomal fraction on ice or store at −90°C, as for the S9 fraction. The microsomal fraction should not be prepared from S9 if the latter has already been frozen.

Metabolism of test chemical using liver S9 of microsomes and calf thymus DNA. The following incubation system can be used:

NADP (12.5 mg/ml in water)	0.1 ml
Glucose-6-phosphate (60 mg/ml in water)	0.1 ml
250 mM MgCl$_2$	0.1 ml
500 mM sodium phosphate buffer pH 7.4	0.6 ml
25% liver S-9 or microsomes	1.0 ml
150 mM KCl	0.5 ml
DNA (20 mg/ml in 150 mM KCl)	0.5 ml
Radiolabelled carcinogen in solvent	x μl
(concentration to be determined, volume of non-aqueous solvent to be <1% of total)	
Glucose-6-phosphate dehydrogenase (for microsomes only)	1 unit

Chemical sources. Nicotinamide adenine diphosphonucleotide disodium salt (NADP), glucose-6-phosphate, glucose-6-phosphate dehydrogenase (from yeast, Grade 1) and calf thymus DNA; all from Boehringer Corporation Limited, Lewes, Sussex. The amounts of liver S-9 or microsomes can be varied remembering to ensure that the above final concentration of KCl is maintained. Carcinogen concentrations can vary between 1 and 10 mM, depending on the optimum concentration for activation. DNA amounts can also be varied.

Conical flasks (25 ml) are used to prepare the incubation mixture and the whole is stood on ice until it is desired to start the experiment. Large-scale incubations are performed by increasing the volumes and ingredients to maintain the same concentrations. To metabolise the chemical, place the flasks in a 37°C shaker/water bath and shake vigorously (100 strokes/min) for the incubation period. It has been shown that for most substrates, metabolism is only linear for some 15 − 20 min. Nevertheless, one is not interested here in enzyme kinetics but in achieving maximum substitution of the DNA.

Recovery of the DNA. DNA is recovered using the procedures outlined later. It may be advantageous to extract any unmetabolised carcinogen using a suitable solvent in order to remove the bulk of the radioactivity. We have found a convenient method for removing most of the unbound radioactivity is to add 2 volumes of methanol to the incubation mixture. This will cause all the proteins and nucleic acids to precipitate. These are recovered by centrifuging at 2000 *g* at 4°C for 10 min. Remove the supernatant containing metabolites and unmetabolised carcinogen with a Pasteur pipette and resuspend the pellet in MUPSE (see later) prior to purifying the DNA.

Either RNA or a protein such as bovine serum albumin can be added to a metabolism incubation in place of DNA if it is desired to determine whether these macromolecules can trap activated metabolites. Both are recovered using similar techniques to those described later for DNA. Indeed it is possible, using the procedures outlined in this chapter, to recover all three macromolecules.

(ii) *In vitro metabolism using liver slices.* Several publications have demonstrated that *in vitro* metabolism using liver S9 or microsomes can lead to the production of metabolites which are not found *in vivo*. This arises, presumably, because the balance between activation and detoxification is altered by disrupting cell structure. In addition, liver microsomal studies only represent one part of the total metabolic pathways for xenobiotics (phase 1). To enable phase 2 reactions (conjugations) to proceed in microsomal incubations it is necessary to add co-factors and substrates. To obtain data more relevant to the *in vivo* situation, therefore, whole cell studies can be performed in which cell metabolism continues to reflect that in the whole animal for a period of time. The simplest way to perform whole cell studies is to utilise liver slices.

Animals. Dose animals of the required species the night before with a few drops of α-tocopherol dissolved in corn oil (50:50 v/v solution). To prepare liver slices, anaesthetise the animals with ether, bleed from the carotid artery and kill by cervical dislocation. Remove the livers and wash with ice-cold 0.9% sodium chloride.

Obtaining liver slices.

(a) Keep the livers on a piece of foil on ice and moisten with ice-cold isotonic saline.

(b) Remove approximately 5 g of liver lobe and place on a piece of filter paper mounted on the stage of a Stadie-Riggs microtome (Arthur Thomas, Philadelphia, USA, obtained through A.R.Horwell, London). Keep the liver constantly moist with ice-cold saline during the cutting of liver slices.

(c) The microtome blades, which should only be used to cut slices from one liver, are covered with a protective layer of grease which should be rinsed off with acetone and dried before use (mind your fingers).

(d) Mount the upper part of the microtome on to the stage, wet the blade with saline and insert it into the guide.

(e) Cut a slice of liver using a sawing movement and remove the slice with a pair of forceps. Always discard the first slice of liver lobe as it consists largely of liver capsule.

Table 1. Composition of Ringer Phosphate Solution.

NaCl	125 mM
KCl	6 mM
MgSO$_4$	1.2 mM
NaH$_2$PO$_4$	1 mM
CaCl$_2$	1.2 mM
Glucose	10 mM
Hepes buffer	15 mM
Adjust pH to 7.4 with NaOH	

(f)   Section the remainder of the lobe into slices as thin as possible, gently drying each slice on a piece of filter paper, weigh and then drop into ice-cold Ringer solution. Slices should weigh approximately 250 mg. Up to three slices can be combined per 10 ml Ringer solution in 50 ml conical flasks. The composition of Ringer phosphate solution is given in *Table 1*.

(g)   Incubate up to three slices per 10 ml of Ringer medium in 50 ml conical flasks at 37°C. Incubation should be under an atmosphere of oxygen using a fitted hood and shaking should be vigorous (100 strokes/min). The thinner the slices that have been cut the less the chance that anoxic cells develop and hence metabolism will remain linear for up to 3 h.

(h)   Add radiolabelled carcinogen at concentrations between 1 and 10 mM, depending on previously determined optima. Liver slices are relatively easy to prepare and are robust. Care should always be taken during their preparation to keep them moist and cold.

Isolation of DNA. Remove slices from the incubation medium and homogenise in MUPSE using a pestle and glass homogeniser. Recover DNA as described later.

(iii) *In vitro metabolism using rat hepatocytes.* One may wish to examine carcinogen activation in a pure population of liver parenchymal cells rather than a mixed population, such as found in liver slices. Various techniques are available for recovering hepatocytes to study the isolated cells (see Chapter 3). We shall describe a technique using *in vivo* perfusion with collagenase.

Animals. Rats are normally used for these studies. Anaesthetise them with nembutal at 50 mg/kg, administered by intraperitoneal injection. Make an incision into the abdomen from the xiphisternum to the pubic bone to expose the liver.

Equipment. Use a variable speed peristaltic pump fitted with 1.5-mm internal diameter flow tubing and a bubble trap to prevent obstruction of the portal vein during perfusion. A 3-way valve is fitted to the tubing on the negative pressure side of the pump. Sterilise the system by circulation of 70% ethanol followed by sterile distilled water. Connect a 21-gauge butterfly needle to the end of the peristaltic tubing for insertion into the portal vein.

Procedure.

(a)   Insert the butterfly needle into the portal vein and clamp with artery forceps.

(b)   Start the perfusion by turning the pump on and pumping sterile solution A (see

Table 2. Media Composition for In Vivo Perfusion with Collagenase.

Solution A		
EGTA	0.5	mM
.NaCl	137	mM
KCl	5	mM
Hepes buffer	10	mM
Glucose	6	mM
Phenol red	0.02 g/l	
NaHCO$_3$	4	mM
Adjust solution pH to 7.2		
Solution B		
Collagenase type 1 (Sigma Chemicals)	100	U/ml
Hepes	10	mM
Williams medium E (WME) obtained		
from Flow Laboratories	Adjust solution pH to 7.4	
EDTA medium	*Weight per litre*	
EDTA	0.25 g	
NaCl	8.0 g	
KCl	0.25 g	
Na$_2$HPO$_4$	1.15 g	
KH$_2$PO$_4$	0.25 g	
Glucose	0.25 g	
Distilled water	to 1 litre	

Table 2), pre-warmed to 37°C, through the liver at a flow-rate of 8 ml/min *via* the 3-way valve.

(c) Immediately sever the subhepatic vena cava to permit perfusate to escape. The liver should lighten in colour as the blood is washed from the circulation. Continue the perfusion for 1.5 min. During this period cannulate the thoracic vena cava by puncturing the right atrium and collect the perfusate using a return cannula.

(d) Clamp the proximate position of the subhepatic inferior vena cava to close the system and increase the pump speed to 40 ml/min for 2.5 min.

(e) Then switch the 3-way valve so that pre-warmed sterile solution B (see *Table 2*) is perfused through the liver. Perfuse the liver for a further 10 min without return of the perfusate to the system. The liver is kept warm at this stage by shining a 40 W light onto it.

(f) Remove the liver to a sterile Petri dish containing 50 ml of solution B. Holding the portis hepatis with forceps, open the capsule of the liver on the inferior surface with small scissors and remove it.

(g) Detach the cells by gently combing with a sterile stainless steel round-toothed comb and shaking them. After complete combing, hepatocytes are dissociated into a suspension leaving a fibrous plug of hepatic connective tissue, which is discarded.

(h) Transfer aliquots of the suspension to 50-ml centrifuge tubes using a wide-bore pipette. Add Williams medium E (WME), supplemented with 10% foetal calf serum and 50 μg/ml gentamycin (WMES), to each tube (25 ml) and sediment the cells at 50 *g* for 5 min. Resuspend the cells in 40 ml WMES per tube by gentle inversion.

(i)   Prepare a 20-fold dilution of cell suspension and add 0.5 ml to 0.1 ml 0.4% trypan blue.

(j)   Determine the viability and concentration of the suspension using a haemocytometer. The expected yield is approximately 2×10^8 cells per 100 g body weight with a viability of at least 90%.

(k)   Adjust the cell suspension to 5×10^5 cells/ml using warm WMES and seed aliquots into 3.5-cm plastic Petri dishes.

(l)   Incubate the cultures at 37°C in a humidified 95% air/5% CO_2 incubator for 2 h. Then remove the medium and add warm fresh WMES containing the radiolabelled carcinogen at the appropriate concentration.

(m)   Incubate the cells for 3 h, remove the medium and wash the cell sheet with sterile phosphate-buffered saline.

(n)   Dislodge the cells from the dish by incubating with warm EDTA medium (see *Table 2*) for 5 min at 37°C, and remove them from the dish by rinsing up and down with a Pasteur pipette, transferring to a centrifuge tube and centrifuging at 1000 *g* for 10 min.

(o)   Remove the supernatant, resuspend the cells in MUPSE and isolate the DNA as described later.

2.3.2 *In Vivo Studies*

In many instances it is desired to study macromolecular binding *in vivo* since the whole animal is more likely to represent man than are single cells or cultures. To obtain the highest extent of binding to macromolecules the radiolabelled carcinogen is administered by either intraperitoneal or intravenous injection. Intubation usually gives lower levels of binding due to factors such as efficiency of absorption from the gastro-intestinal tract. It is important to have radiolabelled carcinogen with a specific activity of at least 100 mCi/mmol. Up to 5 mCi or more of compound may need to be injected, particularly for carcinogens or toxins that bind to macromolecules at low levels. Compound solutions are generally prepared fresh in a suitable solvent such as dimethyl sulphoxide (DMSO), saline or corn oil. It is important to keep injection volumes as small as possible to avoid toxic effects, particularly for DMSO.

Treatment of animals represents the greatest risk in terms of contaminating laboratories, animal rooms, equipment and personnel with radioactivity. Facilities not equipped for such studies should not attempt them.

Once the compound has been administered, animals are killed at time intervals thereafter and organs removed and dropped into ice-cold isotonic saline. After washing, organs are then stored at −90°C until required for processing and analysis, as described below.

2.4 Extraction and Purification of DNA-carcinogen Adducts from Treated Animals

Thaw the organs and wash in ice-cold 150 mM KCl. Excise a small piece of the organ (500 mg) and either homogenise it or dissolve it in Soluene (Packard Instruments Ltd.) prior to scintillation counting to determine the amount of radiolabelled carcinogen present

Table 3. Reagents for the Hydroxylapatite Method of DNA Purification.

MUP — SDS — EDTA (1 litre) (MUPSE)

Chemical	Wt/litre	Final concentration
Urea	480.5 g	8 M
$NaH_2PO_4.H_2O$	33.0 g	0.24 M
SDS[a]	10.0 g	0.035 M
Disodium EDTA.2H$_2$O	3.7 g	0.01 M
Distilled water		500 ml in first instance

Adjust pH to 6.8 with concentrated NaOH and then make up to 1 litre with distilled water.

MUP (Filtered)

Chemical	Wt/litre	Final concentration
Urea	480.5 g	8 M
$NaH_2PO_4.H_2O$	33.0 g	0.24 M
Distilled water		500 ml in first instance

Adjust pH to 6.8 with concentrated NaOH and then make up to 1 litre with distilled water.

Chloroform-isoamyl alcohol-phenol (CIP)

Chemical	Wt/litre
Chloroform	480 ml
Isoamyl alcohol	20 ml
Phenol (crystals)	500 g

The reagents must all be analytical reagent grade.
[a]It is important to use Sigma SDS; we have encountered problems when SDS from other manufacturers was used.

in the organ. As an example, we shall consider the extraction and purification of DNA from the liver of a rat injected with radiolabelled carcinogen 24 h prior to sacrifice.

Numerous protocols have been developed and published for the purification of DNA from living tissue but, in general, they are all modifications of two basic methods; the first being a combination of solvent extraction and enzymic digestion; and the second involving DNA grade hydroxylapatite (Calbiochem) as the final purification step. We shall consider one example of each procedure, and then finally consider modifications required to extract DNA from prokaryotic cells.

2.4.1 *Hydroxylapatite Method* (6)

The reagents needed for this method are given in *Table 3*. The MUPSE solution should be adjusted to pH 6.8 with concentrated NaOH and the volume then made up to 1 litre with distilled water. Filter the solution through Whatman pre-folded, fluted filter paper.

The procedure described below is for a liver weighing approximately 10 g. For differing weights the volumes of reagents should be increased or decreased *pro rata*.

(i)   Transfer the liver to a 150-ml stoppered glass centrifuge tube and add 50 ml of MUPSE. Mince the liver with scissors prior to homogenisation using an Ultra-Turrax blender at full speed for 15 sec.

(ii)   Cool the tube in ice for 2 min and repeat the homogenisation procedure twice more, cooling in ice between each 15-sec period. This procedure should be carried out in a suitable fume cupboard to prevent exposure to radioactive aerosols.

(iii)   Divide the homogenate equally between two glass centrifuge tubes and add an equal volume of MUPSE-saturated CIP to each tube. Stir the mixture at room temperature for 3 min.

(iv)   Centrifuge the tubes at 500 *g* for 3 min at 14°C, after which remove the lower organic layer, containing protein, using a submerged pipette and leaving behind the upper fluffy, white layer.

(v)   Repeat the phenol extraction (MUPSE-CIP) and centrifugation steps twice more.

(vi)   Finally, add an equal volume of diethyl ether to each tube and thoroughly mix the two layers by vigorous stirring for 3 min.

(vii)   After centrifugation, remove and discard the upper organic layer and repeat the extraction. The lower clear layers are then pooled and applied to a hydroxyl-apatite column.

The column is prepared as follows.

(i)   Suspend 15 g of DNA grade hydroxylapatite in 100 ml of 0.014 M sodium phosphate buffer, pH 6.8. Swirl the slurry and leave it to settle for a few minutes.

(ii)   Decant the fines and repeat the process using 100 ml MUP. Then pour the MUP slurry into a 2.6 × 40 cm glass column containing 2 ml MUP.

(iii)   Allow the slurry to settle under gravity and then pump 100 ml of MUP through at 1 ml/min until the u.v. absorbance at 254 nm of the eluent is constant.

(iv)   Remove the liquid head from the column and pump the sample on at 1 ml/min.

(v)   Replace the liquid head and elute the sample with MUP at 1 ml/min.

(vi)   Monitor the u.v. absorbance of the eluent at 254 nm and the first peak observed contains the RNA.

(vii)   When the u.v. absorption again returns to baseline, change the eluent to 0.48 M sodium phosphate, pH 6.8.

(viii) Elute the DNA and collect the u.v.-absorbing peak and dialyse it against 7 litres of distilled water overnight.

(ix)   Flash evaporate the dialysate or for labile adducts use ultrafiltration to reduce the volume to approximately 15 ml and dialyse against 5 mM Bis Tris pH 7.1 for 24 h.

(x)   Measure the volume of the dialysate and add 1/10 of the volume of 1 M NaCl, followed by 2 volumes of ice-cold ethanol. At this stage the DNA should precipitate; however, if this does not happen immediately, store the tubes at −20°C overnight and pellet the precipitated DNA by centrifugation.

(xi)   Wash the DNA with ethanol:acetone 1:1 and finally acetone.

(xii)   Dry the sample under reduced pressure and store at −20°C until required for analysis.

2.4.2 *Solvent Extraction — Enzymic Digestion Method* (7)

Phenol reagent:

Phenol   500 g

Tris-HCl   50 mM added until present in excess

The enzymes quoted below were purchased from Sigma Chemical Co.

(i)   Place the liver in a 150-ml stoppered glass centrifuge tube containing 50 ml 1%

SDS-1 mM EDTA and homogenise in an Ultra-Turrax blender for 15 sec at full speed, cool in ice and homogenise twice more with intermittent cooling. (N.B., to prevent excessive foaming the liver can be homogenised in EDTA and the SDS added afterwards with thorough mixing.)

(ii)   Centrifuge the homogenate at 3500 *g* for 5 min to disperse any foam and then add 25 mg of Proteinase K in 2.5 ml water.

(iii) Incubate the mixture at 37°C for 30 min.

(iv)   Following addition of 2.5 ml 1 M Tris-HCl, pH 8.0, divide the suspension between two centrifuge tubes and extract with 50 ml phenol. (The phenol and all other organic solvents used in the extraction are saturated with 50 mM Tris-HCl pH 8.0.)

(v)   Following centrifugation at 3500 *g* for 5 min, extract the aqueous layer sequentially with 50 ml of phenol:chloroform:isoamyl alcohol (25:24:1) followed by chloroform:isoamyl alcohol (24:1).

(vi)   Add 5 M NaCl (5 ml) to the aqueous layer and precipitate the DNA by the addition of 100 ml of ice-cold ethanol.

(vii) Pool the precipitates, wash twice with 50 ml of 70% ethanol and re-dissolve them in 10 ml of 1.5 mM NaCl:0.15 mM trisodium citrate:1 mM EDTA buffer. Add to this 100 μl of 1 M Tris-HCl, pH 7.4, 750 units of RNase T1 and 1.5 mg of RNase A (in 1.5 ml water, previously heated at 90°C for 10 min).

(viii) Incubate the mixture at 37°C for 15 min, and then add 625 μl of 1 M Tris-HCl, pH 8.0 and extract the whole with 15 ml of phenol:chloroform:isoamyl alcohol (25:24:1).

(ix)   Finally extract the aqueous layer with 15 ml of chloroform:isoamyl alcohol (24:1) and precipitate the DNA by the sequential addition of 1.5 ml of 5 M NaCl and 30 ml of ice-cold ethanol.

(x)   Wash the precipitate with 20 ml of ethanol, then with 20 ml of ethanol:acetone (1:1) and, finally, with 20 ml of acetone.

(xi)   Dry the sample under reduced pressure or in a stream of argon and store at -20°C until required for analysis.

(i) *Extraction of protein.* Protein can be isolated from the phenol layer following the first phenol extraction of the liver homogenate.

(a)   Precipitate the protein by the addition of 2 volumes of methanol.

(b)   After 30 min, centrifuge the suspension at 5000 r.p.m. for 5 min. Then wash the protein pellet sequentially with: 6% TCA (twice); 80% ethanol; absolute ethanol; chloroform:methanol 2:1:6% TCA (three times, once with heating at 100°C).

(c)   Resuspend the pellet in each washing solution by vortexmixing, and then re-sediment by centrifugation at 1000 *g* for 5 min.

(d)   Finally dry the pellet in a stream of argon or digest with 1 M sodium hydroxide. Protein can then be estimated colourimetrically and the extent of binding of radiolabelled carcinogen determined by scintillation counting.

(ii) *Extraction of RNA.*

(a)   Proceed as with the extraction of DNA until the precipitation step. Instead of using ethanol, use 2-ethoxyethanol to precipitate the DNA.

(b)   Then precipitate RNA from the supernatant with excess absolute ethanol.

(c)   Leave the sample overnight at −20°C to allow complete precipitation of RNA which is then sedimented at 10 000 *g* for 15 min.

(d)   Wash the pellet of RNA sequentially with 2 ml of: ice-cold 3 M sodium acetate pH 6.0 (twice); 75% ethanol; and absolute ethanol.

(e)   Redissolve the RNA in 15 mM NaCl/1.5 mM sodium citrate buffer and estimate its concentration colourimetrically. The extent of binding can then be determined from the radioactivity of the solution.

2.5 Preparation of DNA for Quantitation and for Liquid Scintillation Counting

(i)   Dissolve the DNA sample derived from *in vivo* studies in 50 mM Bis Tris buffer, pH 7.1, containing 5 mM $MgCl_2$. At this stage it may be desired to quantitate the extent of carcinogen binding to the purified DNA.

(ii)   Remove an aliquot of DNA solution to a scintillation vial and partially hydrolyse it with DNase I for 30 min at 37°C using 0.1 μg enzyme per μg DNA.

(iii)   Add scintillant and measure the radioactivity of the sample.

(iv)   Remove a second aliquot of the DNA solution and estimate the amount of DNA present using the Burton colourimetric assay (8).

For preparation of diphenylamine reagent, dissolve 2 g of diphenylamine in approximately 80 ml of glacial acetic acid. Add 60% perchloric acid (4.4 ml), followed by 0.5 ml 2% acetaldehyde and make the solution up to 100 ml with the further addition of glacial acetic acid.

For preparation of DNA standards, prepare a stock solution of DNA at 1 mg/ml by dissolving a small, accurately weighed amount of calf thymus DNA in distilled water equivalent to one-half the final volume. Make this up to volume with 10% trichloroacetic acid (TCA) and heat the solution at 90°C for 20 min. It may then be stored in a refrigerator. Dilute 1.0 ml of the above solution with 9.0 ml of 5% TCA to give a working stock solution of 100 μg/ml. From this, prepare a series of standards by dilution in 5% TCA to a final volume of 1.0 ml. Standards are prepared at 0, 10, 20, 40, 60, 80 and 100 μg DNA/ml.

For sample preparation, dilute an aliquot of the DNA solution with 5% TCA to give a final concentration within the standard range. Heat 1.0 ml of this dilution at 90°C for 20 min and allow it to cool.

In order to assay the sample, add 2.0 ml of diphenylamine reagent to all tubes, which are then capped and left at room temperature in the dark overnight. Read all samples against the TCA blank at 600 nm in a spectrophotometer. A standard curve is drawn and the concentration of DNA in the test sample is determined from the curve.

The extent of covalent binding to DNA is then calculated from the ratio of the specific radioactivity of the radiolabelled carcinogen and the specific radioactivity of the DNA.

Returning now to the DNA solution in Bis Tris buffer, the sample is enzymically hydrolysed, as described later for DNA reacted *in vitro*. The volume of buffer and the amounts of enzyme used are scaled down accordingly. It is usually found that the best results are achieved when the final DNA concentration is between 1.5 and 2.0 mg/ml. Following hydrolysis, the adducts are preferentially extracted into n-butanol and prepared for h.p.l.c. analysis, again as described later for *in vitro* reacted DNA samples.

In certain circumstances it may be desirable to analyse the DNA adducts as base adducts rather than as nucleoside adducts. To achieve this, the DNA is dissolved at about 1 mg/ml in a weak buffer such as 15 mM NaCl/1.5 mM sodium citrate. 1 M HCl is then added to give a final concentration of 0.1 M. Addition of the acid causes precipitation of the DNA. The suspension is heated at 70°C for 20 min, during which time the DNA is depurinated leaving acid-soluble oligonucleotides, free guanine, adenine and purine-carcinogen adducts. Following neutralisation, this mixture can be analysed by h.p.l.c.

2.6 High-pressure Liquid Chromatographic Analysis of Hydrolysed DNA

Using as our example DNA which has been isolated and extracted from the liver of a rat treated with a radiolabelled aromatic amine, one should proceed as follows.

(i)   The residue from rotary evaporation of the n-butanol extract of enzymically hydrolysed DNA (see 2.5 and 2.9) will contain carcinogen-nucleoside adducts produced in the DNA. Take this up in 1.0 or 2.0 ml of 30% methanol and determine its radioactivity.

(ii)   Inject an aliquot (containing preferably > 1000 d.p.m.) onto a C_{18} reverse phase column.

(iii)   Elute the column with a gradient of methanol in water, the exact shape of which will depend on the polarity and other properties of the adducts to be identified. In the case of covalently adducted benzidine-DNA, for example, the gradient which proved most satisfactory was 30% methanol for 10 min, 30−70% methanol over 10 min, 70−100% methanol over 5 min and 100% methanol for 5 min (9).

(iv)   During the gradient, collect 30-sec fractions and remove an aliquot from each fraction to determine the radioactivity.

(v)   Finally, plot a graph of fraction number (or time) *versus* radioactivity.

After each run it is advisable to equilibrate the column to the gradient-starting conditions at a rate no greater than 10% per min and, in no case, is it advisable to start at a methanol concentration lower than 5% (see also section 2.1).

2.7 Determination and Synthesis of Reactive Intermediates

In order to determine the structure of *in vivo* DNA adducts following treatment of animals with a carcinogen, it is necessary to generate enough adduct to enable physico-chemical measurements, such as mass- and nuclear magnetic resonance spectra, to be made. This cannot usually be achieved using the *in vivo* system and normally necessitates the synthesis of a presumed proximate reactive intermediate followed by reaction with pure DNA on a large scale *in vitro*. The DNA is then hydrolysed and the adduct purified by repeated semi-quantitative h.p.l.c.

Clearly, at this stage, the likely structure of the presumed reactive intermediate can only be guessed at. However, a knowledge of the activation metabolism of the general chemical class to which the compound under study belongs, will help in more accurately predicting this species. In certain circumstances the intermediate is so reactive that it cannot be isolated (e.g., the olefinic epoxide of aflatoxin B_1) and thus it may be necessary to synthesise a less reactive species which will, nevertheless, produce the same adduct when reacted with DNA (aflatoxin dichloride in the example stated). Another example is the synthesis of N-benzoyloxy-N-methyl-4-aminoazobenzene, which was synthesised by the Millers (10) as an analogue of the less stable N-OH-N-methylaminoazobenzene, though this latter compound was eventually synthesised (11).

To exemplify the approach and practical procedures involved in synthesising a presumptive reactive intermediate we shall describe the case for benzidine (9).

From a knowledge of the metabolism of aromatic amines, it was generally assumed that the likely activation of benzidine would be *via* production of the proximate carcinogenic species N-hydroxybenzidine. It was thought that this would then probably undergo O-esterification to produce the ultimate carcinogenic species which could be N-acetoxybenzidine, N-sulphonyloxybenzidine or similar. However, when rats were injected with either [2,2'-^{3}H]benzidine or [acetyl-^{3}H]acetylbenzidine and the liver DNA analysed 24 h later, the same radiolabelled adduct was detected. Thus it was clear that, prior to probable N-hydroxylation, benzidine was N-acetylated. The fact that N,N'-diacetylbenzidine bound only poorly to rat liver DNA suggested that the presence of an acetyl moiety on the nitrogen atom prior to N-hydroxylation was de-activating. From these data it was assumed that the most likely intermediate in the activation of benzidine to a DNA-binding species would be N-OH-N'-acetylbenzidine. This compound was therefore synthesised from benzidine *via* N'-acetylbenzidine and 4-nitro-4'-acetyl-aminobiphenyl as intermediates.

2.8 Reaction of Postulated Reactive Intermediate with DNA In Vitro

The hydroxylamine produced (N-OH-N'-acetylbenzidine) in the above example is allowed to react with calf thymus DNA.

(i)   Dissolve the DNA (500 mg) in 50 ml of water to which 50 ml of 20 mM potassium citrate-2 mM EDTA buffer, pH 4.6, are added.

(ii)   Saturate the solution with argon, add 2 ml of 0.025 M N-OH-N'-acetylbenzidine in ethanol:DMSO (4:1), and incubate the mixture for 4 h at 37°C. The mixture can then be stored overnight at −20°C if desired.

(iii)   Extract the DNA solution four times with equal volumes of water-saturated n-butanol.

(iv)   Discard the butanol extract and add 10 ml of 1 M NaCl to the aqueous fraction, followed by 500 ml of ice-cold ethanol.

(v)   Spool the precipitated DNA onto a glass rod, squeeze it against the glass container and re-dissolve it in 100 ml of water. The DNA is again precipitated by adding 10 ml 5 M NaCl followed by 500 ml of ice-cold ethanol:acetone (1:1).

(vi)   Collect the DNA as before and wash with acetone and leave to soak for 15 min.

(vii)   Finally pour off the acetone and dry the DNA under reduced pressure or in a stream of argon.

2.9 Hydrolysis of DNA to Deoxyribonucleoside Adducts

(i)   Dissolve the DNA ($\sim$500 mg) from the large-scale reaction in 400 ml of 50 mM Bis Tris buffer, pH 7.1, containing 5 mM $MgCl_2$. (For smaller amounts of DNA, for example samples from rat liver, the amounts quoted are reduced *pro rata* per mg of DNA.)

(ii)   To this add 40 mg of DNase (Sigma DN-Cl) dissolved in 4 ml of 0.9% NaCl.

(iii)   Incubate the solution under argon at 37°C for 3 h. Then add nuclease P1 (16 ml of a 2 mg/ml solution in 1 mM $ZnCl_2$), followed by 200 units of alkaline phosphatase (Sigma type III-S) and 120 units of acid phosphatase (Sigma type I).

(iv)   Incubate the mixture for a further 16 h. Add ammonium sulphate (15 g) to precipitate the enzymic protein. Nucleoside adducts can then usually be recovered by preferential extraction into water-saturated n-butanol (twice with equal volumes).

(v)   Wash the combined butanol extracts once with 400 ml of water containing 15 g ammonium sulphate, and then evaporate to dryness in a rotary evaporator.

(vi)   Take the adduct up in a minimum volume of 20 − 50% methanol (usually between 10 and 50 ml.) Apply the solution to an h.p.l.c. column (semi-prep) in small aliquots using, wherever possible, an isocratic system because repeated injections will be necessary.

(vii)   Collect the adduct peak(s) and reduce it to dryness in a pear-shaped flask. Care should be taken at this stage to use only acid-washed glassware.

(viii) Dissolve the adduct in a minimum volume of suitable re-distilled volatile solvent and transfer to sample tubes for n.m.r. and mass spectral analysis. Reduce the samples to dryness in a stream of argon and place the tubes in a *grease-free* vacuum desiccator until required for analysis.

[N.B. If pure peaks are not obtained from h.p.l.c. then re-chromatography will be required collecting only the central part of the peak for re-application to the column.]

The sample is then analysed by high resolution mass spectrometry to determine its molecular weight. We have found derivatisation with trimethylsilane and analysis by electron impact in-beam desorption has yielded very acceptable mass spectra. Nuclear magnetic resonance spectra are also run on the adduct and compared with spectra obtained from the parent compound (N-acetylbenzidine in this example) and the suspected target nucleoside (deoxyguanosine). It is of course a great help to have experience in interpreting n.m.r. spectra of DNA adducts and to have available spectra of DNA adducts with other carcinogens for comparison. Study of these spectra (e.g., 9,12) in conjunction with data obtained from pH-dependent partitioning experiments (e.g., 9,13,14), pH stability data and u.v. spectral analysis (9,14) should enable an unequivocal determination of the molecular structure of the adduct in question.

2.10 Proof of Structure of Biologically-Produced Adduct

Having synthesised the postulated reactive intermediate of the carcinogen and reacted it with DNA and determined its structure, it is finally necessary to show by as many means as possible that the adduct generated has the same properties and characteristics as the adduct generated *in vivo*. This is usually achieved by demonstrating co-chromatography of the *in vitro* and *in vivo* derived adducts in at least two different chromatographic systems. Clearly, the more ways in which the two samples can be shown to behave similarly the more sure one can be that they are identical. Some of the techniques can be applied to a wide variety of adduct types: such as showing adducts have the same u.v. spectra which behave similarly with change in pH; showing that h.p.l.c. profiles remain similar with changes in pH; and showing that adducts behave similarly in pH-dependent solvent partitioning experiments. Other techniques, however, will be specific for the type of adduct with which one is dealing, e.g., in our benzidine example it was found that the acetyl moiety of the adduct N-(deoxyguanosin-8-yl)-N'-acetylbenzidine could be cleaved by the enzyme carboxylesterase. Thus if the *in vitro* and *in vivo* derived adducts are treated with this enzyme then injection onto reverse-phase h.p.l.c. yields in each case N-(deoxyguanosin-8-yl)benzidine.

3. REFERENCES

1. Miller,J.A. (1970) *Cancer Res.*, **30**, 559.
2. Miller,J.A. and Miller,E.C. (1971) *J. Natl. Cancer Inst.*, **47**, V.
3. Miller,E.C. (1978) *Cancer Res.*, **38**, 1479.
4. Farber,E. (1984) *Cancer Res.*, **44**, 4217.
5. Hemminki,K. and Ludlum,D.B. (1984) *J. Natl. Cancer Inst.*, **73**, 1021.
6. Beland,F.A., Dooley,K.L. and Casciano,D.A. (1979) *J. Chromatogr.*, **174**, 177.
7. Stanton,C.A., Chow,F.L., Phillips,D.A., Grover,P.L., Garner,R.C. and Martin,C.N. (1985) *Carcinogenesis*, **6**, 535.
8. Burton,K. (1956) *Biochem. J.*, **62**, 315.
9. Martin,C.N., Beland,F.A., Roth,R.W. and Kadlubar,F.F. (1982) *Cancer Res.*, **42**, 2678.
10. Poirier,L.A., Miller,J.A., Miller,E.C. and Sato,K. (1967) *Cancer Res.*, **27**, 1600.
11. Kadlubar,F.F., Miller,S.A. and Miller,E.C. (1976) *Cancer Res.*, **36**, 1196.
12. Coles,B.F., Welch,A.M., Hertzog,P.J., Lindsay Smith,J.R. and Garner,R.C. (1980) *Carcinogenesis*, **1**, 79.
13. Moore,P.D. and Koreeda,M. (1976) *Biochem. Biophys. Res. Commun.*, **73**, 459.
14. Kadlubar,F.F., Unruh,L.E., Beland,F.A., Straub,K.M. and Evans,F.E. (1980) *Carcinogenesis*, **1**, 139.

CHAPTER 6

Evaluation of Lipid Peroxidation in Lipids and Biological Membranes

ERIC D.WILLS

1. INTRODUCTION

The study of lipid peroxidation began in the earlier years of this century with investigations of the phenomenon of 'rancidity' in food fats. It was observed that unsaturated fats readily underwent peroxidation in the presence of oxygen and that this process led to the formation of several short chain products which gave the fatty food a very undesirable flavour.

For many years peroxidation was considered in this context but during the past two decades interest in lipid peroxidation in living tissues has received serious attention as a potential toxicological hazard.

1.1 Chemistry of Lipid Peroxide Formation and Products Formed During Peroxidation

Unsaturated fatty acids will undergo peroxidation as free acids, triglycerides or as components of phospholipid. Pure fatty acids are useful as model systems for the study of lipid peroxidation, but in biological systems the major proportion of fatty acids will exist in a combined state either as triglycerides or as phospholipids.

Triglycerides are of major importance in food and in fat depots of the body, whereas the unsaturated fatty acids are important components of tissue phospholipids, especially of membranes. Earlier research indicated that unsaturated fatty acids of triglycerides were oxidised at rates which were slower than equivalent free fatty acids. However, more recent work has shown that the relative rates of metal-catalysed oxidation of trilinolein and linoleate depend largely on the experimental conditions and on the nature of the metals used.

Although few comparisons have been made of the peroxidation rates of phospholipids with those of free fatty acids and triglycerides, it is generally agreed that unsaturated fatty acid components of phospholipids peroxidise very rapidly. The peroxidation of phospholipids of membranes of subcellular components can be readily initiated and is very rapid (1). Similarly, artificial phospholipid micelles in liposomes can undergo rapid peroxidation.

Peroxidation of unsaturated fatty acids is usually initiated by free radicals and the sequence of reactions is illustrated in *Figure 1*. A free radical of the fatty acid is formed as a result of attack by radicals such as $OH^{\cdot}$ which are generated by irradiation, with u.v. or X-rays, or by intervention of catalytic metals such as iron, copper or metal

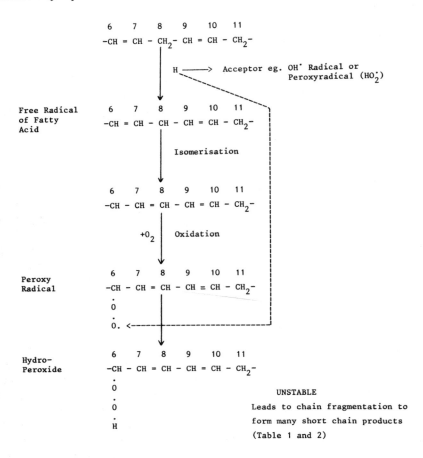

Figure 1. Mechanism of peroxidation of polyunsaturated fatty acids and antioxidant action.

complexes such as haemoglobin. These catalysts initiate the formation of peroxyradicals, R.OO˙ and oxyradicals R.O˙. Other species of active oxygen may be involved in the initiation reaction, as can radicals formed from halogenated hydrocarbons such as CCl_3˙ or CCl_3O_2˙. Isomerisation of the fatty acid free radical then occurs spontaneously followed by oxygen attack to form the peroxy radical. This radical will then attack another molecule of fatty acid, abstracting a H atom to form a fatty acid radical and a molecule of hydroperoxide, so that a chain reaction is initiated. It is thus clear that a very small quantity of OH˙ or other radicals can cause the peroxidation of large quantities of fatty acids.

The nature of the oxygen involved in the initiation of the reaction is still under investigation. The hydroxyl radical OH˙, superoxide anion O_2^-, singlet oxygen O_2˙ and the peroxy radical HO_2˙ have all been considered as possible initiators. The discovery of superoxide dismutase in cells led to the hypothesis that the superoxide radical was likely to be important, but the weight of experimental evidence is generally against this species playing an important role. It has also been suggested that singlet oxygen

($O_2^{\cdot}$) is the very reactive species of oxygen which will form the hydroperoxide. This species could be formed from superoxide anion.

$$O_2^- + O_2^- + 2H^+ > H_2O_2 + O_2^{\cdot}$$
$$O_2^- + H_2O_2 > OH^- + OH^{\cdot} + {}^{\cdot}O_2^{\cdot} \text{ singlet oxygen}$$

Many reactions, such as the autoxidation of haeme, the reaction of oxygen with iron-sulphur proteins, and the autoxidation of thiols are believed to form the superoxide anion. Although several investigations have indicated that singlet oxygen may be involved in lipid peroxidation, its presence during decomposition of superoxide anion or during the peroxidation of linoleic acid cannot always be detected.

Decomposition of hydrogen peroxide, in addition to generating singlet oxygen, will also form $OH^{\cdot}$ radicals which could be involved in the chain initiation and the importance of the role of $OH^{\cdot}$ radicals in lipid peroxidation has not yet been resolved (2). Recently support has been provided for believing that the peroxy radical $HO_2^{\cdot}$ is important in initiating lipid peroxidation. It is thus clear that the real nature of the 'active oxygen' species involved in lipid peroxidation and whether there is one specific species or whether several can play a role is still an open question. Free radicals can be generated by the fission of hydroperoxide catalysed by metals such as iron or copper (3):

$$R.OOH > RO^{\cdot} + HO^{\cdot}$$
$$R.OOH + Fe^{2+} > RO^{\cdot} + OH^- + Fe^{3+}$$
$$2\,R.OOH > R.OO^{\cdot} + R.OH + H_2O$$

All of these species of free radicals ($OH^{\cdot}$, $RO^{\cdot}$ and $R.OO^{\cdot}$) are capable of initiating the peroxidation chains. Only a very small quantity of hydroperoxide is therefore necessary for the initiation of the chain reaction and it is clear that these metals may play an important role in peroxidation by catalysing the decomposition of hydroperoxides and thus generating radicals which enable the chain reactions to proceed. The catalytic activity of metals such as iron in lipid peroxidation is strongly enhanced by ascorbic acid (1,3,4). The precise role which ascorbate plays is uncertain but it may reduce Fe^{3+} to Fe^{2+} to enable the catalysis of the decomposition of hydroperoxides to proceed (3).

Lipid hydroperoxides are very unstable and spontaneously decompose by chain cleavage to form very complex mixtures of aldehydes, ketones, alkanes, alkenes, carboxylic acids and polymerisation products (*Table 1*). The complete analysis of the pro-

Table 1. Products of Lipid Peroxidation[a].

---

1. *Chain cleavage and recurrent oxidation products*
   n-alkanals, 2-alkenals, 2,4-alkadienals, alkatrienals, -hydroxyaldehydes, hydroperoxialdehydes, 4-hydroxyalkenals, 4-hydroperoxialkenals, malonaldehyde, -dicarbonyls, saturated and unsaturated ketones, alkanes, alkenes.
2. *Rearrangement and consecutive products*
   hydroxyacids, keto-acids, keto-hydroxy-acids, epoxi-hydroxy acids, colneleic acid, dihydroxy acids, keto-dihydroxy acids, trihydroxy acids.
3. *Further peroxidation products*
   Cycloendoperoxides (PGG_2) and analogous compounds.
4. *Di- and polymerisation products*
   Di- and polymers linked by ether-peroxi-or C-C-bridges.

---

[a]Esterbauer (21).

Table 2. Analysis of the Products of Peroxidation of Linoleic and Arachidonic Acids[a].

Peroxidation product	nmol aldehyde/mg fatty acid	
	linoleic acid $C_{18:2}$	arachidonic acid $C_{20:4}$
Alkenals (total)	205.7	194.2
Pentanal	6.7	5.2
Hexanal	199.0	189.0
2-Alkenals (total)	25.8	30.9
Pentenal	0	0
Hexenal	1.1	0
Heptenal	10.1	6.2
Octenal	13.1	15.4
Nonenal	1.5	0.3
2,4-Alkadienals	0.44	10.6
Heptadienal	0	0
Nonadienal	0.36	4.6
Decadienal	0.08	6.0
4-Hydroxynonenal ($+20$ μM Fe^{2+})	10.3	10.8

[a]Esterbauer (21).

ducts of peroxidation is a very complex study and the nature and quantity of each product formed are only partially understood. It is very likely that the pattern of the products varies with the conditions of peroxidation and aldehydes formed from peroxidising fatty acids have been analysed and shown to vary with the fatty acid undergoing peroxidation (*Table 2*).

2. METHODS USED TO STUDY LIPID PEROXIDATION

A study of *Figure 1* will show that several methods could theoretically be used for the study of lipid peroxidation. These are:

(i)   measurement of oxygen utilised;
(ii)   measurement of hydroperoxides;
(iii)   measurement of degradation products of hydroperoxides and especially aldehydes.

In addition, direct measurements on the fatty acids involved in the peroxidation process are possible. The initial stage of peroxidation following a free radical attack causes the formation of a conjugated diene and this can be determined spectroscopically. Hydroperoxides usually decompose rapidly and spontaneously to produce many short chain products resulting in the decrease in concentration of unsaturated fatty acids undergoing peroxidation. The concentration of these fatty acids can be followed by gas-liquid chromatographic (g.l.c.) measurements.

In such a complex process as lipid peroxidation it is not possible to select any one method as greatly superior to any other. All have advantages and disadvantages and it is often desirable to apply several methods to a peroxidising system to obtain detailed information about the overall process. Several attempts have been made to correlate the various measurements but in view of the complexity of the peroxidation process these studies have, in general, only been of limited value.

2.1 **Measurement of Oxygen Uptake**

Oxygen uptake by peroxidising unsaturated fats was one of the earliest methods used to study peroxidation and can still be valuable. In biological systems oxygen may be utilised by several reactions and it is important to ensure, using adequate controls, that this is minimal. Oxygen uptake can be measured manometrically or using an oxygen electrode.

2.1.1 *Oxygen Uptake Measured Manometrically*

This method (3) is suitable for the study of the peroxidation of unsaturated fatty acids such as linoleic acid or linolenic acid in the form of aqueous emulsions.

Emulsions must be prepared immediately before use.

(i) Add 9 volumes of a suitable buffer (e.g., 0.025 M phosphate buffer pH 6.0 − 7.4) very rapidly, with gentle shaking to 1 volume of a solution (0.2 M) of the fatty acid dissolved in ethanol. This method of preparation initiates very little lipid peroxidation and produces good stable emulsions. Buffers with pH values less than 5.5 are unsuitable because the fatty acids precipitate out in globules, whereas buffers with a pH greater than 8.5 produce clear solutions of the fatty acids which normally oxidise at a much slower rate than emulsions.

(ii) Introduce 2.0 ml of the emulsion containing 0.02 M fatty acid into the main compartment of the manometer (5) and place a solution of the catalyst under study into the side arm.

(iii) Adjust the total volume of the mixed contents to 3.0 ml by addition of buffer.

(iv) Measure oxygen uptake for 60 min, although this perid can be prolonged for several hours if the rate is slow.

The method is convenient for the study of the catalytic activity of metals, both free and complexed (3) and of the catalytic activity of tissue homogenates and subcellular fractions (4).

2.1.2 *Oxygen Uptake Measured by the Oxygen Electrode*

Oxygen uptake can also be measured accurately on a much smaller scale than is necessary for conventional manometry using the oxygen electrode based on the original design of Clark (see Chapter 9). An Ag/AgCl platinum electrode immersed in a suitable electrolyte conducts an electric current which is proportional to the oxygen activity between the electrodes when a potential (usually −0.8 V) is maintained across the electrodes. It is normally essential to isolate the electrode by means of a thin Teflon membrane which is freely permeable to oxygen. A magnetic stirrer is necessary to help maintain the steady-state of oxygen tensions on both sides of the membrane. The oxygen at the platinum electrode surface is reduced and the current flow induced by these reactions develops a potential across an external resistance which is measured by a recorder. This voltage is proportional to the oxygen activity and to its partial pressure. Before use the recorder must be calibrated by setting the full scale deflection when oxygen-saturated medium is introduced into the electrode. Solutions saturated with a 25% oxygen:75% nitrogen mixture and 50% oxygen:50% nitrogen mixture are subsequently introduced into the chamber and the potential measured. In addition, or alter-

Table 3. Relationship Between Oxygen Uptake and Lipid Peroxide Formation in Incubated Microsomes[a].

Addition to system	Number of experiments	O_2 uptake/malonaldehyde formed molar ratio	
		Mean	S.D.
NADPH	128	15.50	± 4.60
Ascorbate	38	18.68	± 7.95

[a]The medium contained microsomal suspension (1.5 mg/ml) and NADPH (40 μM) in a total volume of 2.0 ml and was measured in the oxygen electrode for $1-2$ min. Sodium phosphate buffer (20 mM) was used in most experiments. Ascorbate (0.5 mM) replaced NADPH and nicotinamide in one series, and sodium phosphate buffer, pH 6.0, was used in most experiments in which ascorbate was present (1).

natively, the medium is made anaerobic (0% oxygen) by the addition of sodium dithionite. Plot the recorder readings against the percentage oxygen saturation of the medium.

The volume of the reaction mixture normally used is $1.0-2.0$ ml. The method has been used to study the oxygen uptake of peroxidising suspensions of a liver microsomal fraction (6). For studies of this type, the incubation mixture should contain $1.5-3.0$ mg microsomal protein/ml, in pH 7.0 sodium phosphate buffer (20 mM), and KCl (90 mM). Oxygen uptake is normally recorded for $0.5-5$ min. Effects of catalysts such as iron/ADP complexes and ascorbate can be studied by introducing suitable quantities into the chamber by means of a microsyringe. The relationship between the oxygen uptake and malonaldehyde produced in peroxidising suspensions of liver microsomes is shown in *Table 3* (6).

2.2 Measurement of Hydroperoxide Formation

The first major product of lipid peroxidation to be formed is a hydroperoxide (R.OOH) (*Figure 1*). Polyunsaturated fatty acids containing several double bonds undergoing peroxidation form several hydroperoxides simultaneously. Hydroperoxides, like all organic peroxides, possess 'peroxide' oxygen which is a powerful oxidising agent and this property may be used to measure true 'lipid peroxide'. Although, theoretically this is a simple procedure, it is technically difficult to obtain accurate and reliable results because hydroperoxides are generally very unstable and break down spontaneously to give complex mixtures of degradation products and, in addition, many other oxidising agents including oxygen itself or species of 'active oxygen' will interfere. Several simple organic peroxides such as benzoyl peroxide can be used as model compounds to test analytical methods for lipid peroxides.

2.2.1 *Oxidation of Fe^{2+} to Fe^{3+} — the Ferric Thiocyanate Method*

This method which is simple in principle, relies on the 'active' peroxide oxygen to oxidise ferrous ions to ferric ions:

$$Fe^{2+} + 2 H^+ + O > Fe^{3+} + H_2O$$

Ferric ions are then detected by addition of ammonium thiocyanate to form the intensely red-coloured ferric thiocyanate which can be determined spectrophotometrically. The method was first introduced in 1921 by Horst, developed by Young *et al.* (7) for the measurement of peroxides in petroleum products, and subsequently for rats by Lea (8). It has not found general favour because erratic results are frequently obtained and oxygen in the reagents often can seriously interfere. The method usually employed is that based on the method of Wagner *et al.* (9) in which peroxide containing a sample is added to a methanolic solution of ferrous thiocyanate.

Procedure.

(i)   Prepare the reagent by dissolving 1.0 g ammonium thiocyanate in methanol deaerated by saturation with nitrogen (200 ml) to which 1.0 ml of 25% (w/v) sulphuric acid has been added.

(ii)   Just before use add 0.2 g finely powdered ferrous sulphate.

(iii) Add the sample (0.4 ml) to 10 ml reagent, and allow to stand in the dark at 20°C for 10 min.

(iv)   Measure the absorbance at 450 nm.

(v)   Calibrate with standard solutions of ferric chloride.

The method is unsuitable for the study of peroxidation in biological systems such as membrane preparations because of the interference by other oxidising contaminants such as hydrogen peroxide. However, it can be used to study peroxide formation in pure unsaturated fatty acids catalysed by metals, or in tissue extracts (3, 4).

2.2.2 *Iodometric Methods*

Iodometric methods originally proposed by Marks and Morrell in 1929 (10) for the determination of hydroperoxides have been used extensively over many years for the measurement of peroxides in fats, and especially in the food industry. The method has also been used for the determination of peroxides in tissues. In principle iodide is oxidised to iodine by the 'active oxygen' of the peroxide to liberate free iodine which is determined.

$$2\ I' + O + 2H^+ \rightarrow I_2 + H_2O$$
$$\text{(Hydroperoxide)}$$

Determination of iodine by titration. The iodine produced was measured in early methods by titration with sodium thiosulphate. Oxygen must be displaced from all solutions. In its original form the method is only suitable for the determination of the hydroperoxide content of relatively large quantities of fat, e.g., $0.2 - 1.0$ g. Although it is possible to displace oxygen from all solutions by passing nitrogen, the following procedure is simple, effective and gives reliable and reproducible results (11).

(i)   Dissolve the fat sample $(0.2 - 1.0$ g) in 25 ml glacial acetic acid in a glass-stoppered conical flask. (Mixtures of solvents e.g., acetic acid:methanol or acetic acid:chloroform may be used if it proves difficult to dissolve the fat in the acetic acid, but it is essential to keep the percentage of acetic acid in the mixture as high as possible.)

(ii)   Add approximately 2 g of solid sodium bicarbonate and replace the stopper as the evolution of carbon dioxide begins to abate.

(iii)   Add 2 ml of a saturated solution of potassium iodide and *immediately* transfer the flask to the dark.

(iv)   Allow to stand for 1 h, add 100 ml of water, and titrate with standard sodium thiosulphate (0.01 − 0.005 N).

The method is standardised using a pure organic peroxide such as benzoyl peroxide, and can be scaled down using smaller quantities of material and smaller volumes of reagents in proportion.

Spectrophotometric determination of iodine. This method developed more recently (12) is based on the same principle as the iodometric methods previously described, but the iodine is determined by measuring the absorption band of the tri-iodide ion in the u.v. region. Excess iodide is converted into the more stable complex, cadmium-iodide-ion.

(i)   Extract the material containing oxidised fat (10 − 100 mg) with chloroform (1 ml/100 mg).

(ii)   Mix 1 ml of the chloroform extract with 8 ml of a glacial acetic acid:chloroform mixture (3:2) and 2 ml of distilled water. To prevent further peroxidation all reagents should contain anti-oxidants, either vitamin E or 2,6-ditert-butyl-p-methyl-phenol (BHT) (1 mg/ml).

(iii)   Shake the mixture thoroughly and separate the phases by centrifugation. Remove 2 ml of the chloroform phase, and add a small quantity of $NaHCO_3$. The acetic acid present in this phase liberates carbon dioxide, and this displaces oxygen from the solution and provides an inert atmosphere.

(iv)   Add 0.1 ml potassium iodide solution (1.2 g/ml), stopper the tubes, and leave in the dark for 30 min at 20°C. After this period has elapsed, add 6 ml of 0.5% cadmium acetate, centrifuge and remove the aqueous layer.

(v)   The absorbance of this layer is read at 350 nm. Calibrate using a solution of pure benzoyl peroxide in chloroform (0.2 − 0.8 μmol/ml).

This method is much more sensitive than the older titration method and thus can be used for the determination of relatively small quantities of fats and of preparations of cellular membranes (13). However, it is not specific for hydroperoxides, and many different oxidising agents will give a positive value. It has also been used to demonstrate hydroperoxide formation after treating unsaturated fats with ionising radiation (14).

More recently a simple micromethod has been described (15). In this procedure great care is taken to exclude oxygen by gassing all solutions with nitrogen which has been washed with alkaline pyrogallol (0.25%).

(i)   Extract the tissue with 19 volumes of a chloroform:methanol solution (2:1 v/v) containing α-tocopherol 0.5 mg/ml. Add methanol (6.3 ml/g tissue) and water (10.0 ml/g tissue).

(ii)   Mix thoroughly and recover the lipid from the chloroform layer by evaporating under nitrogen.

(iii)   Dissolve in a convenient volume of chloroform (e.g., 6.3 ml containing 5 − 20 mg lipid) add 0.5 ml of potassium iodine solution, allow to stand for 1 h in the dark and read the absorbance at 380 nm. 1 μ equivalent of iodine produces an absorbance of 1.226.

2.2.3 *Measurement of Hydroperoxide Formation by t.l.c. and h.p.l.c.*

During the past few years several efforts have been made to separate and determine products of lipid peroxidation using thin-layer chromatography (t.l.c.) and high-performance liquid chromatography (h.p.l.c.). These methods are much more sophisticated than earlier methods in that they require only a very small quantity of material and that several different products can be separated. Unfortunately, however, it is not possible to describe the techniques precisely for a number of reasons: the methods are in a developmental stage; they have been designed mainly to study the products of prostaglandin synthetases, and not primarily those of lipid peroxidation; and reliable standards of the oxidation products are not always available.

Extraction. Aqueous material, for example, a tissue suspension, must first be extracted to remove the lipid oxidation products. All solvents should be treated with an anti-oxidant such as BHT (1 mg/ml) to prevent further peroxidation of fatty acids during the experimental procedure.

(i)   Shake the aqueous sample (e.g., tissue extract) with 2 volumes of acetone, centrifuge and remove the supernatant. Alternatively 4 volumes of ethanol can be used.

(ii) Wash the precipitate with 1 volume of acetone and add to the original extract. Remove the neutral lipids by shaking with 2 volumes of petroleum ether or hexane, centrifuge and discard the upper phase.

(iii) Adjust the pH of the acetone phase to 4.0 − 4.5 by the addition of citric acid or formic acid, and then shake with 2 volumes of chloroform.

(iv) Centrifuge, and remove the lower chloroform layer.

(v)   Repeat the extraction with chloroform, and combine the extracts and then evaporate under vacuum in a rotary evaporator.

Thin-layer chromatography.

(i)   Dissolve the extracted oxidised lipids in chloroform:methanol (2:1 v/v) and spot (50 μl) on to a 20 cm × 20 cm or 20 cm × 5 cm LKD Whatman silicone plate using a Hamilton microsyringe. Some plates required 'activation' by placing in an oven at 110°C for 30 min, but LKD plates are sealed and do not require this treatment.

(ii) Develop the plates using either chloroform-methanol-acetic acid-water (90:8:1:0.8 by vol.) or the organic (upper) phase of a mixture of ethyl acetate-2,2,4-trimethyl-pentane-acetic acid-water (110:50:20:100, by vol.).

(iii) For the detection of hydroperoxides a simple adaptation of the ferric thiocyanate method (Section 2.2.1) is used. Spray the plates with a solution of freshly prepared solution of 0.7 g ferrous ammonium sulphate dissolved in 10 ml 5% ammonium thiocyanate containing 1 ml concentrated sulphuric acid.

Quantitation may be attempted by scanning with a microdensitometer but this is difficult because the spots rapidly fade. The method may be improved to study the peroxidation of ^{14}C-labelled pure fatty acids, or fatty acids incorporated into phospholipids of biological membranes. Proceed as described above, scrape the spots off and measure the radioactivity by scintillation counting after elution with a solvent such as methanol or diethylether.

Two-dimensional chromatography is claimed to give superior separations. Solvent systems which are suitable are ethyl acetate:hexane:acetic acid 75:25:2 (by vol.) or the organic phase of ethylacetate:acetic acid:isooctane:water, 100:10:30:100, (by vol.) followed by chloroform-ether-methanol-acetic acid (45:45:5:2, by vol.).

Separation of hydroperoxides by h.p.l.c. H.p.l.c. is the most sophisticated method available for the separation of peroxidation products but so far methods have not been developed specifically for the study of the formation of hydroperoxides. Methods have, however, been described for the separation of oxidation products formed from arachidonic acid by prostaglandin synthetases and they could be applied to the analysis of lipid hydroperoxides. A typical procedure described in detail (16) for reversed phase h.p.l.c. is summarised below.

(i)   Prepare an extract of the products of lipid peroxidation as described above in methods for t.l.c.

(ii)   Equilibrate an OOS h.p.l.c. column with 26% acetonitrile in water at pH 3.5 at a flow-rate of $1-3$ ml/min depending on the size of the column.

(iii)   Inject the sample and appropriate standards, dissolved in about $50-200$ μl acetone and elute the column isocratically for 23 min. After the initial isocratic period a convex gradient is run from 26% to 80% acetonitrile in water for 12 min. The gradient at 50% acetonitrile is interrupted to allow the separation of several components and after 20 min the gradient is resumed.

(iv)   Collect the samples using a suitable fraction collector.

(v)   Upon completion of the run, purge the column with 100% acetonitrile for 10 min and re-equilibrate with 26% acetonitrile.

(vi)   The eluates are detected by measurement of the absorbance at 234 nm or, if radiolabelled fatty acids have been used, by scintillation counting of the fractions.

Several different procedures have been proposed for h.p.l.c. separation of oxidation products of unsaturated fatty acids and the above is an example of the methods employed. For full details it is essential to consult the original literature.

2.2.4 *Dichlorofluorescein Assay*

Recently, a new method has been described for the detection of lipid hydroperoxides which is claimed to be much more sensitive than methods previously described. The method depends on the oxidation of dichlorofluorescin to dichlorofluorescein by haematin and peroxides. Several peroxides including hydrogen peroxide and lipid peroxides react in the method (17).

(i) *Haematin solution.* Dissolve haematin (0.01 mg/ml) in pH 7.2, sodium phosphate buffer.

(ii) *Dichlorofluorescin solution.* Mix 0.5 ml 1 mM 2,7-dichlorofluorescin acetate dissolved in ethanol with 2.0 ml 0.01 M NaOH. Allow to stand for 30 min and then neutralise with 10 ml of 25 mM sodium phosphate buffer (pH 7.2). Keep at 0°C. Prepare freshly each day.

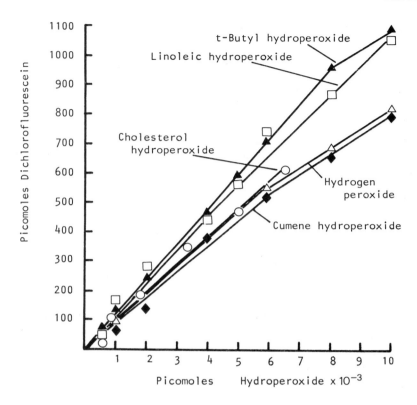

Figure 2. Dichlorofluorescein produced in response to various hydroperoxides. From Cathcart *et al.* (17).

(iii) *Reagent.* Dilute 14 ml haematin solution, by adding 100 ml of pH 7.2 phosphate buffer and boil for 15 min. Purge with argon and add 2 ml of dichlorofluorescin solution.

(iv) *Measurement.* Add 200 μl of solution or suspension for test to 2.9 ml of reagent and incubate at 50°C for 50 min under an argon atmosphere. Measure on a suitable spectrofluorimeter. The excitation wavelength of the spectrofluorimeter should be set to 400 − 470 nm and the emission at 550 nm. The method is not specific for lipid peroxides, but is very sensitive for a number of peroxides (see *Figure 2*).

2.3 Measurement of Degradation Products Formed by Lipid Peroxidation

One of the important consequences of the formation of lipid hydroperoxides by lipid peroxidation is their rapid and spontaneous degradation to large numbers of short-chain products, and especially aldehydes. The exact nature and complete analysis of all these products is still not yet perfected, although recently, through the use of h.p.l.c., it has been possible to separate a large number of the products and identify many of them. The relative proportions of these products vary with the unsaturated fatty acid undergoing peroxidation and a recent analysis is shown in *Table 2*. Despite the fact that analyses for the measurement of lipid peroxidation are not yet complete, some methods for the determination of some of the products have been in use for many years and have prov-ed very valuable for the study of lipid peroxidation.

2.3.1 *The Thiobarbituric Acid (TBA) Test*

This extremely simple method for the measurement of lipid peroxidation was original-ly developed during the 1930s as a test for rancid milk and called the 'Kreiss Test'. The method attracted little attention until 1949, when Wilbur *et al.* (18) showed that the method could be used to demonstrate lipid peroxidation in pure unsaturated fatty acids and in tissue lipids. The method which they described is still used, with only minor modifications, in current research. The procedure is as follows.

(i)   To 2 ml of sample (e.g., solution of fat, tissue homogenate or subcellular frac-tion) add 1 ml of 20% trichloroacetic acid and 2 ml of 0.67% (w/v) thiobar-bituric acid.

(ii) Heat in a lightly-stoppered tube for 10 min in a boiling water bath.

(iii) Centrifuge the precipitated protein if necessary and read the absorbance of the pink colour produced at 530 nm in a spectrophotometer. The coloured product can be extracted with n-butanol if the aqueous solution is turbid.

The method determines aldehydes formed by degradation of hydroperoxide, including malonaldehyde which is used as a standard. The molar extinction coefficient of malonaldehyde is usually quoted as 1.56×10^5 cm^2/mmol (19) and the results of perox-idation experiments are expressed as mmol of malonaldehyde formed.

The great simplicity of the method has encouraged its wide use in investigations of 'lipid peroxidation' and, although often criticised as only an indication of breakdown products, it has proved to be of great value, primarily because of limited interference by other substances found in biological extracts. Furthermore, comparison of the TBA method with other methods of peroxidation measurements frequently show good cor-relation e.g., *Figure 3* (20). There are several aspects of the method which should be noted.

(i)   Trichloroacetic acid is present to provide an acid environment and to precipitate the protein of the samples. The reaction requires a pH of 3 or less to proceed rapidly to completion. Other acids can replace trichloroacetic acid, but trichloroacetic acid is convenient for precipitation of proteins.

(ii) Although many authors have assumed that the method is specific for malon-aldehyde this is not true. Many di-aldehydes react and some mono-enals react but their molar extinction coefficients are usually much less than that of malonaldehyde (21). This fact could be important in interpretation of the results of peroxidation experiments because some of the products, such as hydroxy-alkenals which have important biological activity, react weakly in the TBA method (22).

(iii) Some unsaturated fatty acids such as linoleic acid containing two double bonds form much less malonaldehyde during peroxidation than do unsaturated fatty acids containing three, four or more double bonds. In fact, it has been demonstrated that highly unsaturated fatty acids such as eicosapentaenoic acid ($C_{20:5}$) and docosahexaenoic ($C_{22:6}$) make the main contribution to malonaldehyde formation in incubated tissue fractions (23, 24).

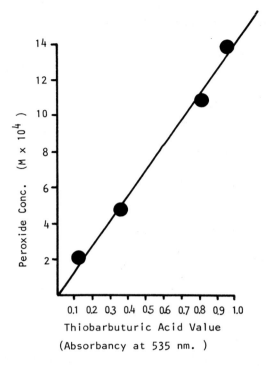

Figure 3. Relationship between the TBA value (absorbancy at 535 nm) and the molar concentration of peroxide in oxidised linolenic acid. The peroxide concentration was determined by measurement of the absorbancy at 232 nm and by using the factor $E^{mol}_{1cm} = 2.17 \times 10^4$. From Wills and Rotblat (20).

(iv)   Several carbohydrates in strong solutions (e.g., 0.25 M sucrose), interfere with the method by producing a yellow-orange colour. Care has therefore to be taken when attempting to apply the method to subcellular fractions prepared in sucrose solutions. It is possible to reduce the interference by carbohydrate by heating at 80°C instead of at 100°C (25, 26).

(v)   Many years ago it was demonstrated that iron had an important effect on TBA values. Although iron is known to be a catalyst of peroxidation of unsaturated fatty acids and of tissue lipids it also enhances the TBA value of peroxidised fatty acids (27) (*Figure 4*). These investigations showed that sufficient iron was present in most samples of trichloroacetic acid to give a pronounced effect and the addition of chelating agents significantly reduced the absorbance values (27). Very few investigations have paid due attention to the role of iron in the TBA method and the problem is difficult to overcome. All iron must be removed from reagents, or alternatively, the solutions may be saturated with iron by adding 1 mM $FeSO_4$ to the trichloroacetic acid (6). Intermediate quantities of iron in the reagents will give unreliable results. The effect of iron in the reaction has never been satisfactorily explained. It is generally assumed that it catalyses rapid breakdown of peroxidised lipids to aldehyde which react in the TBA test. However, the effect of iron during the heating of the reaction to 100°C is always superimposed on the peroxide value of tissue suspension or pure fats incubated with iron at 37°C for a prolonged period, so that if the proposed explanation is correct the products of peroxidation formed initially must be stable to iron at 37°C but rapidly decomposed at 100°C.

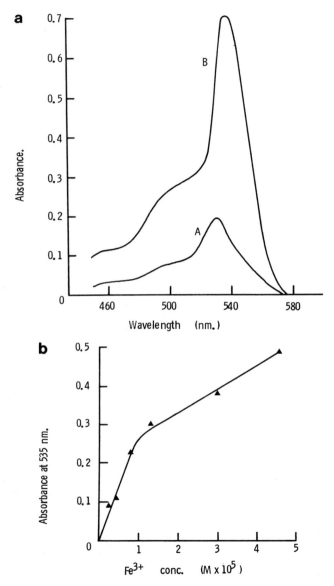

Figure 4. (a) Effect of Fe^{3+} on the spectrum of the colour produced in the TBA method by an oxidised linolenic acid emulsion. 1-cm light path. **A,** oxidised linolenic acid (1.0 ml, 2×10^{-3} M); **B,** oxidised linolenic acid (1.0 ml, 2×10^{-3} M) + Fe^{3+} (1.0 ml, 2×10^{-4} M). **(b)** Effect of Fe^{3+} on the TBA value of a partially oxidised emulsion of rat liver lipids. 1-cm light path. Lipid emulsion (1 ml, 2.83 mg lipid/ml, 0.175% SDS) made up to 2.0 ml with a solution of Fe^{3+} for each determination. From Wills (27).

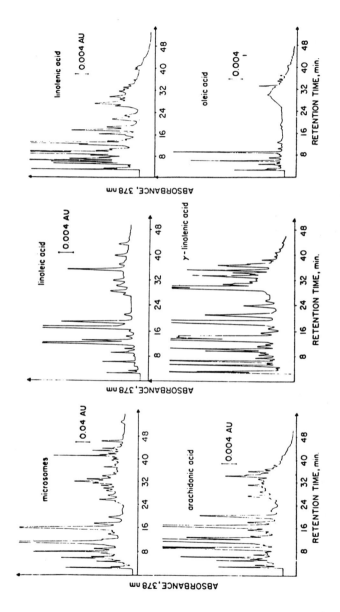

Figure 5. H.p.l.c. separation of the dinitrophenylhydrazones of the carbonyls derived from peroxidation of unsaturated fatty acids and microsomes. Separation conditions: spherisorb ODS, 4.6 × 200 cm, methanol/water 8/2, detector wave length 378 nm, the injected volume was 20 μl; this volume contained the carbonyls produced by 0.28 mg protein, 0.16 mg linoleic, linolenic, γ-linolenic, oleic or 0.08 mg arachidonic acids. From Esterbauer (21).

2.3.2 *Separation of Aldehydes Formed by Lipid Peroxidation by h.p.l.c.*

Recently, Esterbauer *et al.* (22) have described a method for separation of the dinitrophenyl hydrazines of aldehydes by h.p.l.c.

(i)   Centrifuge the sample (for example of liver microsomes) at 100 000 *g* for 40 min.

(ii)   Mix the supernatant with an equal volume of dinitrophenyl hydrazine reagent (50 mg recrystallised 2,4-dinitrophenyl hydrazine after extraction of impurities with hexane, dissolved in 100 ml 1.0 M HCl). Stand for 12 h in the dark.

(iii)   Extract with chloroform (2 × 3 ml) and centrifuge to separate the phases.

(iv)   Remove traces of water by freezing to −20°C and filtering.

(v)   Remove excess dinitrophenyl hydrazine reagent and contaminating hydrazone of acetone, formaldehyde and acetaldehyde by t.l.c. Use dichloromethane as the first developer on a 20 × 20 cm gel plate and benzene as the second developer. Scrape off all hydrazones with Rf values greater than 0.12, except for contaminants, and extract with methanol (2 × 5 ml). These are the carbonyls of 'low and medium polarity'. Carbonyls with Rf values less than 0.12 are also extracted and are termed 'polar carbonyls'. Further t.l.c. may be necessary with this group to remove some of the contaminating aldehydes of the reagent.

(vi)   Separate the carbonyls by h.p.l.c. Inject 50 μl of a methanol solution of purified products onto a Zorbax octadecyl silicate (ODS) column (4.6 mm × 25 cm). Carry out isocratic elution for 20 min with methanol:water mixture (4:1) followed by a linear gradient of 80 − 100% methanol for 20 min. Set the flow-rate at 0.9 ml/min and the detection wavelength at 378 nm. A typical separation is shown in *Figure 5*.

2.3.3 *Formation of Ethane and Pentane*

The discovery by Riely *et al.* (28) that volatile hydrocarbons such as ethane and pentane produced by peroxidation of unsaturated fats could be detected and measured in the exhaled gases of animals was an important advance because it was the first method which could be used to measure lipid peroxidation *in vivo*. Subsequently the method has been extensively applied to the study of culture cells. The method has an important advantage in that it is non-invasive.

A typical apparatus is illustrated in *Figure 6* which is suitable for the study of a mouse. Circulate air through H_2SO_4 to trap exhaled NH_3, 10% KOH to trap CO_2 and dry ice/propanol to condense water vapour. Regulate the flow by means of a pump. A flow-rate of 100 − 200 ml/min is needed for a 0.5 litre chamber. The oxygen reservoir allows pure oxygen to replace the extracted carbon dioxide which is absorbed in the KOH. Samples (~50 ml) are removed by the syringe (A) and the oxygen inlet is closed by a stopcock (B) whilst the sample is removed.

The samples are analysed by gas chromatography, but concentration is often necessary before analysis. This may be achieved using a Chemical Data Systems Model 310 concentrator (29). Several different columns have been described as being suitable for separation of hydrocarbons by g.l.c., such as Poropak N (29) or Poracil C (30). It is suggested, however, that several proprietary g.l.c. columns would be suitable, and it is advisable to investigate the performance of columns which are currently available for g.l.c. in the laboratory.

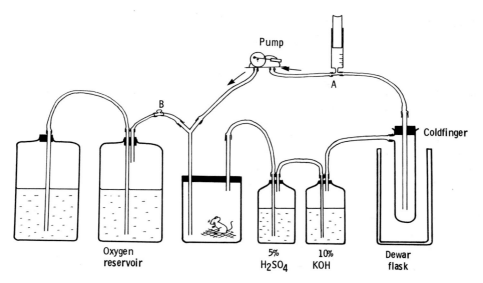

Figure 6. Ethane and lipid peroxidation. The breath collection chamber and air circulating system are described in the text. Air flow is the direction shown by the arrow on the pump. The pump is turned off and the oxygen supply is closed at stopcock **(B)** when a sample is removed from stopcock **(A)** for analysis. An equal volume of purified air is added back to the chamber *via* stopcock **(B)** to maintain constant oxygen tension. From Lawrence and Cohen (29).

Calibration gases containing mixtures of alkanes in nitrogen are obtainable and are used to calibrate the columns. The method can, in addition to the study of small animals such as mice or rats, be used to study alkane formation in cultured cells and in perfused organs. The principle of the method is identical to that used for animals, in that the cell suspension (e.g., isolated hepatocytes) in a flask replaces the chamber occupied by the animal. For perfused tissues a special perfusion table is required (30).

The results of a typical experiment demonstrating the effect of CCl_4 on lipid peroxidation, and measurement by alkane formation, are shown in *Figure 7*.

2.3.4 *Measurement of Fluorescent Products*

It was observed by Chio and Tappel (1969) (31), that when lipid peroxidation occurred in the presence of compounds containing free amino groups such amino acids, proteins or nucleic acids, the aldehydes formed by peroxidation reacted with the amino groups to form fluorescent products. The chromophore group is believed to be $N - C = C - C = N$. Amino groups are nearly always present during peroxidation of lipids in biological system and thus fluorescent products are normally obtained.

(i)   Homogenise the tissue or tissue fraction (~ 0.2 g) which is to be analysed in 20 volumes of chloroform:methanol (2:1).

(ii)   Add an equal volume of water, mix thoroughly and centrifuge at 3000 r.p.m. (2600 RCF) for 1 or 2 min, or until the phases are separated.

(iii)   Remove 1 ml of the chloroform layer and add 0.1 ml methanol for fluorometric analysis.

The lipid-soluble fluorescent products which arise as a result of lipid peroxidation

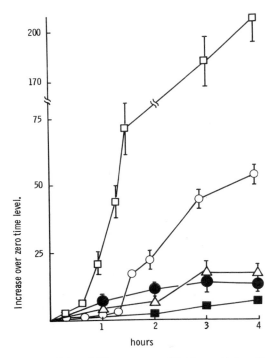

Figure 7. Comparison of the sensitivity of different assays of lipid peroxidation. Hepatocytes (10^6 cells/ml) were incubated at 37°C in the presence of 5 mM carbon tetrachloride. The extent of lipid peroxidation was determined by the five different methods at various times and the results expressed in terms of the increase over the zero time level. (●), Malondialdehyde production; (■), fluorescent products; (○) chemiluminescence; (□), ethane production; and (△), pentane production. Results are expressed as the mean values (± S.E.M.) of four separate experiments. The zero time values were: 1.56 nmol malondialdehyde/10^6 cells; 154 arbitrary fluorescence units (fluorescent products); 206 c.p.s. (chemiluminescence); 0.3 pmol ethane/10^6 cells; and 1.67 pmol pentane/10^6 cells. From Smith *et al.* (34).

have an excitation maximum of 340-370 nm and an emission maximum of 420 − 470 nm and any spectrofluorometer with this capability will be suitable for the analysis. A typical study on peroxidised microsomes is shown in *Figure 8*.

The method is relatively simple and of some biological importance because it measures the binding of aldehydes to proteins, which could clearly cause significant changes in their biological activity. It suffers from the disadvantage that there are no suitable standards. Values must be expressed as fluorescent values and only comparative experiments are possible (cf. *Figure 8*).

2.4 **Measurements of Unsaturated Fatty Acids**

Some indirect methods of investigating lipid peroxidation which do not depend directly on the measurement of the hydroperoxide formed, or the aldehydes produced are available. In the first tube described, the determination of the conjugated diene is an early event in lipid peroxidation, whilst the second method described involves the analysis of fatty acids which are degraded as a result of peroxidation.

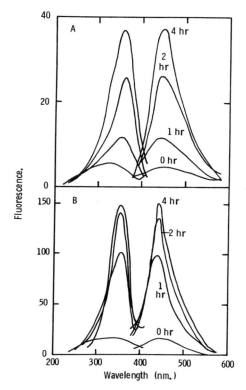

Figure 8. Fluorescence spectra of (**A**) mitochondria (1.9 mg protein) and (**B**) microsomes (1.6 mg protein) peroxidised over a 4 h time period. The 0.5 ml sample was extracted into 3 ml of chloroform:methanol, 2:1. The sensitivity setting was 0.03; 1 μg of quinine sulfate/ml 0.1 N H₂SO₄ had a relative fluorescence intensity of 60 at a sensitivity setting of 0.3. From Dillard and Tappel (35).

2.4.1 *Measurement of Conjugated Dienes*

It will be noted from *Figure 1* that a very early event in the process of peroxidation, before hydroperoxides are produced, is the formation of a conjugated diene system − CH = CH − CH =. Many years ago it was observed by the famous organic chemist R.B.Woodward that conjugated dienes always produced an intense absorption in the u.v. range, 215 − 250 nm. Peroxidised lipids give an intensive absorption at 233 nm which may be easily measured.

The procedure adopted depends on the nature of the lipid undergoing peroxidation. Cyclohexane is the preferred solvent and if a peroxidised, pure, unsaturated fatty acid is under study it may merely be necessary to extract the acid with the solvent. Alternatively an aqueous alkaline solution (pH 9.0) will enable a clear solution of the fatty acid to be obtained and read directly. When the method is used for biological systems, however, a more complex procedure and extraction is essential.

(i)   Extract 1 volume of tissue suspension or homogenate with 20 volumes of chloroform:methanol mixture (2:1 v/v). Mix thoroughly and allow to stand for 10 min.

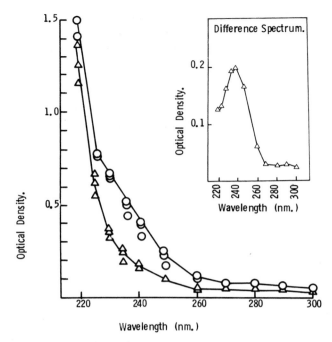

Figure 9. Conjugated diene absorption in rat liver microsomal lipids 5 min after intragastric administration of CCl₄. The lower rate is end absorption of non-peroxidised microsomal lipids from normal untreated rats. From Recknagel and Glende (36).

(ii) Centrifuge at 260 g for 10 min. Decant the organic layer and adjust to 20 volumes with the chloroform:methanol mixture.

(iii) Add 1 volume of water and mix *gently*. Remove the top aqueous methanol layer; centrifuge if necessary for a short period to separate the phases.

(iv) Remove the chloroform under a stream of nitrogen or in a rotary evaporator.

(v) Dissolve the chloroform-free lipid in pure cyclohexane and scan the spectrum between 220 nm and 300 nm.

The 233 nm peak is normally only shown clearly when pure unsaturated fatty acids such as linoleate are peroxidised. When tissue samples are used, it is necessary for a control sample to be compared with a peroxidised sample. Typical results using liver microsomes are shown in *Figure 9*.

2.4.2 *Analysis of Fatty Acids*

Lipid peroxidation causes fragmentation of unsaturated fatty acids, so that analysis for fatty acids during lipid peroxidation will show the disappearance of the fatty acids involved in peroxidation. The methods suffer from the disadvantage that many metabolic reactions other than lipid peroxidation can cause disappearance of fatty acids. However, it offers the advantage over methods which measure hydroperoxides or degradation products, that the unsaturated fatty acids undergoing peroxidation are identified precisely.

The methods can be used for natural pure fatty acids, natural fats or for tissue extracts.

(i) *Extraction of lipids.* In order to prevent peroxidation of the fatty acids, particularly unsaturated fatty acids, during the extraction add BHT (5 mg/ml) to the fractions to be analysed. Also add BHT to the solvents used to extract the lipids, and all solvents used in handling them subsequently. Add 2 ml of chloroform:methanol (2:1 v/v) containing BHT (5 mg/100 ml) to 1 ml of tissue homogenate or fraction. For pure oils take $10-20$ mg of the fat and add 2 ml of the chloroform:methanol mixture containing BHT. Add a further 1 ml of chloroform:methanol mixture to all samples followed by 1 ml of distilled water. Mix well and centrifuge at 2000 r.p.m. for 10 min to separate the phases. Remove the upper phase. Evaporate the lower chloroform layer, containing the extracted lipids, to dryness under reduced pressure in a rotary evaporator at 40°C.

(ii) *Preparation of fatty acid methyl esters.* Add 2 ml of the transmethylation system (25% of a 14% boron trifluoride-methanol complex, 20% benzene and 55% methanol) to $5-10$ mg of the dry lipid sample in a thick walled borosilicate glass tube. Place inside a screw-top brass reaction tube and flush with nitrogen. Place the sealed tubes in a boiling water bath for 30 min and stop the reaction by placing the tubes in ice. Add hexane (4 ml) containing BHT (5 mg/100 ml) to extract the fatty acid methyl esters and then drop 5 N NaOH (1 ml) slowly through the hexane while cooling the tubes on ice. Extract the fatty acid methyl esters a second time with 2 ml of hexane. Combine the two extracts and dry by shaking with a small quantity of solid anhydrous sodium sulphate. Evaporate to dryness under vacuum and re-dissolve the residue in a small volume of hexane/BHT. Store the samples in sealed tubes under nitrogen at $-20°C$ until analysis by g.l.c. Between 90 and 95% of the original lipid sample is transmethylated using this method, and each fatty acid is methylated and recovered with equal efficiency.

(iii) *Analysis of fatty acid methyl esters by gas liquid chromatography.* Analyse the methyl esters of fatty acids using a suitable g.l.c. such as a Pye Unicam Series 204 gas-liquid chromatograph fitted with a CDP1 integrator. The column (150 cm × 4 mm i.d.) is packed with 10% diethylene glycol stearate on Chromosorb W $80-100$ mesh. Set the column temperature to 190°C and the injector and detector temperatures at 250°C. Adjust the flow-rate of the carrier gas (nitrogen) to 60 ml/min. A hydrogen flame ionisation detector is used.

Inject the fatty acid methyl esters, dissolved in a small quantity of hexane, onto the top of the column in a volume of $0.5-1.5 \mu l$ using a Hamilton microsyringe. The standards used are: lauric, $C_{12:0}$; myristic, $C_{14:0}$; palmitic, $C_{16:0}$; palmitoleic, $C_{16:1}$; stearic, $C_{18:0}$; oleic, $C_{18:1}$; linoleic, $C_{18:2}$; linolenic, $C_{18:3}$; arachidic, $C_{20:0}$; arachidonic, $C_{20:4}$; eicosapentaenoic, $C_{20:5}$; and erucic, $C_{22:1}$. Calculate the peak areas manually or using a Pye Unicam CDP1 integrator.

Each peak area is expressed as a percentage of the total area of the methyl ester peaks in a particular sample and the quantity of each acid as a percentage by weight of the total fatty acids is calculated for each sample. A typical analysis showing the effect of the destruction of polyunsaturated fatty acids as a consequence of peroxidation is shown in *Table 4*. It will be noted that the highly unsaturated fatty acids, containing $4-6$ double bonds, are much more liable to peroxidation that linoleic acid, containing two double bonds.

Table 4. Comparison of Losses of Polyunsaturated Fatty Acids from Pure Fats and Membrane Phospholipids after Peroxidation.

Fatty acid	% Loss of fatty acid				
	Incubated suspensions			γ-Ray irradiation	
	Liposomes[a]	Liver[b] microsomes	Human[c] sperm	Herring oil[d]	
				400 Krad 13 days	1000 Krad 3 days
ω-6 { $C_{18:1}$	0	8.7	0	0	0
$C_{18:2}$	12	19.2	0	0	26
$C_{20:4}$	40	26.7	33	–	–
ω-3 { $C_{20:5}$	–	–	–	100	100
$C_{22:6}$	49	62	43	89	93

[a]Konings *et al.* (39).
[b]Eichenberger *et al.* (40).
[c]Jones *et al.* (41).
[d]Hammer and Wills (42).

3. APPLICATION OF METHODS TO PURE LIPIDS, TISSUE HOMOGENATES, SUBCELLULAR FRACTIONS AND WHOLE ANIMALS

3.1 Pure Fatty Acids and Fats

Nearly all the methods described in the preceding section can be used to measure lipid peroxidation; a notable exception is the determination of fluorescent products (Section 2.3.4), which depends on the binding of degradation products of peroxidation to proteins. Several points should however be noted:

(i)   Dilute aqueous solutions of pure fatty acids may be easily prepared if the pH is adjusted to an alkaline value (pH 9). This enables many of the analyses to be readily performed in aqueous solution. Fats must be dissolved in organic solvents and miscibility with aqueous reagents can present problems.

(ii)   When measuring degradation products, for example by the TBA method, it must be realised that some of the products reacting with this reagent are readily water soluble, whilst others are lipophilic and remain with the lipid in the organic solvent phase. When removing samples for the determination of degradation products this fact must be taken into account.

3.2. Tissue Homogenates and Subcellular Fractions In Vitro

Oxygen uptake by tissue homogenates or fractions may be due to many reactions and it is important to ascertain that lipid peroxidation is the major reaction which is consuming oxygen. The method has been successfully applied to the study of lipid peroxidation in the microsomal fraction of liver (*Figure 10*) (6).

Measurement of hydroperoxides, although important, is usually unsatisfactory in tissue samples on account of the large variety of active oxygen species which react in the methods described. Limited success however, has been achieved using these methods (13, 25).

Degradation products of lipid peroxidation in tissue homogenates and subcellular frac-

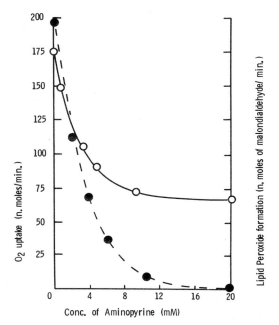

Figure 10. Comparison of the effects of aminopyrine on lipid peroxide formation and O_2-uptake by microsomes in the presence of NADPH. The experiments were carried out in the oxygen electrode for a period of $1-2$ min. Peroxide determinations were carried out on the whole suspension by the TBA method (2.0 ml) immediately after removal from the reaction vessel. O, O_2-uptake; ●, lipid peroxide formation. From Wills (37).

tions have been measured frequently by the TBA test which is simple and reliable. When using the TBA method it is important to note the important role which iron may play in the test and that varying concentrations of iron in the sample may cause erratic results (27). Sophistication of the method using h.p.l.c., which enables separation of the products to be achieved, now gives a much more reliable indication of the actual degradation products formed.

Fluorescent products may also be measured but determinations of ethane and pentane formation requires elaborate equipment and are not generally used for studies of peroxidation in homogenates.

Fatty acids of tissues undergoing peroxidation may be readily extracted and analysed, and conjugated dienes may be measured after extraction. Many substances can however interfere with the spectrum of the conjugated diene and careful control measurements are necessary.

3.3 Measurement of Peroxide Concentration of Tissue In Vivo and in Homogenates

Numerous efforts have been made to measure lipid peroxides in tissues rapidly removed from animals after death in an attempt to establish the peroxide concentration in the tissues *in vivo*. These have nearly all ended in failure, or given erroneous results, because handling and manipulation of the tissues always induces peroxidation. Only measurements of ethane and pentane production in living animals or cultured cells have so far given a reliable indication that lipid peroxidation can occur *in vivo*. However, caution must be exercised in the interpretation of the results because both these gases can be produced by bacterial action in the digestive tract.

4. SIGNIFICANT CONSEQUENCES OF LIPID PEROXIDATION

Lipid peroxidation can cause cellular damage in several ways. Firstly, many very reactive free radicals are formed during the process and these can react with, and cause damage

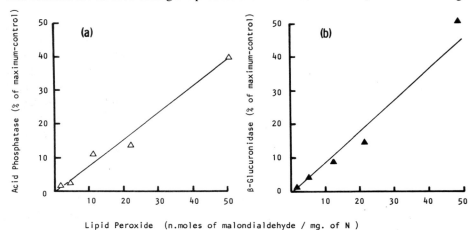

Figure 11. Relationship between the increase of lipid 'peroxide' and the release of acid phosphatase **(a)** and β-glucuronidase **(b)** from liver lysosomes irradiated with doses in the range 0 – 50 Krad. Lipid 'peroxide' (malonaldehyde) values are expressed as increases over non-irradiated controls. From Wills and Wilkinson (26).

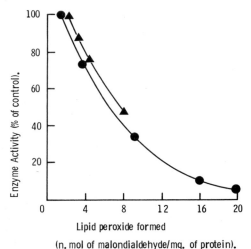

Figure 12. Comparison of effect of peroxidation on oxidative demethylation and glucose-6-phosphate activity of microsomal suspensions (containing 3.0 mg protein/ml). Peroxidation was induced by incubation with ascorbate or NADPH. For oxidative demethylation of aminopyrine, microsomal suspensions were irradiated with doses (5 – 50 Krad) of γ-rays from a ^{60}Co source before determination of lipid peroxide and oxidative demethylation ● , glucose-6-phosphatase; ▲ , oxidative demethylation. From Wills (38).

to, cellular macromolecules. Secondly, the decomposition of hydroperoxides formed from polyunsaturated fatty acids is likely to be the major cause of the cellular toxicity caused by lipid peroxidation. Cellular damage could arise in two ways. The fragmentation of the unsaturated fatty acid chains of phospholipid membrane components will cause disruption of plasma or subcellular membranes. This has been demonstrated to occur in membranes of the red blood cell (32), lysosomes (26) (*Figure 11*) and endoplasmic reticulum (14). Damage to the erythrocyte membranes causes leakage of cell components; damage to the lysosomal membrane causes leakage of enzymes; and damage to the endoplasmic reticulum causes reduction in the activity of enzymes associated with the membrane (for example, glucose-6-phosphatase and enzymes involved in oxidative drug metabolism) (14) (*Figure 12*). Alternatively, or in addition, cellular damage can be caused by decomposition products of peroxidation, and, during recent years, 2-alkenals and 4-hydroxyalkenals have been shown to react at neutral pH with many biomolecules of the cell, especially those possessing sulphydryl groups. It is therefore likely that some of the cytotoxic effects of products of lipid peroxidation can be explained by the production of these aldehydes, particularly 4-hydroxynonenal (21, 33), in combination with disintegration of membranes.

5. REFERENCES

1. Wills,E.D. (1969a) *Biochem. J.*, **113**, 315.
2. Morehouse,L.A., Tien,M., Bucker,J.R. and Aust,S.D. (1983) *Biochem. Pharmacol.*, **32**, 123.
3. Wills,E.D. (1965) *Biochim. Biophys. Acta*, **98**, 238.
4. Wills,E.D. (1966) *Biochem. J.*, **99**, 667.
5. Umbreit,W.W., Burns,R.H. and Stauffer,J.F. (1957) *Manometric Techniques*, published by Burgess Publ. Co., Minneapolis.
6. Wills,E.D. (1969b) *Biochem. J.*, **113**, 325.
7. Young,C.A., Vogt,R.R. and Nieuland,J.A (1936) *Ind. Eng. Chem. (Anal.)*, **8**, 198.
8. Lea,C.H. (1945) *J. Soc. Chem. Ind.*, **64**, 106.
9. Wagner,C.D., Clever,H.L. and Peters,E.D. (1947) *Ind. Eng. Chem. (Anal.)*, **19**, 980.
10. Marks,S. and Morrell,R.S. (1929) *Analyst*, **54**, 503.
11. Skellon,J.H. and Wills,E.D. (1948) *Analyst*, **73**, 78.
12. Swoboda,P.A.T. and Lea,C.H. (1958) *Chem. Ind.*, 1090.
13. Wills,E.D. (1971) *Biochem. J.*, **123**, 983.
14. Wills,E.D. (1980) *Int. J. Radiat. Biol.*, **37**, 403.
15. Bunyan,J., Murrell,E.A., Green,J. and Diplock,A.T. (1967) *Br. J. Nutr.*, **21**, 475.
16. Eling,T., Tainer,B., Ally,A. and Warnock,R. (1982) in *Methods in Enzymology*, Vol. **86**, Academic Press, New York and London, p. 511.
17. Cathcart,R., Schweiss,E. and Ames,B.N. (1984) in *Methods in Enzymology*, Vol. **105**, Packer,L. (ed.), Academic Press, New York and London, p. 352.
18. Wilbur,K.M., Bernheim,F. and Shapiro,O.W. (1949) *Arch. Biochem. Biophys.*, **24**, 305.
19. Sinnhuber,R.O., Yn,T.C. and Yu,T.C. (1958) *Food Res.*, **23**, 626.
20. Wills,E.D. and Rotblat,J. (1964) *Int. J. Radiat. Biol.*, **8**, 551.
21. Esterbauer,H. (1982) in *Free Radicals, Lipid Peroxidation and Cancer*, McBrien,D.C.H. and Slater,T.F. (eds.), Academic Press, London and New York, p. 101.
22. Esterbauer,H., Cheeseman,K.H., Dianzani,M.U., Poli,G. and Slater,T.F. (1982) *Biochem. J.*, **208**, 129.
23. Hammer,C.T. and Wills,E.D. (1978) *Biochem. J.*, **174**, 585.
24. Wills,E.D. (1984) in *Oxidative Stress*, Sies,H. (ed.), Academic Press, London and New York.
25. Fortney,S.R. and Lynn,W.S. (1964) *Arch. Biochem. Biophys.*, **104**, 241.
26. Wills,E.D. and Wilkinson,A.E. (1966) *Biochem. J.*, **99**, 657.
27. Wills,E.D. (1964) *Biochim. Biophys. Acta*, **84**, 475.
28. Riely,C., Cohen,G. and Lieberman,M. (1974) *Science (Wash.)*, **183**, 208.
29. Lawrence,G.D. and Cohen,G. (1984) in *Methods in Enzymology*, Vol. **105**, Packer,L. (ed.), Academic Press, New York and London, p. 311.

30. Muller,A. and Sies,H. (1984) in *Methods in Enzymology*, Vol. **105**, Packer,L. (ed.), Academic Press, New York and London, p. 305.
31. Chio,K.S. and Tappel,A.L. (1969) *Biochemistry (Wash.)*, **8**, 2821.
32. Tsen,C.C. and Collier,H.B. (1960) *Can. J. Biochem. Physiol.*, **38**, 957.
33. Schauenstein,E. (1982) in *Free Radicals, Lipid Peroxidation and Cancer*, McBrien,D.C.H. and Slater,T.F. (eds.), Academic Press, London and New York, p. 159.
34. Smith,M.T., Thor,H., Hartzell,P. and Orrenius,S. (1982) *Biochem. Pharmacol.*, **31**, 19.
35. Dillard,C.J. and Tappel,A.L. (1984) in *Methods in Enzymology*, Vol. **105**, Packer,L. (ed.), Academic Press, New York and London, p. 352.
36. Rechnagel,R.O. and Glende,E.A. (1984) in *Methods in Enzymology*, Vol. **105**, Packer,L. (ed.), Academic Press, New York and London, p. 331.
37. Wills,E.D. (1969c) *Biochem. J.*, **113**, 333.
38. Wills,E.D. (1971) *Biochem. J.*, **123**, 983.
39. Konings,A.W.T., Damen,J. and Trieling,W.B. (1979) *Int. J. Radiat. Biol.*, **35**, 343.
40. Eichenberger,K., Bohni,P., Winterhalte,K.H., Kawato,S. and Richte,C. (1982) *FEBS Lett.*, **142**, 59.
41. Jones,R., Mann,T. and Sherins,R. (1979) *Fertil. Steril.*, **31**, 531.
42. Hammer,C.T. and Wills,E.D. (1979) *Int. J. Radiat. Biol.*, **35**, 323.

CHAPTER 7

Preparation and Use of Renal and Intestinal Plasma Membrane Vesicles for Toxicological Studies

M.IQBAL SHEIKH and JESPER V.MØLLER

1. INTRODUCTION

The opportunity to prepare plasma membranes in relatively large yield offers promising prospects in renal toxicology, namely to study the mechanism of tissue damage at the subcellular level. The proximal tubule of the nephron stands out as one of the prime targets for many toxicological agents. Examples are poisoning with heavy metals like mercury, lead, bismuth, and the antitumor drug cis-platinum (1,2). Administration of many antibiotics like aminonucleosides (2) and cephalosporins (2,3) also may lead to degenerative or necrotic changes of the proximal tubule. The hepatotoxic effect of halogenated hydrocarbons like chloroform is often accompanied by necrosis of the proximal tubule (1,4). In addition, compounds have been used in the laboratory to create selective damage of particular nephron segments or functions of the proximal tubule in animal experiments. These compounds include dichromate (5,6), uranylnitrate (5) and maleic acid (7,8).

The selectivity of the proximal tubule of the nephron for many nephrotoxic agents may, to a large extent, reflect the well developed capacity of this segment of the nephron for transport of a variety of substances by specific mechanisms, pinocytotic activity, and possibly also the presence of many cellular binding sites. For instance, the toxic effect of cephalosporins appears to be related to a high level of intracellular accumulation, due to tubular uptake across the basolateral membrane by the organic anion (hippurate) system. Part of the accumulation of lead (9,10) and mercurial diuretics (11,12) occurs by active transport processes. Pinocytosis may be important for intracellular accumulation of gentamycin (2) and inorganic mercury (13). However, there is in general a paucity of data on the renal handling of nephrotoxic substances as well as on the role of transport defects in the development of degenerative processes. Membrane vesicles are potentially useful for examining the transport and interaction of nephrotoxic agents with membranes. The use of membrane vesicles would also permit the assessment of transport of normal metabolites and other properties of the plasma membranes as part of the biochemical characterisation of degenerative changes observed during administration of nephrotoxic agents.

The primary aim of this chapter is to discuss in detail the preparation and properties of transport-competent brush-border (luminal) and basolateral membrane vesicles from pars recta and pars convoluta of the proximal tubules. The methods described are in routine use in our laboratory and do not require the use of specialised equipment beyond what is available in any laboratory with biochemical facilities. We also describe methods for characterisation of the membrane preparations, including transport assays. In addition, we comment on the adaptation of the present methods to the preparation of luminal vesicles from intestinal mucosa.

2. BASIC TECHNIQUES

2.1 Principles of the Procedures

The methods described here represent a further development of procedures previously reported (14,15). In short, they are characterised by the following features: (i) an efficient, but gentle, homogenisation step, which aims at a breakage of most of the renal tubular cells, without disruption of intracellular organelles; (ii) removal of intracellular organelles by low-speed and medium-speed centrifugation; (iii) the use of divalent cations to separate luminal membranes from basolateral and other cellular membranes; (iv) a self-generating Percoll gradient to further purify basolateral membranes; and (v) dissection of discrete regions of the kidney to provide populations of proximal tubules, predominantly derived from pars convoluta and pars recta.

Addition of divalent cations in the separation of luminal from other membranes was pioneered by Booth and Kenney (16) and is also used in procedures described by Kinsella *et al.* (17) and Kippen *et al.* (18). The separation is based on a higher density of negative electrostatic charges on luminal than on basolateral membranes. In effect separation by divalent cations can be seen as the chemical equivalent of separations based on differences in electrophoretic mobility (19,20). Due to a high content of sialic acid the surface charge density of luminal membranes is sufficiently high to shield bound divalent cations from interaction with other luminal vesicles, whereas the divalent cations are able to interact with the negative charges on two adjoining basolateral membranes.

The use of a self-generating Percoll gradient (instead of a sucrose gradient) to separate slower sedimenting basolateral membranes from luminal membranes has been described by various authors (21−24). In the present context, Percoll is used as a further purification step after treatment with divalent cations and sedimentation of membraneous material by low-speed centrifugation. The procedures described below include:

(i)   a method for preparation of luminal and basolateral membranes from whole cortex which is summarised in Scheme 1;

(ii)   a short method for preparation of luminal membranes alone (Section 2.9);

(iii)   a method for preparation of luminal and basolateral membrane vesicles from pars convoluta and pars recta of the proximal tubule (25) by dissection from selected regions of the kidney (Section 2.10);

(iv)   a method for preparation of luminal membranes from small intestine.

2.2 Equipment

The separations are based on differential centrifugations in an angle rotor. These are performed at 4°C in a refrigerated centrifuge (Sorvall SS-34 rotor in a Sorvall R-2CB centrifuge), except for the last step in the preparation of basolateral membranes where we use a Beckman preparative centrifuge with a Ti-50 angle rotor to remove Percoll.

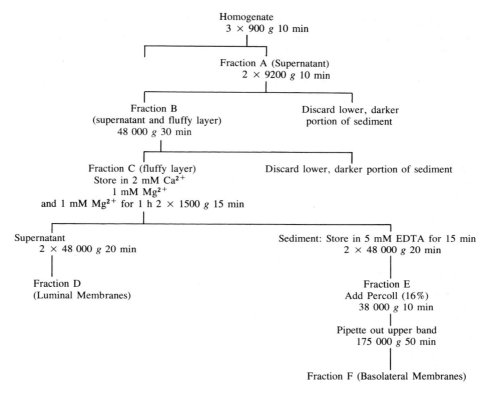

Scheme 1. Purification of luminal and basolateral membranes

However, this step can be substituted by a more lengthy centrifugation in the Sorvall centrifuge, if a preparative centrifuge is not available. It should be noted that all *g*-forces given are those calculated at the bottom of the tubes. Homogenisation of tissue is performed in a 30-ml Potter-Elvehjem glass tube with a Teflon pestle (clearance 0.2 mm) attached to a motor-driven homogeniser (Braun, Melsungen, FRG).

2.3 **Media**

(A)   Physiological saline (0.15 M NaCl).
(B)   0.30 M sucrose, 25 mM Hepes, titrated with Tris to pH 7.0 at 20°C.
(C)   Medium (B) in which is dissolved 0.2 mM of the proteinase inhibitor phenyl-methylsulfonyl fluoride (PMSF) (Sigma Chem. Co., St. Louis, MO).
(D)   0.30 M mannitol, 25 mM Hepes, titrated with Tris to pH 7.0 at 20°C.
(E)   Medium (D) in which is dissolved 5 mM EDTA and 0.2 mM PMSF.
(F)   Medium (D) in which is dissolved 2 mM $CaCl_2$, 1 mM $MgCl_2$ and 1 mM $MnCl_2$.

2.4 **Animals**

Adult rabbits have been used predominantly as the source of kidneys in our standard procedure. However, successful preparations by the same method have been obtained with other biological starting materials such as calf and pig kidney. The same procedure can also be used for kidneys of newborn rabbits (14). These experiences suggest that the methods are of general utility.

2.5 Preparation of Kidney Homogenate

(i)   Adult rabbits are killed by bleeding from the jugular vein after anaesthesia with Nembutal (20 mg/kg body weight, injected into the ear-vein) or after a blow on the neck.

(ii)   After opening the abdomen ligate the renal artery and vein and gently free the kidneys from perirenal fat, taking care to preserve the renal artery and vein.

(iii) Transfer the isolated kidneys to ice-cold medium (A).

(iv)   Insert a hypodermic needle into the renal artery and perfuse the kidneys with approximately 15 ml medium (B) until they are bleached. All subsequent operations should take place at $0-4°C$.

(v)   Cut the kidneys into halves by a section through the upper and lower pole, and cut out small pieces of cortex.

(vi)   Weigh the cut tissue on a pan and mince suitable portions ($\sim 4-5$ g) with a pair of scissors before transferring to a 30 ml Potter-Elvehjem tube.

(vii) Add medium (C) in the proportion of 5 ml/g tissue.

(viii) Homogenise with 30 up-and-down strokes performed at a gentle speed with the Teflon piston at maximal speed (1500 r.p.m.), the tube being immersed in an ice-water mixture to prevent overheating.

2.6 Preparation of a Crude Plasma Membrane Fraction

2.6.1 *First Centrifugation Step*

(i)   Pour portions of the homogenate ($20-30$ ml) into 38-ml centrifuge tubes and centrifuge in a SS-34 rotor at 2750 r.p.m. (900 g) for 10 min.

(ii)   Collect the supernatant and store on ice.

(iii) Add 1/4 of the original homogenate volume of medium (C) to the sediment, transfer to the Potter-Elvehjem tube and homogenise again by performing 30 strokes at full speed.

(iv)   Centrifuge the homogenised sample as before. Repeat the homogenisation and centrifugation of the pellet once more and combine the three supernatants (the combined supernatant is designated as *Fraction A*).

2.6.2 *Second Centrifugation Step*

(i)   Homogenise fraction A with 10 strokes at maximal speed and centrifuge for 10 min at 8750 r.p.m. (9200 g).

(ii)   Suspend the white and fluffy upper part of the sediment by whirling and gentle detachment with a Pasteur pipette to separate it from the denser, dark-brown bottom part of the pellet.

(iii) Pour off the suspended supernatant and homogenise with five strokes (this time performed by hand).

(iv)   Centrifuge homogenised supernatant again for 10 min at 8750 r.p.m., and remove the supernatant and fluffy layer as before.

(v)   Homogenise the suspended sample with 10 strokes by hand. The resulting preparation, representing the crude plasma membrane fraction, is designated as *Fraction B*.

2.7 **Preparation of Luminal Membranes**

(i)   Centrifuge fraction B at 20 000 r.p.m. (48 000 *g*) for 30 min.

(ii)   Discard the clear, red-coloured (haemoglobin) supernatant and add 50 ml of medium (F), containing divalent cations, to each tube.

(iii)   Suspend the fluffy layer with the aid of a Pasteur pipette, and homogenise the suspension by 10 strokes at maximal speed (*Fraction C*).

(iv)   Store the suspension in the cold room at 2°C for 1 h with gentle magnetic stirring and then centrifuge for 15 min at 3500 r.p.m. (1500 *g*).

(v)   Carefully pour off the bulk of the supernatant so as to avoid admixture from the loosely-packed sediment.

(vi)   Pour off and discard the remaining part of the supernatant which is contaminated with sediment.

(vii)   Add 50 ml of medium (F) to the sediment and repeat the procedure.

(viii) Combine the two supernatants and centrifuge for 20 min at 20 000 r.p.m. (48 000 *g*).

(ix)   Suspend the sediment in 8 ml of medium (E) and centrifuge for 20 min at 20 000 r.p.m., resuspend by homogenising with 10 strokes at maximal speed, followed by centrifugation for 20 min at 20 000 r.p.m.

(x)   Finally suspend the sediment in 6 ml medium (E). This fraction represents the purified luminal vesicles and is designated as *Fraction D*. The yield is approximately 60 mg protein per 20 g kidney cortex.

2.8 **Preparation of Basolateral Membranes**

(i)   The sediment from the 3500 r.p.m. step described above contains the membraneous material precipitated with divalent cations. Treat this sediment with EDTA by suspension in 8 ml of medium (E) and homogenisation (10 strokes by hand).

(ii)   After standing for 15 min, centrifuge the suspension for 20 min at 20 000 r.p.m. (48 000 *g*). Repeat this step once more, but this time using medium (D) (i.e., an EDTA-free medium) for resuspension. The resultant preparation is designated as *Fraction E*.

(iii)   In the next step mix fraction E (3.0 ml) with 28 ml of medium (D) and 6.04 ml Percoll (Pharmacia, Uppsala, Sweden) and centrifuge the sample at 16 000 r.p.m. (38 000 *g*) for 10 min. After the centrifugation two main bands are formed, at the bottom and upper parts of the tube (see *Figure 1*).

(iv)   Remove the upper band carefully with a Pasteur pipette and transfer to 8-ml tubes for centrifugation in a Beckman Ti-50 rotor at 45 000 r.p.m. for 50 min (175 000 *g*). After the centrifugation the sedimented membranes are found to be layered on top of a dense, glassy pellet of Percoll.

(v)   Remove the sedimented membranes with a Pasteur pipette, suspend in approximately 1.5 ml of medium (D), and homogenise with 10 strokes by hand.

Figure 1. Appearance of tubes after centrifugation of fraction E in 16% Percoll for 10 min at 38 000 *g*. Arrows indicate the upper and lower bands during preparation of basolateral membrane vesicles from the pars convoluta and pars recta, respectively.

This preparation predominantly contains basolateral membranes and is designated as *Fraction F*. The yield is approximately 40 mg protein per 20 g of kidney cortex.

2.9 Short Procedure for Preparation of Luminal Membranes Alone

2.9.1 *Media*

(G)   310 mM sorbitol, 15 mM Hepes titrated with Tris to pH 7.5 at 20°C.
(H)   Medium (G) to which $CaCl_2$ is added to a concentration of 100 mM.

2.9.2 *Homogenisation*

Prepare the kidneys in the same way as described under Section 2.5 and perfuse with medium (G). Homogenise in 9 ml medium (G) per gram of cortex by 3 x 15 strokes at maximal speed.

2.9.3 *Centrifugation*

First centrifugation. Fill the homogenate (20 − 30 ml) into 38 ml tubes and centrifuge at 3500 r.p.m. (1500 *g*) for 15 min in the SS-34 rotor.

Second centrifugation. Pour off the supernatant and suspend the fluffy layer of the pellet with the aid of a Pasteur pipette. Add medium (G) to 1/4 of the original volume and homogenise the suspension with 10 strokes at 700 r.p.m. Then add medium (H) to the homogenate in the proportion of 1 volume to 9 volumes of homogenate. Stir the mixture gently on a magnetic stirrer in the cold room for 20 min.

Third centrifugation. Centrifuge the mixture at 3500 r.p.m. (1500 *g*) for 15 min.

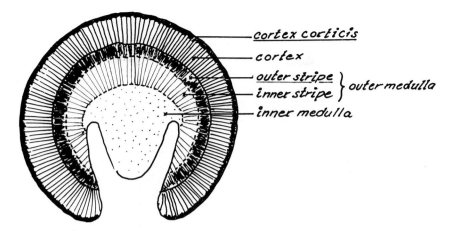

Figure 2. Diagram showing the various zones of the rabbit kidney. Tissue for preparation of membranes from tubuli contorti proximalis is taken from cortex corticis, and tissue for preparation of tubuli rectae proximalis is taken from outer stripe of medulla adjoining the cortex.

Fourth centrifugation. Centrifuge the supernatant from the third centrifugation step for 30 min at 14 500 r.p.m. (25 000 *g*). Suspend the pellet, which represents the luminal membrane preparation, in medium (G) and homogenise with 10 strokes at 700 r.p.m.

2.10 Preparation of Membranes from Tubuli Contorti and Rectae Proximalis

The preparative procedures are identical to those described above, and the separation is solely based on dissection of specific regions of the kidney (see *Figure 2*). For the preparation of membranes derived from the convoluted part of the proximal tubule, cut slices to a maximum of 0.8 mm below the kidney surface (cortex corticis) with a razor blade.

For the preparation of tubules derived from pars recta first remove the cortex and then dissect out $1 - 1.5$ mm from the outer stripe of the outer medulla. Approximately $1 - 1.5$ g outer cortex is obtained from an adult rabbit kidney and slightly less from the outer medulla. The yield of luminal membrane vesicles prepared by the method described in 2.10 is approximately $1 - 1.5$ mg/g of outer cortex and $0.5 - 1$ mg/g of outer medulla. Usually 12 kidneys are used for a preparation of vesicles from pars convoluta and pars recta. Both preparations are minced and homogenised as described in Section 2.5.

2.11 Storage of Preparations

The preparations deteriorate quite rapidly if stored at $0°C$, especially with respect to transport properties. However, they can be kept intact for many months after quick freezing in liquid N_2. The samples ($10 - 20$ mg protein/ml) are frozen in aliquots of around 300 μl in plastic vials. It is recommended that thawing of the vial content is performed rapidly in a $40°C$ water bath, with manual mixing and that the vial is transferred to $0°C$ immediately after the complete thawing of vial contents.

159

2.12 **Isolation of Brush Border Membrane Vesicles from Small Intestine**

It is to be expected that methods similar to those described here for the isolation of luminal and basolateral plasma membranes from renal proximal tubules will be useful in the study of the transport function of epithelial cells from other organs. Mucosal cells of the small intestine have a similar plasma membrane morphology and transport function as those of the proximal cells of the kidney tubule. Transport studies on brush border membranes, isolated from mucosal scrapings, have been performed since the 1960s. During this period the methodology used in the isolation of intestinal plasma membranes has changed considerably (26 – 29). To those interested in the study of transport function in other organs than the kidney, we wish to point out the potential applicability of the same kind of methods as employed here by describing in outline a method developed by Burkhardt *et al.* (30) to isolate highly purified luminal membrane vesicles from the rat small intestine. The method is based on two precipitations with 10 mM $MgCl_2$ (instead of one precipitation with 10 mM $CaCl_2$ as in the case of kidney, cf. Section 2.9).

2.12.1 *Procedure*

(i)   Kill male Wistar rats, weighing 180 – 200 g by a blow on the head. All subsequent procedures take place at 4°C.

(ii)   Remove the entire small intestine and divide into ten segments of equal length, numbered 1 – 10 from the proximal to the distal direction. Segments 2 – 4 and 7 – 9 can be used to isolate membrane vesicles from jejunum and ileum, respectively.

(iii)   Scrape the mucosal cells after eversion of the intestinal pieces and thorough rinsing with 155 mM NaCl. The scraping is performed by gentle sliding of a small glass plate over the mucosal surface.

(iv)   Homogenise the isolated cells using a Waring blendor in 300 ml solution containing 60 mM mannitol, 1 mM EGTA and 2.4 mM Tris/HCl (pH 7.1).

(v)   Add to the homogenate solid $MgCl_2$ to a concentration of 10 mM.

(vi)   After solubilisation and standing for 15 min, centrifuge the homogenate at 3000 *g* for 15 min.

(vii)   Centrifuge the supernatant at 27 000 *g* for 30 min.

(viii) Suspend the resulting pellet in 60 ml solution containing 60 mM mannitol, 5 mM EGTA and 12 mM Tris/HCl (pH 7.4) and homogenise in a Potter-Elvehjem tube by 20 strokes at 900 r.p.m. using a tight-fitting Teflon pestle.

(ix)   Add to the homogenate solid $MgCl_2$ to a concentration of 10 mM.

(x)   After 15 min, centrifuge the sample at 3000 *g* for 15 min.

(xi)   Centrifuge the supernatant at 27 000 *g* for 15 min.

(xii)   Suspend the pellet in 30 ml of buffer containing 300 mM mannitol and 10 mM potassium phosphate (pH 7.4) and homogenise in the Potter Elvehjem tube by 20 strokes at 900 r.p.m.

(xiii) Centrifuge the homogenate at 47 000 *g* for 30 min.

(xiv)   Suspend the pellet, which represents the luminal membrane preparation, in 310 mM mannitol and 15 mM Hepes/Tris (pH 7.4) to the desired volume.

3. CHARACTERISATION OF THE PREPARATIONS

3.1 Enzyme Activities

Measurements of enzyme activities are used, not only in the assessment of purity of the final preparations, but also as a guideline during the preparation (14). Markers of luminal membranes include a number of hydrolases: alkaline phosphatase, glycosidases (maltase, sucrase, trehalase), and peptidases (leucine aminopeptidase, γ-glutamyl-transpeptidase). Alkaline phosphatase has been widely used for this purpose. There is some doubt concerning the exclusive location of alkaline phosphatase in the luminal membranes (31,32) and it is advisable to include an assay for other types of hydrolases, e.g., trehalase and leucine aminopeptidase. At the present time Na,K-ATPase is used as the marker for basolateral membranes, but it should be noted that appreciable Na,K-ATPase activity is present in all segments of the nephron (33,34). The practical details of the measurement of enzyme activities that are particularly useful as markers of luminal and basolateral membranes and mitochondria are described in Section 3.4.

Table 1 shows measurements of enzyme activities, protein and DNA content in homogenate and fractions A − F. In addition to enzymes characteristic of luminal and basolateral membranes, we have used succinate dehydrogenase and acid phosphatase as markers of mitochondrial and lysosomal membranes, while DNA content serves to monitor the presence of nuclei. It can be seen from the table that the preparation of a crude plasma membrane fraction (C) leads to an approximately 5-fold enrichment both in luminal and basolateral membrane fragments, while mitochondrial and nuclear content is appreciably reduced. The enrichment of fraction D (luminal membranes), after divalent cation precipitation of the basolateral membranes, is 13- to 14-fold, compared with the original homogenate, but 25 − 30 times with respect to lack of Na,K-ATPase as compared with fraction F. Contamination with mitochondria and nuclei at this stage is exceedingly low, although some acid phosphatase activity remains. However, the indication of lysosomal presence on the basis of remaining acid phosphatase activity may be artefactual, since it may represent occluded or membrane-associated enzyme

Table 1. Distribution of Marker Enzymes During Purification of Basolateral and Luminal Membrane Vesicles.

Fractions	(Na,K)-ATPase	Alkaline phosphatase	Trehalase	Succinate dehydrogenase	Acid phosphatase	DNA (mg per g homogenate protein)
	sp. act.	sp. act.	sp. act.x10	sp. act.x10²	sp. act.x10²	
Homogenate	0.20	0.12	0.36	2.94	1.3	16.3
Fraction A	0.24	0.15	0.55	4.52	1.5	11.57
Fraction B	0.32	0.22	0.95	1.87	1.6	0.86
Fraction C	1.11 (5.6)	0.52 (4.3)	1.94 (5.4)	0.65 (0.22)	1.2 (0.9)	0.37
Fraction D	0.12 (0.60)	1.57 (13)	4.79 (13)	0.05 (0.02)	0.7 (0.5)	0.004
Fraction E	2.15	0.24	0.74	0.07	0.9	0.009
Fraction F	3.18 (16)	0.18 (1.5)	0.64 (1.8)	0.03 (0.1)	0.1 (0.8)	0

Specific activities are given as μmol of substrate hydrolysed/min per mg of homogenate protein. Numbers in parentheses refer to purification factors, compared with homogenate, of crude plasma membranes, luminal membranes and basolateral membranes (Fraction C, D and F, respectively). References to assay methods are described in the text (Section 3.1).

Table 2. Enrichment Factors for Various Marker Enzymes in Preparations of Luminal Membrane Vesicles from Tubuli Contorti Proximalis and Tubuli Rectae Proximalis.

Enzyme	Enrichment factor	
	Pars convoluta	Pars recta
Alkaline phosphatase	8	12
Maltase	9	9
Leucine aminopeptidase	16	31
(Na,K)-ATPase	0.1	0.4
Succinate dehydrogenase	0.04	0.04
Yield (mg protein/g cortex protein)	6 – 9	3 – 6

Enrichment factors are calculated as the ratio between specific activity in the luminal membrane fractions and the homogenate.

that is released from the lysosomes during the preparative procedure.

Table 1 shows that the material precipitated by divalent cations (fraction E) consists mainly of basolateral membranes, but with an admixture of luminal membranes. The Percoll step leads to preferential removal of luminal membranes (from fraction E) so that the final purification in fraction F with respect to the original homogenate is 15-fold as estimated by measurements of Na,K-ATPase activity.

Table 2 shows the enrichment factors for enzyme activities in preparations of luminal membranes from the convoluted and straight part of the proximal tubule by the short method. Enrichment factor for alkaline phosphatase is slightly lower than that obtained for whole cortex (*Table 1*). The enrichment of leucine aminopeptidase activity is considerably higher than that for alkaline phosphatase and maltase especially in tubuli recti. The reason for the higher enrichments obtained with leucine aminopeptidase is not quite clear at present and raises the question as to whether some markers traditionally used for luminal membranes may not be quite as specific as previously considered.

3.2 Transport Properties

The isolated membrane fractions show excellent transport properties, and appropriate transport assays are useful to detect cross-contamination (i) between the luminal and basolateral membranes and (ii) between membrane vesicles derived from pars convoluta and pars recta. Transport can be detected either by the Millipore technique, using radioactively labelled transport substrate, or with the aid of a potential-sensitive dye, if transport is electrogenic. The rationale and performance of these techniques are described in Sections 3.5 and 3.6.

Figure 3A shows Na^+-dependent overshoot of L-malate and D-glucose by luminal membranes as demonstrated after addition of 155 mM Na^+. Note that addition of the same concentration of K^+ does not produce transport of either compound. For the basolateral membrane preparation (*Figure 3B*) Na^+ induces electrogenic transport of L-malate, but not of D-glucose, indicating the absence of transport-competent luminal vesicles in this preparation. *Figure 4* shows that Na^+-dependent, active transport of *p*-aminohippurate is a specific property of the basolateral membranes.

Studies on plasma membranes derived from pars recta and pars convoluta are still

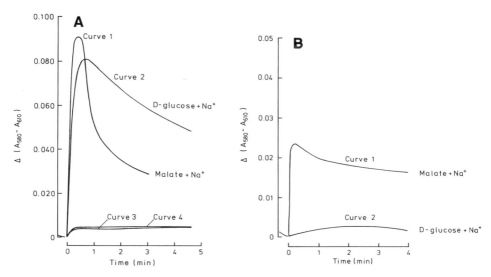

Figure 3. A. Uptake of L-malate and D-glucose by luminal vesicles. The uptake occurs in co-transport with Na$^+$ and electrogenic transport is monitored with the aid of a potential sensitive dye, 3,3'-diethyldi-carbocyanine iodide. This is done by measuring changes in the absorption difference of the dye at 580 and 610 nm on an Aminco-Chance dual beam spectrophotometer (see Section 3.6). The vesicles were prepared in 310 mM mannitol and 15 mM Hepes/Tris buffer (pH 7.5), and were added to a cuvette containing 3.0 ml of 15 mM Hepes/Tris buffer (pH 7.5) and dye (7 μg/ml) together with either 155 mM NaCl (Curves 1 and 2) or 155 mM KCl (Curve 3), or 155 mM choline chloride (Curve 4). **B.** Electrogenic transport by basolateral membrane vesicles. The experimental set-up was the same as described above. Curve 1 shows depolarisa-tion caused by addition of 5 mM L-malate in the presence of 155 mM NaCl in the medium, while Curve 2 demonstrates the absence of Na$^+$-dependent D-glucose transport in the basolateral membrane preparation. Taken from data published by Sheikh *et al.* (14).

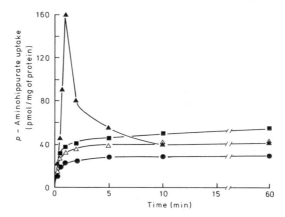

Figure 4. Na$^+$-dependent uptake of *p*-aminohippurate by basolateral membrane vesicles, measured by the Millipore filtration technique (Section 3.5). In this experiment basolateral and luminal vesicles were prepared in 300 mM sucrose and 25 mM Hepes/Tris buffer (pH 7.0). The reaction was started by addition to a basolateral membrane preparation of 25 mM Hepes/Tris (pH 7.0), 0.0075 mM plus [^{14}C]-labelled *p*-aminohippurate, and 150 mM NaCl (▲), or 150 mM KCl (△), or 150 mM NaCl and 1 mM probenecid (●). The lack of a Na$^+$-dependent overshoot of *p*-aminohippurate uptake by a luminal membrane vesicle preparation is demonstrated by ■—■. Taken from data published by Sheikh and Møller (35).

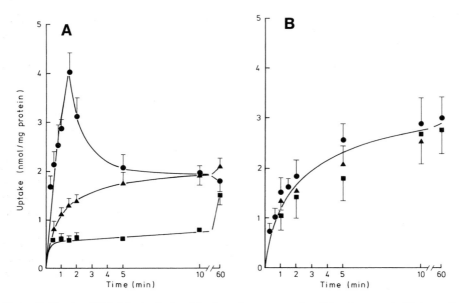

Figure 5. Uptake of $^{22}Na^+$ by luminal vesicles from the pars convoluta (**A**) and pars recta (**B**) to monitor H^+-Na^+ exchange. At zero time, 100 μl of incubation medium was added to 20 μl of membrane vesicle suspension (12 – 13 mg protein/ml). After varying incubation periods at 20°C, the uptake of $^{22}Na^+$ was stopped by adding 1 ml ice-cold incubation medium (without radioactive Na^+). The vesicles were pre-loaded in mannitol (179-191 mM), 1 mM Na^+, and Mes/Hepes/Tris buffer systems to produce pH 5.9 (●) and 7.4 (▲,■) the external medium contained mannitol (187 mM), 1 mM Na^+ with $^{22}Na^+$ and Mes/Hepes/Tris buffer to produce pH 7.4 (●,▲) and pH 6.1 (■). Note that only in the case of the pars convoluta is there evidence of Na^+-H^+ exchange as indicated by an overshoot of Na^+ uptake in the presence of a H^+-gradient ($pH_i > pH_o$). Taken from data published by Kragh-Hansen *et al.* (36).

in their infancy. *Figure 5* demonstrates that in our preparations Na^+-H^+ exchange is localized exclusively to pars convoluta. Other studies indicate that D-glucose has a high affinity and is co-transported with 2 Na^+ in pars recta, while the Na^+:D-glucose stoichiometry is 1:1 in pars convoluta. In the case of D- or L-lactate, transport is Na^+-specific in pars recta, while it is stimulated both by Na^+ and K^+ in pars convoluta. The locus of action of nephrotoxic agents may show specific features, e.g., dichromate tends to cause degeneration of tubuli contorti cells, while mercuric chloride preferentially causes necrosis of pars recta cells (5). On this basis we emphasise the need in future studies seriously to consider that pars recta and pars convoluta are to be considered as two structural entities with quite different functional properties. These will not be revealed in preparations derived from whole cortex, but require the isolation of luminal membranes from discrete regions of the kidney as explained in Section 2.10.

3.3 Electron Microscopy

Electron microscopy is useful in the characterisation of the preparations. The techniques employed can only be dealt with here in outline. For detailed procedures the reader should consult a standard treatise on the subject. Reference 37 is a good introduction to the subject.

Electron microscopic examination of negatively stained specimens of vesicle preparations is useful to assess to what extent the brush border organisation in luminal membrane fractions is maintained after vesicularisation. For this technique, grids are employed which have been coated with carbon and Formvar. The technique is as follows:

(i)   Apply to the grids a droplet of membrane suspension, diluted to about 0.05 mg protein/ml with 155 mM NaCl.

(ii)   After 30 sec, soak excess fluid from the grid with filter paper.

(iii)   Immediately add a drop of staining solution (potassium phosphotungstate, pH 7.2) and after 40 sec remove excess stain by soaking with filter paper.

After 10 − 15 min the grid is ready for electron-microscopic examination. *Figures 6C* and *7C* show electron micrographs of luminal membrane vesicles from adult rabbit kidney. It is seen that rod-like structures, reflecting the constrained organisation of the original brush border, are a prominent feature of the preparation. In addition the vesicles are characterised by a coat of small particles on the outer surface, indicating a right-side-out orientation.

Electron microscopy on sectioned and stained specimens may serve as an additional check on the purity and state of the preparation (absence of intracellular organelles, size distribution of vesicles, presence of open membrane fragments). In brief the procedure is as follows:

(i)   Spin down vesicles (0.5 ml of suspension containing ~5 mg protein/ml) in a rapid bench centrifuge (e.g., Beckman microfuge at 15 000 r.p.m.).

(ii)   Fix the specimen with a Karnowsky-type medium (2.5% glutaraldehyde, 2% paraformaldehyde, dissolved in 0.1 M phosphate buffer, pH 7.3), followed by washing (suspension and centrifugation) in 0.1 M sodium cacodylate buffer (pH 7.3), and then with 1% OsO_4, dissolved in 50 mM veronal buffer (pH 7.3).

(iii)   After dehydration with increasing concentrations of ethanol (40, 60, 75, 90, 96 and 100%), exchange ethanol with propylene oxide and embed in Epon (38).

(iv)   Section the specimen and stain with both uranyl acetate and lead acetate. An example of the results obtained by this technique for basolateral membranes is shown in *Figure 8*.

Microscopic section of the intact tissue is useful to characterise the nature of the starting material. For this purpose we have used the following procedure (36):

(i)   Anaesthetise rabbits with ethylurethan 1 h prior to fixation.

(ii)   After giving a small supplement of Mebumal anaesthesia in an ear vein, fix the kidneys by vascular perfusion of the whole body. The fixative, 2% glutaraldehyde in 0.1 M phosphate buffer, (pH 7.4 and 37°C), is infused into the abdominal aorta 2 cm distal to the origin of the renal arteries. Immediately after the start of the perfusion cut the inferior vena cava.

(iii)   Continue perfusion for 15 min under a hydrostatic pressure of about 150 cm.

(iv)   Take small blocks of cortex corticis and outer medulla tissue from the kidneys (see Section 2.10) and fix for another 45 min.

(v)   Rinse in 0.1 M cacodylate buffer, pH 7.4, and postfix in 2% OsO_4.

(vi)   After embedding in Epon, cut sections of 3 μm thickness, stain with 0.1% toluidine blue and study by light-microscopy.

(vii)   Cut sections of 50 − 70 nm thickness from the same tissue blocks, double-stain

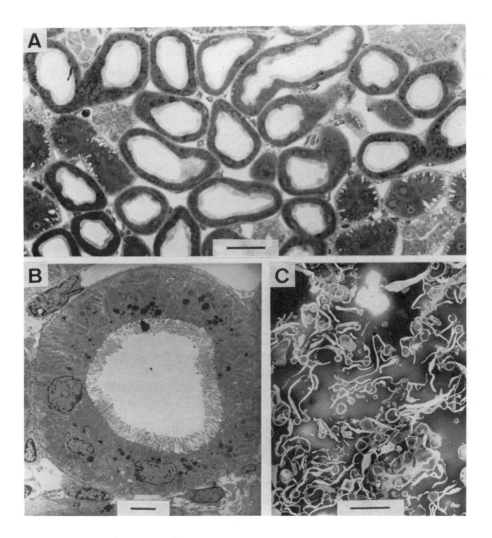

Figures 6 and 7. Uses of microscopy in the evaluation of the origin of membrane vesicles.
Figure 6. A. Light-microscopy of tissue taken from outer cortex, showing transverse sections of patent, convoluted proximal tubules. The bar: 30 μm. **B.** Electron microscopy of a transverse section of a proximal convoluted tubule (S_1-cell). Note morphology typical of S_1-cells well preserved brush-border, numerous mitochondria and lysosomes. Extensive interdigitation of tubule cells. The bar: 3 μm. **C.** A negatively stained preparation of luminal membrane vesicles prepared from the same tissue as shown in micrograph A. Note the presence of rodlike structures characteristic of luminal vesicles. The bar: 1 μm.

with uranyl acetate and lead citrate and examine in a Zeiss 10B electron microscope at 60 KV.

The result of such procedures for cortex corticis is shown in *Figure 6*. At low magnification (*Figure 6A*) the tissue is seen to contain numerous tubular structures with open lumina which are derived from proximal convoluted tubules, and only relatively

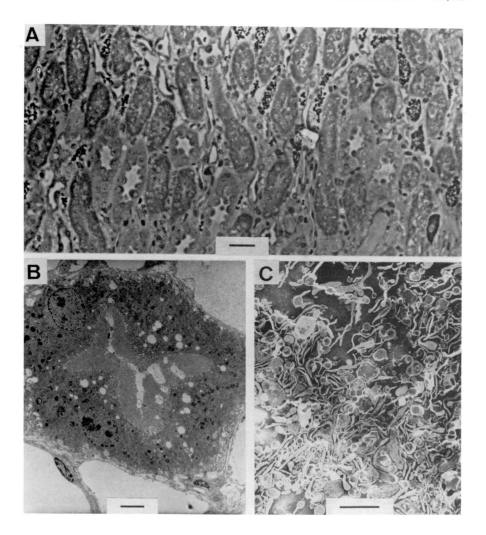

Figures 6 and **7.** Uses of microscopy in the evaluation of the origin of membrane vesicles.
Figure 7. A. Light-microscopy of longitudinal section through tissue taken from outer strip of outer medulla. Note the presence of pars recti proximal tubules in conjunction with other tubular structure (derived from ascending limb of Henlé, collecting ducts). The bar denotes 30 μm. **B.** Electron microscopy of a transverse section of an S_3-cell. The bar: 3 μm. **C.** A negatively stained preparation of luminal membrane vesicles prepared from the same tissue as shown in micrograph A. The bar: 1 μm. Taken from data published by Kragh-Hansen *et al.* (36).

few other tubular structures, which probably represent cortical collecting ducts and distal convoluted tubules. *Figure 6B* shows the morphology of the cells of the proximal tubule by electron-microscopic observation. It is seen that the cells are characterised by (i) numerous infoldings of the basal (peritubular) membrane, in which region the mitochondria are predominantly localised, (ii) pronounced 'brush border' organisation

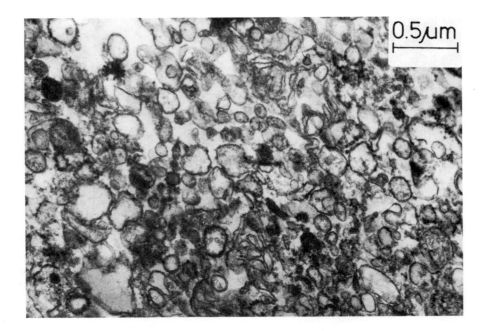

Figure 8. Thin section of basolateral membrane vesicles. A basolateral membrane preparation from rabbit whole renal cortex was treated as described in the text and the stained specimen was observed at a 36 000 × magnification. Taken from Reference 14.

of the luminal cell membrane, (iii) a multitude of subcellular organelles including mitochondria and lysosomes, and (iv) extensive interdigitation. These findings are characteristic features of proximal convoluted tubule cells. *Figure 6C* shows a negatively stained preparation of membrane vesicles derived from the same type of tissue as shown in *Figure 6A*. The picture shows that the preparation predominantly contains luminal membrane vesicles.

Figure 7 shows the tissue and membrane vesicles derived from outer medulla. At low magnification (*Figure 7A*) the tissue is seen to contain many proximal tubules (~50%) in conjunction with other tubular structures which probably represent collecting ducts, loops of Henle and blood capillaries. At high magnification (*Figure 7B*) the proximal tubular cells are seen to have a simpler shape and to interdigitate less than the cells from cortex corticis (*Figure 6B*). They contain smaller, less numerous and more randomly distributed mitochondria, which do not show a close association to the plasma membrane. Furthermore, a greater number of large vacuoles are seen, and the small endocytotic vesicles are decreased in number compared to the cells shown in *Figure 6B*. These features are typical for pars recta of the proximal tubule of rabbit kidney (39).

As is apparent from Section 3.1 there are only subtle differences in the enzyme activity profile of vesicles derived from pars convoluta and pars recta. Therefore, the demonstration of the origin of the membrane vesicles by electron microscopy is an essential step in the characterisation in this case.

3.4 Procedures for Measurement of Enzyme Activities

3.4.1 *Maltase and Trehalase*

Principle. Maltase and trehalase catalyse the hydrolysis of the disaccharides, maltose and trehalose, respectively, to D-glucose. The concentration of D-glucose liberated is measured by enzymatic spectrophotometry.

1. D-glucose + ATP $\xrightarrow{\text{hexokinase}}$ Glucose-6-P + ADP

2. Glucose-6-P + NAD^+ $\xrightarrow{\text{glucose-6-P dehydrogenase}}$ D-gluconolactone-6-P + NADH + H^+.

Reagents.

(A) Substrate: 25 mM maltose or 20 mM trehalose, dissolved in 30 mM potassium phosphate buffer (pH = 6.3).

(B) Perchloric acid: 3.4% (w/v).

(C) Reagent kit from Diagnostica, Roche, Basle, Switzerland. The contents of bottle No. 1 (containing ATP (>50 μmol), NAD^+ (>30 μmol), hexokinase (>7 U) and glucose-6-phosphate dehydrogenase (>8 U) are dissolved in 1.5 M triethanolamine buffer (pH 7.5) which is obtained by diluting the stock triethanolamine solution in bottle 2 with 650 ml de-ionised H_2O.

Procedure.

(i) Dilute membrane vesicles from pars convoluta or pars recta or corresponding homogenates with de-ionised H_2O to an approximate final protein concentration as given below:

A.	Pars convoluta homogenates:	12 mg protein/ml
B.	Pars convoluta vesicles:	6 mg protein/ml
C.	Pars recta homogenates:	12 mg protein/ml
D.	Pars recta vesicles:	6 mg protein/ml

(ii) Add 100 μl of renal tissue preparation of 100 μl of maltose or trehalase solution (A), thermostated at 20°C. Stop the reaction after exactly 5 min by addition of 500 μl perchloric acid (B). Centrifuge the samples for 15 min at 1000 g. To 150 μl of supernatant add 1500 μl of reagent kit solution (C). After 20 min read the absorbance at 340 nm versus H_2O as a reference. Calculate liberated D-glucose on the basis of an absorption coefficient of NADH of 6.22 mM^{-1} cm^{-1}.

3.4.2 *Determination of Leucine Aminopeptidase (LAP) Activity*

LAP catalyses the following reaction:

(i) L-leucyl-naphthylamide + H_2O $\xrightarrow{\text{LAP}}$ L-leucine + β-naphthylamine

   β-naphthylamine is measured by a modified Bratton-Marshall procedure;

(ii) β-naphthylamine + $NaNO_2$ $\xrightarrow{H^+}$ Diazo reagent;

(iii) Diazo reagent + N-(1-naphthyl)ethylenediamine $\rightarrow$ Blue Azo dye.

Reagents. A kit for the determination can be purchased (kit No. 251 from Sigma Chem. Corp., St. Louis, MO, USA), consisting of the following solutions:

(A) LAP substrate (Stock No. 251-1), containing L-leucyl-β-naphthylamide, 200 mg/l, dissolved in phosphate buffer (pH 7.1).

(B)   2 N HCl
(C)   Sodium nitrite, 2 mg tablets (Stock No. 251-4). Before use prepare a fresh 0.2% solution by dissolving 1 tablet in 1 ml de-ionised H_2O.
(D)   Ammonium sulfamate, 0.5% (w/v), (Stock No. 251-3).
(E)   Coupling reagent: 0.5 mg/ml N-1-naphthylethelenediamine, dihydrochloride (Stock No. 251-5), dissolved in 95% ethylalcohol.
(F)   LAP calibration solution: β-naphthylamine, 1.8 mg/dl (Stock No. 251-5).

Procedure.

(i)   Dilute membrane vesicles from pars convoluta or pars recta or corresponding homogenate with de-ionised H_2O to an approximate final protein concentration as given below:

A.	Pars convoluta homogenates:	0.8 mg protein/ml
B.	Pars convoluta vesicles:	0.03 mg protein/ml
C.	Pars recta homogenates:	1.5 mg protein/ml
D.	Pars recta vesicles:	0.08 mg protein/ml

(ii)   For each determination prepare 3 tubes as follows:

Tube 1 (reagent blank): 0.5 ml H_2O + 0.5 ml LAP substrate (A)
Tube 2 (vesicle blank): 0.5 ml H_2O + 0.5 ml of renal tissue preparation
Tube 3 (test sample): 0.5 ml LAP substrate (A) + 0.5 ml renal tissue preparation.

(iii)   Incubate all tubes for 1 h at 37°C. Stop reaction with 0.5 ml HCl (B) and then add 0.5 ml nitrite solution (C). Mix quickly and wait exactly 3 min. Add to each tube 1 ml ammonium sulfamate (D) (to destroy unreacted $NaNO_2$), mix quickly and wait for exactly 3 min. Add to each tube 2 ml coupling reagent (E). Mix and wait for 45 ± 10 min (room temperature).

(iv)   Read absorbance at 580 nm versus H_2O as a reference.

Calculations. The absorbances of the reagent blank and vesicle blank are subtracted from that of the test sample. To convert absorbances to concentrations prepare a standard curve by carrying LAP calibration solution (F), and suitable dilutions thereof with 2 N HCl, through the same colour development procedure (addition of nitrite etc.).

3.4.3 Na^+, K^+-ATPase Activity

Principle. Na^+, K^+-ATPase catalyses the hydrolysis of ATP to ADP and inorganic phosphate in presence of Na^+, K^+, and Mg^{2+}. The production of inorganic phosphate can be measured by spectrophotometry after formation of a reduced phosphomolybdate complex. The presence of other ATPases in the preparation is corrected for by addition of ouabain which acts as a specific inhibitor of Na^+, K^+-ATPase.

Reagents.

(A)   Histidine buffer. Dissolve 75 mM histidine, 325 mM NaCl, 5 mM KCl and 16 mM $MgCl_2$ in de-ionised H_2O.
(B)   2-amino-2-methyl-1,3-propanediol: 1 M, dissolved in de-ionised H_2O.
(C)   ATP buffer: Dissolve 7 μmol Na_2ATP in 10 ml of solution (A), and adjust the pH of the solution to 7.79 at 20°C (corresponding to pH = 7.4 at 37°C) with

solution (B) ($\sim$0.15 ml). Dilute the mixture to a final volume of 20 ml. This buffer should be used on the day it is prepared.

(D)   Histidine (30 mM), dissolved in de-ionised H_2O, and adjusted to pH = 7.4 with HCl.

(E)   0.5 N HCl.

(F)   Ammonium molybdate: 10% (w/v), in de-ionised H_2O.

(G)   Dissolve 3 g ascorbic acid in 100 ml 0.5 N HCl at 4°C, and add 5 ml ice-cold ammonium molybdate (F) dropwise with magnetic stirring. This solution can only be used for up to 2 days when stored at 4°C.

(H)   Dissolve 10 g sodium arsenate and 10 g sodium citrate in 490 ml de-ionised H_2O. After solubilisation add 10 ml glacial acetic acid.

(I)   Ouabain (10 mM): Dissolve in de-ionised H_2O.

(J)   Inorganic phosphate stock solution. 1 mM NaH_2PO_4 in de-ionised H_2O. This solution is diluted to 0.1 and 0.2 mM for the standard curve.

(K)   Deoxycholate. Solubilise sodium deoxycholate in de-ionised H_2O at a concentration of 5% (w/v).

(L)   Dilution buffer. Dissolve 310 mM mannitol and 25 mM Hepes/Tris (pH 7.5) in de-ionised H_2O.

Procedure.

(i)   Dilute membrane vesicles from pars convoluta or pars recta or corresponding homogenates with dilution buffer (L) to an approximate final protein concentration as given below:

      A.   Pars convoluta homogenates: 12 mg protein/ml
      B.   Pars convoluta vesicles: 6 mg protein/ml
      C.   Pars recta homogenates: 12 mg protein/ml
      D.   Pars recta vesicles: 6 mg protein/ml

(ii)   To 250 μl protein sample, thermostated at 37°C, add 5 μl deoxycholate (K).

(iii) Transfer 50 μl into each of two tubes (a and b), also thermostated at 37°C, containing:

    (a) 450 μl ATP buffer (C) + 50 μl ouabain (J)
    (b) 450 μl ATP buffer (C) + 50 μl H_2O.

(iv)   After 5 min add 1 ml of ice-cold solution (G). Exactly 6 min later add 1.5 ml solution (H).

(v)   Incubate the mixture for 10 min at 37°C and read the absorbance at 850 nm. Also carry a reagent blank, consisting of 450 μl ATP buffer (C), 50 μl H_2O, and 50 μl solution (L), through the same procedure. The concentration of inorganic phosphate produced by ATPase activity is calculated from the standard curve, after subtraction of the ATP blank value. The standard curve is made from 450 μl standard solutions (J) to which is added 50 μl H_2O and 50 μl solution L.

3.4.4 *Succinate Dehydrogenase Activity*

Principle. Succinate dehydrogenase activity is determined by measuring the transfer of 2[H] from succinate to an artificial receptor, namely 2,4-dichloro-phenol-indophenol (DCPIP).

Reagent. Dissolve 33 mM sodium succinate, 50 mM KH_2PO_4, 0.04 mM DCPIP, and 1 mM KCN in H_2O. Adjust the pH of the solution to 7.6 during solubilisation.

Procedure. All membrane fractions are diluted to 6 mg protein/ml with de-ionised H_2O. Pipette 3 ml of solution into a glass cuvette and thermoequilibrate the cuvette for 5 min at 37°C in the cuvette compartment of the spectrophotometer. Then add 50 μl of membrane fraction and mix. Follow for 10 min the change in absorbance at 600 nm, resulting from the reduction of DCPIP. An absorbance change of 1 is produced by reduction of 0.0524 μM DCPIP.

3.5 Millipore Filtration Technique

3.5.1 *Principle*

The reaction is usually started by addition to a vesicle suspension of the compound whose transport is to be studied in radioactively labelled form. After timed intervals, transport is stopped in aliquots of the reaction mixture by addition of ice-cold quench solution, which may contain a specific inhibitor of the transported compound. The vesicles are then separated from the suspension medium by suction through an appropriate filter.

3.5.2 *Equipment*

Filtration apparatus. *Figure 9* shows a schematic diagram of a convenient filtration apparatus which is frequently used for rapid separation of membrane vesicles from the incubation medium. By the use of this apparatus, simultaneous vacuum filtration of up to twelve samples can be achieved for comparative analysis of membranes retained on the filter surface. The keyed top plate contains 12 sample cups which rest on a filter support plate with screens on which to place the filters. Sealing is accomplished by turning the handwheel nut to tighten the filters against the O-rings of the top plate and connect the plates with the reservoir. By insertion of a test tube rack, the reservoir may also serve as a collection vessel for filtrate, which passes through the filters after establishment of vacuum. If all twelve cups are not needed for operation, stoppers are plugged into unused sample cups to maintain vacuum.

Millipore filters. These filters are made from chemically derivatised cellulose. A filter type must be chosen which retains all vesicles in a suspension, but which is not so fine as to decrease flow rates to low values. Usually filters with a nominal pore size of 0.45 μm are a good compromise. It should be ascertained that the filter material does not adsorb the transported compound and is chemically compatible with the medium to be filtered. Our own choice is to use cellulose acetate (SM 11106) and cellulose nitrate (SM 11306) filters (Sartorius GmbH, Göttingen, FRG) to study the transport of various carboxylic acids and amino acids, respectively.

Automated apparatus for short time uptake measurements. The procedures involved in quenching transport may be performed manually. However, for quick time-interval determinations (1 – 2 sec) we recommend the use of a relatively simple, automated apparatus, originally described by Kessler *et al.* (40), which may be obtained from Innovac Labor, Adliswill, Switzerland. Short time uptake measurements are often needed for precise determination of initial rates, due to efflux of substrate and/or dissipation

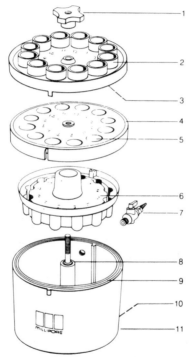

Figure 9. Sampling manifold for Millipore filtration, assembled and exploded view. **1**: Handwheel nut. **2**: Top plate. **3**: Location of filter sealing O-rings. **4**: Filter support plate. **5**: Support screens. **6**: Test tube rack. **7**: Ball valve with hose connector. **8**: Inner O-ring. **9**: Outer O-ring. **10**: Drain plug. **11**: Reservoir Chamber. Manufactured by Millipore Corporation, Bedford, MA, USA.

of the driving force (Na^+-gradient in many cases) during the uptake period. The apparatus (*Figure 10*) consists of an electronic control unit, a vibrator for starting the reaction, and a stop solution injector. A schematic diagram of the stop solution injector is shown in *Figure 11*. Stop solution in reservoir C is expelled through the injection assembly E at the set time by a shot of compressed air. The siphon reservoir is replenished with stop solution from reservoir A at the end of each shot. This action is controlled by valve B which remains open until the siphon reservoir is filled to the

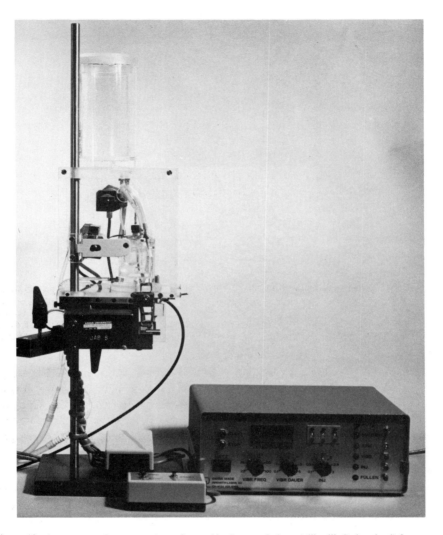

Figure 10. Appearance of apparatus (manufactured by Innovac Labor, Adliswill, Switzerland) for automatic processing of samples for transport measurements, consisting of stop solution injector, vibrator with sled for insertion of reaction test tube, and electronic console.

level of Pt-electrode D. Both reservoirs A and C are kept cooled by a surrounding mantle, percolated with ice-cold water.

3.5.3 *Procedure for Transport Measurements*

Transport measurements are usually carried out on samples, thermostated at around 20°C and stored in 5 ml clear polystyrene test tubes (diameter 13 mm, length 70 mm). For the manual procedure:

(i)   Transfer 50 μl membrane vesicle suspension (15 − 25 mg protein/ml) to the polystyrene test tube.

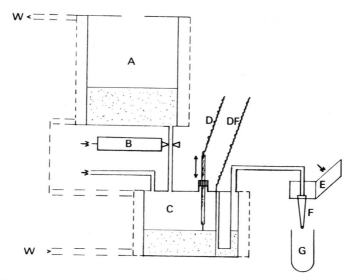

Figure 11. Schematic diagram of stop solution injector of the automatic apparatus for transport measurements. **A**: Reservoir for stop solution. **B**: Valve, on when siphon is empty, off when siphon is filled, controlled by two Pt-electrodes **D** and **DF**, of which the height of electrode **D** is adjustable so that the ejection volume can be varied. **C**: Siphon. **D** and **DF**: Pt-electrodes. **E**: Injection assembly. **F**: Tip. **G**: Test tube, fitted in the sled of the vibrator. **W**: Cooling water. **B**: Compressed air, by one main line to valve B and valves for injection assembly and for ejection of stop solution (pressure ~1.8−1.0 bar).

(ii)　At time zero quickly add 100 μl substrate medium, containing radioactive isotope (0.25−0.5 μCi for ^{14}C), with rapid mixing on a Whirlimixer. After a suitable time interval, add 850 μl of ice-cold stop solution and proceed with Millipore filtration as described below. Usually transport is measured with 15 sec intervals initially, while time periods may be extended to several minutes at later stages, see *Figures 4* and *5*.

Measurements with the automated apparatus are performed as follows:

(i)　Place two 10 μl drops of vesicle suspension and substrate medium respectively close to each other at the bottom of the polystyrene test tube. Take care that the position of the two drops is in the same direction as the subsequent shaking action.

(ii)　Place the test tube in a sled connected to the vibrator (variable frequency 20−100 Hz). Set the desired reaction time on the electronic control unit.

(iii)　Mix the drops by switching the vibrator to the ON position. (The frequency and the amplitude of the shaking have to be properly adjusted in order to get rapid mixing of the drops without dispersion of the fluid). At the set time the reaction is stopped automatically by the addition of stop solution from the injector.

After the addition of the stop solution, manually or automatically:

(i)　Filter the contents of the test tube through Millipore filter.

(ii)　Rinse the filter and test tube by washing twice with 2.5 ml ice-cooled stop solution. Continue to process the next sample in the same way. Normally to complete a transport assay requires $6-10$ different samples. At the end, treat two blanks containing transport medium and membranes, denatured by boiling for 2 min, in the same way.

After the assay, transfer the filters to a mini-scintillation vial, let the filters stand overnight to dry, and add 2 ml Lumagel (Lumac, The Netherlands) for counting by lipid scintillation. Determine specific radioactivity by counting 50 and 100 μl of samples of substrate medium diluted $250-500$ times with unlabelled medium. Calculate transport from the radioactive counts of each filter after subtraction of the blank.

3.5.4 *Examples of Media Composition*

(A)   Membrane vesicle preparations, suspended in 310 mM mannitol and 15 mM Hepes buffer (pH 7.5), are used at a protein concentration of $15-25$ mg/ml.

(B)   Substrate media. Transported substrate (e.g., 0.1 mM D-glucose, 0.05 mM L-proline, or 0.05 mM p-aminohippurate, together with ^{14}C-labelled isotopes of these compounds ($25-50$ μCi/ml), dissolved in 310 mM mannitol and 15 mM Hepes (pH 7.5). To test cation dependency it is convenient to substitute mannitol with iso-osmotic (155 mM) NaCl or KCl. To create a H^+-gradient substitute the Hepes buffer with 15 mM Mes/Tris buffer (pH 5.5), cf. *Figure 12*.

(C)   Stop solution (0°C). Same composition as substrate medium, except for absence of radioactive isotope, and addition of competitive transport inhibitor (0.2 mM phlorizin to stop D-glucose transport, 10 mM L-proline to stop L-proline transport, and 1 mM probenecid to stop p-aminohippurate transport).

3.5.5 *Comments on the Technique*

Millipore filtration is the most commonly used technique for transport measurements, but requires great care, if precise results are to be obtained. During separation of vesicles from the suspending medium, transported substrate may leak from the vesicles by a time-dependent process, and some vesicles may be damaged, resulting in release of vesicle contents. In addition, the vesicle volume (~ 2 μl/mg protein) is small, compared to the volume of fluid that is needed to wet the filter ($40-50$ μl for a 25 mm diameter filter); hence the need for thorough rinsing of the filters. Often the existence of specific transport systems in membrane vesicles is inferred from comparisons of uptake of structurally related compounds when the required radioactive labelled substrates are unavailable.

3.6. Spectrophotometric Registration of Electrogenic Transport

3.6.1 *Principle*

The light absorption of many carbocyanine dyes is sensitive to changes in transmembrane potential (41). If transport of a compound is associated with the movement of electric charges across a membrane, this causes a partial depolarisation of a membrane, that can be registered by an absorbance change of the added carbocyanine dye.

3.6.2 *Apparatus*

A dual wavelength spectrophotometer (DW2, American Instrument Co.) is used to

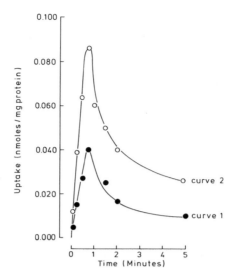

Figure 12. Na$^+$ and H$^+$-dependent uptake of L-proline by luminal membrane vesicles prepared from pars convoluta of the proximal tubule from rabbit kidney, as measured by Millipore filtration technique in the presence of a Na$^+$ or a H$^+$ gradient (extravesicular > intravesicular). 20 μl of a vesicle suspension (20 mg/ml) was incubated at different time intervals in 100 μl incubation mixture consisting of 155 mM NaCl, 50 μm L-proline, 10 μM L-[^{14}C]-proline in 15 mM Hepes/Tris, pH 7.5 (curve 1) or in 15 mM Mes/Tris pH 5.5 (curve 2). The reaction was stopped by adding 1 ml ice-cold stop solution consisting of 155 mM NaCl and 10 mM L-proline in 15 mM Hepes/Tris buffer pH 7.5 or in 15 mM Mes/Tris pH 5.5, respectively. After termination of the reaction the suspension was rapidly filtered through a Sartorius membrane filter (0.45 μM, type SM 11306). The filter was washed twice with 2.5 ml ice-cold stop solution, dried overnight and the radioactivity was counted in a liquid scintillation counter after addition of Lumagel. Taken from Røigaard-Petersen et al. (unpublished data).

compensate for changes in light absorption, caused by the turbid membrane suspension. This is done by measuring changes in the absorption difference of two different wavelengths. The spectrophotometer is used in conjunction with an accessory for magnetic stirring of the cuvette in the cuvette compartment. In addition the cuvette compartment should be thermostated (20°C) with the aid of a circulating thermostat bath. Addition of small volumes to the sample during the experiment is performed with Hamilton syringes.

3.6.3 Media

(A)   Membrane vesicle preparation at a protein concentration of 15−25 mg/ml, suspended in 15 mM Hepes/Tris buffer (pH 7.5) and 310 mM mannitol.

(B)   3,3′-diethyloxydicarbocyanine iodide (Eastman Kodak, Rochester, NY, USA) is dissolved at a concentration of 14 μg/ml in degassed 15 mM Hepes/Tris buffer (pH 7.5). The solution is bubbled with N$_2$ during solubilisation and stored at 2°C in a dark bottle that is covered with aluminium foil. Fresh dye solution is made after 3−4 days.

(C)   Medium for creation of a salt gradient. Dissolve 310 mM NaCl, or 310 mM KCl, or 310 mM choline chloride in 15 mM Hepes/Tris buffer (pH 7.5).

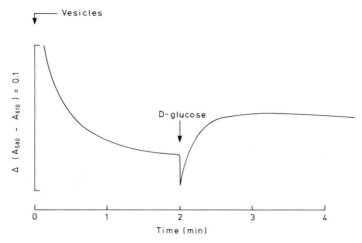

Figure 13. Changes in absorbance of potential-sensitive carbocyanine dye after addition of vesicles at time zero and addition of 5 mM D-glucose at the time of maximal hyperpolarisation. The experiment was performed as described in the text, in the presence of a NaCl gradient. The deflection at the time of D-glucose addition represents dilution caused by the substrate addition. For further details see reference 42.

(D)   Transported substrate which is solubilised at a high concentration (e.g., 200 − 400 mM) in medium (C).

3.6.4 *Procedure*

(i)   Turn on the spectrophotometer thermostat. When the spectrophotometer has warmed up (~ 30 min) transfer 1.5 ml each of media (B) and (C) to the cuvette and place it in the cuvette compartment.

(ii)   Set Monochromator *1* at 580 nm and Monochromator *2* at 610 nm. Balance the beams from the two monochromators and adjust the pen to the desired level on the recorder, using an absorption scale of $0 - 0.2$.

(iii)   After these adjustments have been performed, start the experiment by adding 60 μl of the vesicle suspension (A). Immediately after the addition an increase in $\Delta(A_{580} - A_{610})$ occurs as a consequence of binding of part of the dye to the vesicular membranes. If at this step the pen deflects outside the scale, adjust the pen with the beam balance knob so that changes in dye absorption can be followed approximately 10 sec after the vesicle addition. After this time a *decrease* in $\Delta(A_{580} - A_{610})$ occurs which is caused by the low electrolyte concentration of vesicular fluid, compared to that of the suspending medium. Since the vesicles are more permeable to Cl^- than to the cation added, a hyperpolarisation of the vesicles (inside negative) is produced which is registered as a downward deflection of dye absorption (*Figure 13*). At the time of maximal deflection, add a small volume $(25 - 50\ \mu l)$ of transported substrate. Electrogenic transport (of positive charge) results in partial depolarisation and an increase in $\Delta(A_{580} - A_{610})$. The amplitude and initial rate of the response is approximately proportional to transport rate (43).

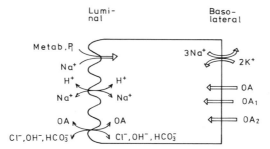

Figure 14. Transport functions of the proximal tubule. The figure demonstrates Na^+-dependent transport metabolites and inorganic phosphate, Na^+-H^+ exchange and anion-exchange of organic (OA) and inorganic anions at the luminal membrane. The active transport of Na^+ and K^+ by Na^+-K^+-ATPase and the tubular secretory system for organic anions (OA), the hippurate system, are located at the basolateral membrane. In addition, the basolateral membranes contain specific systems for tubular uptake of acidic amino acids (OA_1) and di- and tricarboxylic acids (OA_2)

3.6.5 Comments on the Technique

The technique requires the use of a specialised spectrophotometer, but is very simple to perform. Once it has been set up, a large number of experiments can be performed within a working day to test for properties such as cation dependency, affinity for transport and competition between different transported substrates. But note that a response requires that a *net* electrical charge is carried across the membrane in connection with each transport cycle. For instance, in the case of citrate transport across the basolateral membrane, the number of Na^+ co-transported (2 or 3) matches the anionic (divalent or trivalent) charge of citrate. This results in electrically silent transport that does not give rise to any changes in dye absorption (43).

4. APPLICATIONS OF THE PLASMA MEMBRANE PREPARATIONS

The main function of renal plasma membranes is to modify and regulate the composition of the glomerular filtrate by a series of transcellular transport processes that require oxidative energy. In turn, the transport functions are important for cell homeostasis, and interference with these processes is therefore potentially important in the development of degenerative changes in renal cells by administration of nephrotoxic compounds. Let us first briefly consider the roles played by the luminal and basolateral membranes in transcellular transport by the proximal cells (*Figure 14*). Many metabolites, including sugars, amino acids and other organic acids are transported across the luminal membrane by Na^+-dependent co-transport systems, and in some cases by H^+-dependent co-transport (*Figure 12*). This results in intracellular accumulation of metabolites, due to the imbalance of inorganic cation composition between the extracellular and intracellular fluid (low Na^+ and high K^+ inside the cell). This imbalance is created by the Na^+,K^+-ATPase, situated at the basolateral membrane. The further transfer of reabsorbed solutes from the intracellular space to the blood is considered to occur by facilitated carrier systems in the basolateral membrane. The basolateral membrane also contains systems for active uptake of organic anions by the hippurate system (44), and specific systems for active uptake of glutamate (24), malate

(14), citrate (43) and other dicarboxylic acids (45). The luminal membrane contains exchange systems for anions like urate, salicylate, hippurate (46,47), a Na^+-H^+ exchange system (48), and is the site for pinocytotic activity (39).

The fact that the Na^+,K^+-ATPase is the ultimate driving force for most other transport activities via cation-dependent transport systems makes it possible in vesicle experiments to simulate transport conditions as they exist *in vivo*, by the imposition of cation gradients across the membranes. It is therefore possible to study the effect of nephrotoxic agents on transport function, independent of other effects that these agents may have on cell function. The effect of maleic acid on tubular function provides a case in point. Administration of this compound has been shown to lead to the development of a Fanconi-like syndrome, characterised by increased excretion of bicarbonate, glucose, amino acids, phosphate, Na^+, and Cl^- (7,8) which is accompanied by histological changes (49,50). Reynolds *et al.* (51) found that substantial amounts of maleic acid interact with the luminal membrane preparations of the proximal tubule. However, the Na^+-dependent transport of sugars, amino acids and phosphate by isolated luminal membranes isolated from kidneys exposed to maleic acid was unchanged (51,52). The failure to detect an effect of maleic acid on subcellular transport was considered to support other evidence (53) that maleic acid causes a permeability defect in the distal nephron, leading to a non-discriminatory loss of solutes with the urine (51). However, a recent study emphasises that the main site of action of renal tubular alkalosis, associated with maleic acid administration, is located to the proximal tubule (54). Hence it appears more reasonable to assume that maleic acid affects tubular transport by interference with renal metabolism (54), a conclusion which is supported by studies on the effect of nucleic acid on the oxidative energy metabolism of mitochondria (55). However in our opinion, the published studies are not sufficiently detailed to rule out the possibility of a functional effect of maleic acid on renal plasma membranes (e.g., on Na^+-H^+ exchange) since alkalosis is the most prominent manifestation of maleic acid administration on renal function (7).

From another point of view, the use of plasma membranes as compared to *in vivo* conditions, is also helpful in the study of transport function since after administration of nephrotoxic agents *in vivo* evaluation of kidney function is complicated by factors such as changes in hemodynamics and glomerular filtration rate. Indeed it has been suggested that hemodynamic changes, leading to a curtailment of renal cortical blood supply, are the major cause of tubule necrosis and decrease of kidney function in acute mercury poisoning (56). In principle, transport function as an intrinsic property of plasma membranes could be studied by the use of isolated membrane vesicles after administration of nephrotoxic agents in animal experiments (in particular under conditions of chronic administration which do not lead to extensive cell necrosis), or by addition of the nephrotoxic compound directly to the vesicles.

Vesicles may also be used to study the interaction (transport, binding) of the nephro-toxic compounds themselves with the membrane. In the case of cephalosporins, there is evidence from studies *in vivo* and kidney slices of active uptake across the basolateral membrane by the hippurate system (57), but vesicle experiments to study the details of the transport are not available. Aminoglycosides are bound to luminal membranes (58), and a high affinity site interaction with acidic phospholipids has been demonstrated (59). Binding of aminoglycoside possibly represents the initial step in cellular uptake by pinocytosis. This binding is inhibited by other cationic compounds (58) and by Ca^{2+}, which may help to explain a beneficial effect of Ca^{2+} administration of gentamycin nephrotoxicity *in vivo* (59).

Other possible uses of the preparations are to test the enzymatic function and intactness of membrane function (leakiness of vesicles to solutes). For the basolateral membrane, studies on Na^+, K^+-ATPase activity (60) and the adenylate cyclase system are important functional attributes. The activity of the latter may be assayed by the use of parathyroid hormone to stimulate formation of c-AMP from $\alpha[^{32}P]ATP$ (61).

6. REFERENCES

1. Maher,J.F. (1976) *The Kidney*, Brenner,B.M. and Rector,F.C. (eds.), W.B. Saunders Co., Philadelphia, Vol. **II**, Chapter 31.
2. Porter,G.A. and Bennett,W.M. (1981) *Am. J. Physiol.*, **241**, F1-F8.
3. Tune,B.M. and Fravert,D. (1980) *Kidney Int.*, **18**, 591.
4. Oliver,J., Mac Dowell,M. and Tracy,A. (1951) *J. Clin. Invest.*, **30**, 1307.
5. Biber,T., Mylle,M., Baines,A.D., Gottschalk,C.W., Oliver,J.R. and Mac Dowell,M.C. (1968) *Am. J. Med.*, **44**, 664.
6. Bräunlich,H., Fleck,C., Weise,C. and Stopp,M. (1979) *Exp. Pathol.*, **17**, 486.
7. Berliner,R.W., Kennedy,T.J. and Hilton,J.G. (1950) *Proc. Soc. Exp. Biol. Med.*, **75**, 791.
8. Harrison,H.E. and Harrison,H.C. (1954) *Science*, **120**, 606.
9. Vander,A.J., Mouw,D.R., Cox,J. and Johnson,B. (1979) *Am. J. Physiol.*, **236**, F373.
10. Craan,A.G., Nadou,G. and P'An,A.Y.S. (1984) *Am. J. Physiol.*, **247**, F773.
11. Campbell,D.E.S. (1960) *Acta Pharm. Toxicol.*, **171**, 213.
12. Cafruny,E.J. and Gussin,R.Z. (1967) *J. Pharm. Exp. Therap.*, **155**, 181.
13. Madsen,K. (1982) Effects of chronic mercury exposure and ageing on the lysosomal system of the rat proximal tubule, Thesis, Aarhus.
14. Sheikh,M.I., Kragh-Hansen,U., Jørgensen,K.E., and Røigaard-Petersen,H. (1982) *Biochem. J.*, **208**, 377.
15. Kragh-Hansen,U., Røigaard-Petersen,H., Jacobsen,C. and Sheikh,M.I. (1984) *Biochem. J.*, **220**, 15.
16. Booth,A. and Kenney,A.J. (1974) *Biochem. J.*, **142**, 575.
17. Kinsella,J.L., Holohan,P.D., Pessah,N. and Ross,C.R. (1979) *Biochim. Biophys. Acta*, **552**, 468.
18. Kippen,I., Hirayama,B., Klinenberg,J.R. and Wright,E.M. (1979) *Biochim. Biophys. Acta*, **556**, 161.
19. Murer,H. and Kinne,R. (1980) *J. Membr. Biochem.*, **55**, 81.
20. Kinne,R. and Kine-Safran,E. (1981) *Eur. J. Clin. Biochem.*, **25**, 436.
21. Scalera,V., Huang,Y.-K., Hildmann,B. and Murer,H. (1981) *J. Membr. Biochem.*, **4**, 49.
22. Inui,K.-I., Kokano,T., Takano,M., Kitazawa,S. and Hori,R. (1981) *Biochim. Biophys. Acta*, **647**, 150.
23. Windus,D.W., Cohn,D.E., Klahr,S. and Hammerman,M.R. (1984) *Am. J. Physiol.*, **246**, F78.
24. Sacktor,B., Rosenbloom,I.L., Liang,C.T. and Cheng,L. (1981) *J. Membr. Biochem.*, **60**, 63.
25. Turner,R. and Moran,A. (1982) *Am. J. Physiol.*, **242**, F406.
26. Miller,D. and Crane,R.K. (1961) *Biochim. Biophys. Acta*, **52**, 293.
27. Eichholz,A. and Crane,R.K. (1965) *J. Cell Biol.*, **26**, 678.
28. Forstner,G.G., Sabesin,S.M. and Isselbacher,K.J. (1968) *Biochem. J.*, **106**, 381.
29. Schmitz,J., Preiser,H., Maestracci,D., Ghosh,B.K., Cerda,J.J. and Crane,R.K. (1973) *Biochim. Biophys. Acta*, **323**, 98.
30. Burkhardt,G., Kramer,W., Kurz,G. and Wilson,A.F. (1983) *J. Biol. Chem.*, **258**, 3618.
31. Mamelock.R.D., Solomon,S.T., Newcomb,K., Bildstein,C.L. and Liu,D. (1982) *Biochim. Biophys. Acta*, **692**, 115.
32. Colad,B. and Maroux,S. (1980) *Biochim. Biophys. Acta*, **600**, 406.
33. Jørgensen,P.L. (1980) *Physiol. Revs.*, **60**, 864.
34. Knepper,M. and Burg,M. (1983) *Am. J. Physiol.*, **244**, F579.
35. Sheikh,M.I. and Møller,J.V. (1982) *Biochem. J.*, **208**, 243.
36. Kragh-Hansen,U., Røigaard-Petersen,H. and Sheikh,M.I. (1985) *Am. J. Physiol.*, in press.
37. Wischnitzer,S. (1970) *Introduction to Electron Microscopy*, Pergamon Press, New York.
38. Luft,J. (1961) *J. Biophys. Biochem. Cytol.*, **9**, 409.
39. Maunsbach,A. (1973) in *Handbook of Physiology, Section 8: Renal Physiology*, Orloff,J. and Berliner,R.W. (eds.), *Am. Phys. Soc.*, *Washington*, p. 31.

40. Kessler,M., Tannenbaum,V. and Tannenbaum,C. (1978) *Biochim. Biophys. Acta*, **509**, 348.
41. Sims,P.J., Waggoner,A.S., Wang,C.-H. and Hoffman,J.F. (1974) *Biochemistry*, **13**, 3315.
42. Kragh-Hansen,U., Jørgensen,K.E. and Sheikh,M.I. (1982) *Biochem. J.* **208**, 359.
43. Jørgensen,K.E., Kragh-Hansen,U., Røigaard-Petersen,H. and Sheikh,M.I. (1983) *Am. J. Physiol.*, **244**, F686.
44. Møller,J.V. and Sheikh,M.I. (1983) *Pharm. Revs.* **34**, 315.
45. Burckhardt,G. (1984) *Europ. J. Physiol.*, **401**, 254.
46. Blomstedt,J.W. and Aronson,P.S. (1980) *J. Clin. Invest.*, **65**, 931.
47. Manganel,M., Ginsau,B., Murer,H., Roch-Ramel,F. (1985) *V Eur. Colloq. Renal Physiol.*, p. 22.
48. Kinsella,J.L. and Aronson,P.S. (1980) *Am. J. Physiol.*, **238**, F461.
49. Christensen,E.I. and Maunsbach,A.B. (1980) *Kidney Int.*, **17**, 771.
50. Verani,R.R., Brewer,E.D., Ince,A., Gibson,J. and Bulger,R.E. (1982) *Lab. Invest.*, **46**, 79.
51. Reynolds,R., McNamara,P.D. and Segal,S. (1978) *Life Sci.*, **22**, 39.
52. Silverman,M. (1981) *Membr. Biochem.*, **4**, 63.
53. Bergeron,M., Bubord,L. and Hausser,C. (1976) *J. Clin. Invest.*, **57**, 1181.
54. Al'Bander,H.A., Weiss,R.A., Humphreys,M.H. and Morris,R.C.,Jr. (1982) *Am. J. Physiol.*, **243**, F604.
55. Rogulski,J., Pacanis,A., Adamovicz,W. and Angielsky,S. (1974) *Acta Biochim. Pol.*, **21**, 403.
56. DiBona,G.F., McDonald,F.D., Flamenbaum,W., Dammin,G.J. and Oken,D. (1971) *Nephron*, **8**, 205.
57. Tune,B.M. and Fernholt,M. (1973) *Am. J. Physiol*, **225**, 1114.
58. Just,M. and Haberman,E. (1977) *Nauyn-Schmiedebergs Arch. Pharm.*, **30**, 67.
59. Sastrasinh,M., Knauss,T.C., Weinberg,J.M. and Humes,H.D. (1982) *J. Pharm. Exp. Therap.*, **222**, 350.
60. Cronin,R.E., Nix,K.L., Ferguson,E.R., Southern,P.M. and Heinrich,W.L. (1982) *Am. J. Physiol.*, **242**, F477.
61. Scholer,D.W. and Edelman,I.S. (1979) *Am. J. Physiol.*, **237**, F350.

CHAPTER 8

Preparation and Characterisation of Microsomal Fractions for Studies on Xenobiotic Metabolism

BRIAN G.LAKE

1. INTRODUCTION

The microsomal fraction of a cell may be defined as that portion of the post-mitochondrial supernatant which is sedimented by a centrifugal force of $100\ 000 - 250\ 000\ g$ for $60 - 120$ min (1). The morphological nature of the pellet obviously depends on the cell type from which it is derived, but with hepatocytes consists of fragments of rough and smooth endoplasmic reticulum. Unlike intracellular organelles (e.g., mitochondria, lysosomes) the endoplasmic reticulum is extensively disrupted by homogenisation and membrane fragments 'pinch off' to form closed vesicles or 'microsomes' (2). These vesicles may contain ribosomes on their outer surface (i.e., if derived from rough endoplasmic reticulum), the outside of the microsomal membrane corresponding to the cytoplasmic surface of the endoplasmic reticulum with the luminal surface inside. Allowing for the recovery and purity of the microsomal fraction the endoplasmic reticulum of rat hepatocytes contains approximately 20%, 50% and 60% of the total cellular protein, phospholipid and RNA, respectively (2).

Although the microsomal fraction is thus an artefact of tissue homogenisation it contains several enzymes involved in xenobiotic metabolism including cytochrome P450 dependent mixed function oxidase activities (3), the mixed function amine oxidase described by Ziegler (4), epoxide hydratase (5) and UDP-glucuronyltransferase (6). A complicating factor is that multiple forms exist of several xenobiotic metabolising enzymes including microsomal cytochrome P450 (7) and UDP-glucuronyltransferase (6) activities.

The purpose of this chapter is to describe basic techniques for the preparation of microsomal fractions and the assay of a range of xenobiotic metabolising enzyme activities. Whilst the procedures described have been primarily developed for rodent liver they may be applied, with appropriate modifications as required, both to other species and to extrahepatic tissues.

2. BASIC TECHNIQUES

2.1 Equipment

2.1.1 Homogenisers

The classic homogeniser for hepatic tissue is the Potter-type, Teflon-glass, motor driven homogeniser (A.H.Thomas Company, Philadelphia, PA, USA; Arnold R.Horwell Ltd.,

London, UK). This homogeniser is available in various sizes of which size C is suitable for rat liver, size B for mouse liver or resuspending microsomal pellets and size A for even smaller (e.g., biopsy) samples. The glass homogenising vessel is kept in a plastic container (e.g., a plastic 500 g chemical bottle packed with ice). This keeps the sample cool during homogenisation and also protects the operators hands should the homogeniser fracture during use. The Teflon pestle is mounted vertically in a suitable motor, such as a domestic power drill, which can be rotated between 1000 and 2500 r.p.m. For convenience and safety the motor is best operated by a foot switch. As homogenisation is an important determinant of the nature of the final preparation obtained, workers are advised to maintain a constant set of conditions between experiments. These include the source of the homogeniser, the motor speed and the number of return strokes (i.e., passes of the pestle from the top to the bottom of the grinding vessel and back again) to effect adequate tissue disruption.

2.1.2 *Centrifuges*

To prepare the post-mitochondrial supernatant fraction a 'high speed' centrifuge (maximum speed up to 18 000 − 24 000 r.p.m.) is required and for the microsomal fraction an 'ultracentrifuge' (maximum speed 50 000 − 75 000 r.p.m.). Post-mitochondrial supernatant fractions can be prepared with ultracentrifuges, although this is usually less convenient. Generally refrigerated centrifuges and appropriate angle rotors from any of the major centrifuge manufacturers may be employed. With respect to centrifuge tubes polycarbonate tubes are preferable to polypropylene tubes as they permit easier visualisation of sedimented pellets. For ultracentrifuges threaded tubes with low cost sealing cap assemblies are easier and quicker to use than tubes with conventional sealing cap assemblies.

2.1.3 *Other Apparatus*

In addition to homogenisers and centrifuges the following will be required:
(i)   Surgical equipment for removal of tissue from experimental animals.
(ii)   Stop clocks or timers.
(iii)   Various pipettes.
(iv)   Parafilm and tissues.
(v)   Glass beakers and measuring cylinders for storage of liver samples and homogenates. As with centrifuge tubes these should be kept cooled in ice.

2.2 Homogenising Media

The literature contains a variety of homogenising media, of which media based on either isotonic sucrose (0.25 M, 85.6 g/l) or KCl (0.154 M, 11.5 g/l) are the most popular. Whilst sucrose is generally preferable for classical subcellular fractionation studies, KCl is more appropriate for work on xenobiotic metabolising enzymes as it is more efficient in removing haemoglobin from microsomal preparations and thus reduces interference in spectral assays. Homogenising media should be adjusted to around pH 7.4 prior to use or ideally should contain 5 − 50 mM of a suitable buffer such as Tris or Hepes. Some workers also advise the addition of 1 − 5 mM EDTA to prevent lipid peroxidation and additional homogenising media are given by Guengerich (8). However, for most studies a medium comprising 0.154 M KCl containing 50 mM Tris-HCl, pH 7.4 [i.e. 11.5 g KCl and 6.055 g Tris base in 1 litre (adjusted to pH 7.4 with HCl)] should be suitable. This medium is stable for at least 12 weeks at 4°C.

2.3 **Biological Starting Material**

Small rodents, such as rats, mice and hamsters, should be killed as quickly and humanely as possible (e.g. by cervical dislocation). Anaesthetics such as ether and barbiturates are known to affect enzyme activities and hence should be avoided. However, larger animals may have to be killed by exsanguination under a suitable anaesthetic. In all instances the livers should be immediately excised into beakers containing ice-cold homogenising medium. As a wide variety of factors are known to affect enzyme activities care should be taken to reduce inter-experiment variation as much as possible. These factors include the choice of animal age, sex and strain together with health status, the type of diet fed and environmental conditions. The known circadian rhythm in xenobiotic metabolism can be minimised by maintaining animals under an appropriate light/dark cycle and killing them for tissue preparation at the same time of the day. Finally, it is common practice to starve rodents overnight before using them for tissue preparation. This procedure depletes hepatic glycogen which otherwise would be recovered in the microsomal pellet and also improves the recovery of microsomes (9).

2.4 **Preparation of Whole Homogenate, Post-mitochondrial Supernatant and Microsomal Fractions**

Throughout this procedure keep tissue and tissue fractions cold (0−4°C) by using pre-cooled homogenisation medium and centrifuge rotors, and by keeping tissue homogenisers, measuring cylinders and centrifuge tubes in ice.

(i) Remove excess blood from the liver sample by washing in homogenising medium, dissect away any pieces of connective tissue etc., and gently blot the liver sample dry with tissue paper.

(ii) Weigh the liver sample on a top pan balance, record the weight, and then quickly transfer the liver to a homogenising vessel which is kept in ice. Add an appropriate amount of homogenising medium (the volume is dictated by the desired concentration of the whole homogenate, e.g. for a 25% homogenate add approximately 3 ml/g tissue), and homogenise the liver sample with five return strokes of the motor-driven pestle. For large liver samples (>12 g), cut them into suitable portions and pool the resulting homogenates. With rodent liver it is usually not necessary to scissor mince the sample prior to homogenisation but with liver samples from some other species (e.g., certain primates and man), which contain larger amounts of connective tissue, the samples may need to be cut into small pieces in order to obtain adequate tissue disruption. This can be effected either by scissor mincing the tissue or by cutting up the liver sample with a scalpel on a glass plate or other suitable surface. Homogenates which contain large amounts of connective tissue may require to be filtered through nylon mesh or surgical gauze before centrifugation.

(iii) Carefully decant the homogenate, whilst examining it for pieces of undisrupted tissue, into a measuring cylinder and make up the desired volume with washings of homogenising medium from the homogeniser tube. In the case of rat liver adjust the final homogenate to approximately 0.25 g fresh tissue/ml (i.e., a 25% homogenate) although microsomes can also be readily prepared from 10%, 20% and 33% homogenates; more dilute homogenates often being prepared when only small liver samples are available (e.g., from a single mouse). Record the final volume of the homogenate and cover the top of the measuring cylinder with 'Parafilm'. Mix the homogenate well by gentle inversion of the cylinder and transfer a desired amount immediately to pre-cooled centrifuge tubes.

(iv)   Cap and balance the tubes according to the centrifuge manufacturers instructions and prepare a post-mitochondrial supernatant fraction by centrifugation at 10 000 g average for 20 min at $2-4°C$. Carefully remove the tubes from the centrifuge and decant the post-mitochondrial (10 000 g average) superantant into a pre-cooled measuring cylinder or other suitable receptacle. The presence of a floating lipid layer on top of the supernatant fraction is often observed and unless all of the supernatant is required remove this with the aid of a Pasteur pipette and discard it.

(v)   Separate the microsomal fraction from the cytosol by further centrifugation of the post-mitochondrial supernatant. Although the literature contains many different combinations of rotor speed and run times a combined relative centrifugal force/time of 105 000 g average h^{-1} is generally sufficient to sediment the microsomal pellet. The choice of ultracentrifuge rotor is dictated by the volumes of post-mitochondrial supernatant fractions to be centrifuged but generally any angle rotor of nominal tube capacity of $10-50$ ml which can be spun at 40 000 or preferably 50 000 r.p.m. is suitable. In the authors laboratory MSE (Crawley, West Sussex, UK) 8 × 14 ml angle aluminium or titanium ultracentrifuge rotors are regularly used at 50 000 r.p.m. For a centrifuge rotor capable of producing say 160 000 g average the centrifuge run time at full speed (i.e., ignoring rotor run up and run down times) will thus be approximately 40 min (to provide an overall relative centrifugal force/time of 105 000 g average h^{-1}).

(vi)   Transfer known volumes of post-mitochondrial supernatant into appropriate sized pre-cooled ultracentrifuge tubes which are then capped and balanced according to the manufacturers instructions. Carefully load the tubes into the rotor and centrifuge at $2-4°C$ at an appropriate speed/time combination as outlined above. After centrifugation, carefully remove the tubes from the rotor and if required remove the cytosol fraction (i.e., the 105 000 g average h^{-1} supernatant) with a Pasteur pipette whilst discarding any floating lipid layer.

(vii)   Although the microsomal pellet can be used at this stage most workers normally used 'washed' microsomal fractions for experimental purposes. Washing the preparation reduces the haemoglobin content and thus is to be particularly recommended for spectral studies. To prepare washed microsomes simply resuspend the pellet in fresh homogenising medium with the motor-driven homogeniser (normally size B or even A) and centrifuge again [as in step (vi)] to obtain a microsomal pellet. Decant the supernatant and resuspend the washed microsomes by homogenisation in fresh homogenising medium to an appropriate tissue concentration and record the volume of the suspension.

(viii) Calculate the tissue concentration (i.e., g fresh tissue/ml) of whole homogenate, post-mitochondrial supernatant and microsomal fractions as follows:

Concentration (g/ml) of whole homogenate and post-mitochondrial supernatant

$$= \frac{\text{Weight of liver (g)}}{\text{Volume of whole homogenate (ml)}}$$

Concentration (g/ml) of microsomal fraction

$$= \frac{\text{Concentration of}}{\text{post-mitochondrial}} \times \frac{\text{Volume of post-mitochondrial supernatant spun (ml)}}{\text{Volume of microsomal fraction (ml)}}$$

(ix) The microsomal fraction is now ready for use although for some studies the protein content (see Seciton 2.8) is first determined and the microsomal fraction diluted to a known protein concentration before use.

(x) For most studies microsomal preparations should be used on the day of preparation. A typical preparation time from tissue to resuspended washed microsomes being some 3.25 h, with approximately an additional hour if protein is to be assayed before the sample is used. In some instances it may not be possible to use the preparation immediately and thus the preparation will have to be stored. Various storage procedures are described in the literature of which the best method appears to be the storage of microsomal suspensions in 0.154 M KCl containing 10 mM HEPES-HCl buffer pH 7.6 containing 1 mM EDTA and 20% glycerol (10) at −20 to −70°C (8) or preferably in liquid nitrogen (10). However, it should be stressed that the various microrosomal enzyme activities, spectral characteristics etc. are differentially affected by storage (11) and thus the storage stability of any particular microsomal parameter should be elucidated before routinely assaying activities in stored material.

2.5 Subfractionation of Microsomal Preparations

The microsomal fraction as prepared above consists of a mixture of disrupted rough (i.e., with attached ribosomes) and smooth endoplasmic reticulum fragments. It is possible to separate rough and smooth microsomal fractions by various procedures of which the method of Dallner (9,12) is the most convenient. Briefly this involves preparing a post-mitochondrial supernatant fraction in 0.25 M sucrose (see Section 2.4), adding caesium chloride to a final concentration of 15 mM and then layering the sample over 1.30 M sucrose containing 15 mM caesium chloride. As caesium ions have a higher affinity for rough microsomes they cause this material to aggregate and hence centrifugation of the sample will result in a pellet of rough microsomes with the smooth microsomes remaining around the interface of the two sucrose layers. The smooth microsomal fraction is collected either by aspiration with a Pasteur pipette or else by puncturing the bottom of the centrifuge tube. The fraction is then diluted with 0.25 M sucrose and centrifuged as in Section 2.4 to obtain a microsomal pellet. The preparation of rough and smooth microsomes can be achieved in a variety of angle and swing-out rotors (9,12), and if required the preparations obtained can be further subfractionated

(13). Cytochrome P450 content and mixed function oxidase enzyme activities are generally higher in smooth than in rough microsomes from rat liver, although this depends on the particular species examined (12,14).

2.6 Preparation of Microsomal Fractions from Extrahepatic Tissues

The general principles of preparation of hepatic microsomal fractions described above may also be applied to extrahepatic tissues. With soft tissues (e.g., kidney and testis) no particular modifications may be required, but tissues containing large amounts of connective tissue (e.g., lung and skin) may require additional procedures (15). The preparation of intestinal microsomal fractions can be particularly troublesome as the epithelial cells (which contain xenobiotic metabolising enzymes) need to be separated from the connective tissue cells; enzyme activity varies in different regions of the intestinal tract and enzyme activity can be rapidly denatured by the presence of intestinal proteases. Intestinal epithelial cells can be prepared either by scraping the cells from the inner surface of opened intestinal segments (15) or by liberating the cells by mechanical vibration from segments of intestine everted over stainless steel rods (16). The yield of enzymatically active intestinal microsomes may be improved by various procedures including the addition of a trypsin inhibitor during homogenisation, the incorporation of glycerol and dithiothreitol in the homogenising medium, and also by adding heparin to prevent agglutination and protein aggregation (17,18).

Generally cytochrome P450 content and associated enzyme activities are lower in extrahepatic tissues than in the liver, but by employing the more sensitive assays (e.g., radiometric and fluorimetric procedures) a range of activities should still be easily measurable. In preparations containing high amounts of haemoglobin, such as lung microsomes, various procedures have been developed for reducing interference in the spectral estimation of cytochrome P450 (15,19).

2.7 Preparation of Microsomal Fractions by Procedures other than Ultracentrifugation

Apart from ultracentrifugation other techniques including gel filtration (20), calcium aggregation (21) and isoelectric precipitation (22) have been developed for the preparation of microsomes from post-mitochondrial supernatant fractions. While the gel filtration method is not suitable for processing several samples the other two methods have been employed in studies of xenobiotic metabolism. In the calcium aggregation method 8 mM $CaCl_2$ (final concentration) is added to the post-mitochondrial supernatant fraction resulting in a calcium-dependent aggregation of endoplasmic reticulum fragments which may then be harvested by a comparatively low speed centrifugation (2000−25 000 g for 10−15 min) (21−24). Similarly the acid precipitation method relies on the isoelectric precipitation of microsomal membrane when the pH of the post-mitochondrial supernatant is adjusted to pH 5.5 with acetate buffer, the microsomal fraction again being obtained by low speed centrifugation (10 000 g for 10 min) (22). Both the calcium aggregation (23) and isoelectric precipitation (16) techniques have also been used to isolate microsomal fractions from extrahepatic tissues. Although both these procedures remove the requirement for an expensive ultracentrifuge they do have disadvantages. For example the calcium aggregation method appears to lead to a loss of ribosomes

(24), whereas the acid precipitation method is reported to result in an inactivation of certain xenobiotic metabolising enzymes (such as cytochrome P-450 and NADPH-cytochrome c reductase) (21).

2.8 **Estimation of Protein**

An estimation of the protein content of the microsomal preparation is almost always required as cytochrome content, enzyme activities etc. are normally expressed per unit of protein. Of the methods available the Lowry procedure (25) is the most commonly used. The advantages of this method are that it is reliable, relatively rapid and that tissue samples when diluted with 0.5 M NaOH may be stored at +4°C for several days prior to analysis. The coloured complex is thought to arise predominantly from reaction between the alkaline copper-phenol reagent and the tyrosine and tryptophan residues of the protein sample. As such there is the possibility of an estimation error should the content of these two amino acids differ markedly between the test sample and protein standard. Bovine serum albumin (e.g., Fraction V from Sigma Chemical Company, Poole, Dorset, UK) is generally employed as the protein standard. Finally, care should be taken to avoid erroneous results by allowing for the presence of interfering compounds in the sample such as detergents or glycerol (e.g. for stored microsomal preparations).

2.8.1 *Apparatus*

(i)   Single beam spectrophotometer (e.g. Cecil, Cambridge, UK; CE272) equipped with either 1-cm cuvettes or a 1-cm flow-through cell).
(ii)   Stop-clock or timer.
(iii) Various pipettes.
(iv)   Vortex mixer.
(v)   Glass or plastic disposable 10 ml nominal capacity test tubes and racks.

2.8.2 *Reagents*

(i)   0.5 M NaOH. Dissolve 20.0 g of NaOH in 1000 ml in distilled water.
(ii)   2% (w/v) Na_2CO_3. Dissolve 40.0 g of anhdyrous Na_2CO_3 in 2000 ml of distilled water.
(iii) 1% (w/v) Copper sulphate. Dissolve 2.5 g of $CuSO_4.5H_2O$ in 250 ml of distilled water.
(iv)   2% (w/v) Potassium sodium tartrate. Dissolve 5.0 g of $KNaC_4H_4O_6.4H_2O$ in 250 ml of distilled water.

The above reagents are all stable for at least 12 weeks at room temperature.

(v)   500 μg/ml Bovine serum albumin. Dissolve 50 mg of bovine serum albumin in 100 ml of 0.5 M NaOH. This reagent may require prolonged stirring to dissolve the protein and is stable for 12 weeks at +4°C.
(vi)   Folin and Ciocalteu's phenol reagent. The stock commercial reagent (e.g., from Fisons plc, Loughborough, Leics., UK) is stored at +4°C and any unused reagent discarded after 24 weeks. Great care should be taken to avoid contamination of the stock reagent. The working reagent is prepared by diluting one volume of stock reagent with one volume of distilled water immediately before use.

(vii)   Lowry reagent C. This is prepared by mixing 1 volume each of the copper sulphate and sodium potassium tartrate solutions with 100 volumes of the sodium carbonate solution. This reagent is prepared immediately before use. Note that excess Lowry reagent C and diluted Folin and Ciocalteu's phenol reagent are prepared over that required for the total number of assays, to permit samples with poor duplicates or requiring additional dilution etc. to be re-assayed without the need to prepare an additional standard curve.

2.8.3 *Procedure*

Test samples and standards are assayed in duplicate and reagent blanks are assayed in triplicate.

(i)   Prepare a suitable dilution (i.e., dilute to between 250 and 1000 μg protein/ml) of the sample in 0.5 M NaOH. For example, with rat liver microsomes (0.25 g of fresh tissue/ml) add 0.5 ml of the sample to 9.5 ml of 0.5 M NaOH and mix with a vortex mixer.

(ii)   Set up a triplicate reagent blank by pipetting 1 ml aliquots of 0.5 M NaOH into three glass or plastic tubes.

(iii)   Set up duplicate test samples by pipetting a suitable aliquot (0.1−0.3 ml of the diluted sample) and making the volumes up to 1 ml with 0.5 M NaOH.

(iv)   Set up duplicate 50−200 μg, in 25 μg increments, bovine serum albumin standards. This is achieved by pipetting 0.10, 0.15, 0.20, 0.25, 0.30, 0.35 and 0.40 ml of aliquots of the stock standard and making the volumes up to 1 ml with 0.5 M NaOH.

(v)   To all tubes add 5 ml of freshly prepared Lowry reagent C and mix tube contents with a vortex mixer.

(vi)   After at least 10 min add 0.5 ml of freshly diluted Folin and Ciocalteu's phenol reagent to each tube and immediately mix tube contents with a vortex mixer.

(vii)   After 30 min mix the tube contents again and measure and record the absorbance at 720 nm of all the tubes employing a suitable spectrophotometer.

2.8.4 *Calculation of Results*

(i)   Subtract the mean value of the triplicate reagent blanks from the mean values of the standard and test samples. Re-assay samples with poor duplicates (e.g., which differ by >0.03 absorbance units).

(ii)   Construct a graph of absorbance at 720 nm (y axis) against μg bovine serum albumin (x axis) and by interpolation convert the absorbance values of the test samples into μg of protein. A more convenient method is to employ a suitable calculator equipped with a program for linear regression analysis. As judged either graphically or by linear regression analysis (correlation coefficient $r \geq 0.995$) there should be a linear relationship between absorbance and protein concentration over the 50−200 μg range assayed.

(iii)   Samples with absorbances either less than the 50 μg standard or greater than the 200 μg strandard must be re-assayed after appropriate adjustment of sample volume selected for assay.

(iv) Calculate the protein concentration (mg/ml) of the samples by the formula:

$$\frac{\mu\text{g protein (from graph)}}{1000} \times \frac{1}{\text{volume of dilution assayed (ml)}} \times \frac{\text{volume of microsomal sample (ml) plus volume of 0.5 M NaOH (ml)}}{\text{volume of microsomal sample (ml)}}$$

For example, for a microsomal fraction where 0.3 ml aliquots of a dilution of 0.5 ml of the sample plus 9.5 ml of 0.5 M NaOH were used the equation will be:

$$\frac{\mu\text{ protein}}{100} \times \frac{1}{0.3} \times \frac{(0.5 + 9.5)}{0.5} = \frac{\mu\text{ protein}}{15}$$

(v) Calculate the microsomal protein content of the original tissue (mg microsomal protein/g fresh weight tissue) by the formula:

$$\frac{\text{Microsomal fraction protein (mg/ml)}}{\text{Microsomal fraction tissue concentration (g fresh tissue/ml)}}$$

2.8.5 *Additional Comments*

(i) The literature contains many variations of the Lowry (25) procedure and the observed protein linearity range depends on the particular assay conditions employed. When using any set of assay conditions protein contents should only be derived from sample dilutions which fall within the standard curve used.

(ii) If only a small sample or a very dilute sample is available then it may be more convenient to construct the standard curve over the range $0-50$ μg bovine serum albumin.

(iii) If applying the method to whole homogenate fractions it is usually necessary to digest the dilutions in 0.5 M NaOH prior to analysis by placing them in stoppered tubes in a 37°C water bath for approximately 3 h.

3. CHARACTERISATION OF PREPARATIONS

The microsomal fraction contains a variety of Phase I and II xenobiotic metabolising enzyme activities. Phase I enzymes include cytochrome P450 dependent mixed function oxidase enzymes together with components (3) of the microsomal electron transport chain [e.g., cytochrome b_5, NAD(P)H-cytochrome c reductase], the mixed function amine oxidase described by Ziegler (4) and non-specific esterase enzymes (26). Phase II enzymes include epoxide hydratase (5), UDP-glucuronyltransferase (27) and GSH S-transferase (28) activities. Because of the wide diversity of microsomal xenobitic metabolising enzymes the methods described below will focus on the determination of microsomal haemoproteins, additional spectral properties of cytochrome P450 and assays for NADPH-cytochrome c reductase and some selected mixed function oxidase enzyme activities.

3.1 **Determination of Cytochrome P450**

The principle of the spectral determination of cytochrome P450 is that the reduced haemoprotein combines with carbon monoxide to give a characteristic absorption spectrum (29) with a maximum at 450 nm (hence the name cytochrome P450). The procedure described below permits the rapid determination of the total cytochrome P450 content [i.e., the sum of all the various isoenzymes (7)] in the sample.

3.1.1 *Equipment*

(i)   Dual beam recording spectrophotometer. Because microsomal suspensions are turbid, spectral measurements are best performed in instruments with provision to place the cuvettes close to the photomultiplier in order to minimise light scattering by the sample. Although not absolutely necessary for work with liver microsomes, instruments equipped with automatic baseline correction systems [e.g. Perkin Elmer (Beaconsfield, Bucks, UK) model 557 or Shimadzu MPS-2000 (available from V.A.Howe and Company Ltd., London, UK)] are very useful for studies with either whole cell homogenates or microsomal preparations from extrahepatic tissues.

(ii)   Matched glass or quartz 1 cm pathlength cuvettes.

(iii) Stop-clock or timer.

(iv)   Various pipettes.

(v)   Parafilm and tissues.

(vi)   Glass or plastic disposable test tubes.

3.1.2 *Media*

(i)   Homogenising medium.

(ii)   0.2 M Phosphate buffer, pH 7.4. This is stable for several weeks at room temperature.

(iii) Carbon monoxide (house cylinder in an efficient fume cupboard).

(iv)   Sodium dithionite (general laboratory reagent grade).

3.1.3 *Procedure*

(i)   Prepare a suitable dilution of the microsomal fraction in 0.2 M phosphate buffer, pH 7.4. For example, with rat liver microsomes (0.25 g of fresh tissue/ml) add 2 ml of suspension to 4 ml of buffer mix and keep in ice until analysis.

(ii)   Add a small quantity of solid sodium dithionite and using parafilm mix gently by inversion.

(iii) Divide the sample between two matched 1 cm spectrophotometer cuvettes and with a dual beam spectrophotometer scan the cuvettes between 400 and 500 nm. Ideally this should result in a perfectly flat baseline or at least in a baseline with no appreciable absorbance difference between 450 and 490 nm. Should a poor baseline be obtained check that the cuvettes are a matched pair and are clean. Certain spectrophotometers (e.g. see Section 3.1.1) have facilities for automatic baseline correction should this be necessary.

(iv)   In an efficient fume cupboard, gently bubble the contents of the test cuvette with carbon monoxide for approximately 1 min and then record the difference spectrum between 400 and 500 nm of the test and reference cuvettes. The spectrum should contain a prominent absorption maximum at around 450 nm and with samples where the haemoglobin content is either low or absent (e.g., rat hepatic washed microsomes) there should not be a significant absorption maximum at 420 nm. A significant absorption maximum at 420 nm in such samples represents the presence of cytochrome P420 (29), which is the inactivated form of cytochrome P450, and is indicative of poor technique. However, in certain instances a significant absorption at 420 nm will be observed (see Section 3.1.5).

3.1.4 *Calculation of Results*

(i) From the spectrophotometer chart determine $\Delta A(450-490)$nm, that is the difference between the absorption maximum at around 450 nm and the baseline at 490 nm, correcting if necessary for any absorption difference between these wavelengths prior to bubbling the test cuvette with carbon monoxide.

(ii) The extinction coefficient of cytochrome P450 has been determined to be 91 cm^2/mmol (29). Hence the cytochrome P450 concentration (nmol/ml) of the sample is calculated by the formula:

$$\frac{\Delta A(450-490)nm}{1} \times \frac{1000}{91} \times \frac{\text{volume of microsomal sample (ml) plus volume of phosphate buffer (ml)}}{\text{volume of microsomal sample (ml)}}$$

The cytochrome P450 concentration of the sample may be expressed as either nmol/mg microsomal protein or nmol/g liver by simply dividing by either the sample protein (mg/ml) or tissue (g liver/ml) concentration, respectively.

For example, a microsomal sample (7.4 mg protein/ml) diluted 2 ml plus 4 ml phosphate buffer has a $\Delta A(450-490)$nm value of 0.19.

$$\text{Cytochrome P450 content (nmol/mg microsomal protein)} = \frac{0.19}{7.4} \times \frac{1000}{91} \times \frac{(2+4)}{2} = 0.85$$

3.1.5 *Additional Comments*

(i) Cytochrome b$_5$ may be determined in the same sample prior to estimation of cytochrome P450. After determining cytochrome b$_5$ (see Section 3.2.3) simply reduce both cuvettes with sodium dithionite and then proceed as described above.

(ii) The wavelength maximum of cytochrome P450 can be diagnostic of the effect of treatment with various compounds (see Section 3.5) and may be determined in the same sample employed for estimation of cytochrome P450 content. The method is adequately described by Mailman *et al.* (30) and it is recommended that the spectrophotometer is calibrated employing the absorption maximum of a Holmium oxide filter at 453.2 nm.

(iii) Cytochrome P450 may also be determined in whole homogenates of liver (31) and isolated hepatocytes (32). In such studies the method of estimation is modified so that both cuvettes are bubbled with carbon monoxide but only the test cuvette is reduced with sodium dithionite. According to Guengerich (8) the limit of detection of cytochrome P450 in extrahepatic tissues is influenced more by the presence of haemoglobin than the sensitivity of the spectrophotometer. Various procedures are available for minimising the effect of haemoglobin interference in the lung and other extrahepatic tissues (15,19).

3.2 **Determination of Cytochrome b_5**

Cytochrome b_5 is the other haemoprotein present in microsomal fractions from mammalian tissues and is readily estimated (29) from its redox spectrum of NADH-reduced versus oxidised cytochrome. The reduction of cytochrome b_5 is catalysed by the presence of a microsomal flavoprotein enzyme namely NADH-cytochrome b_5 reductase (EC 1.6.2.2).

3.2.1 *Equipment*
See Section 3.1.1.

3.2.2 *Media*
(i)   Homogenising medium.
(ii)   0.2 M Phosphate buffer, pH 7.4.
(iii)   NADH.

3.2.3 *Procedure*
(i)   Prepare a suitable dilution of the microsomal fraction in 0.2 M phosphate buffer, pH 7.4 (see Section 3.1.3).
(ii)   Divide the sample between two matched 1 cm spectrophotometer curvettes and with a dual beam spectrophotometer scan the cuvettes between 400 and 500 nm (see Section 3.1.3).
(iii)   Add a small quantity of solid NADH [or if preferred 25 μl of a freshly prepared 2% (w/v) solution in buffer] to the test cuvette and using parafilm mix gently by inversion.
(iv)   After approximately 2 min record the redox spectrum between 400 and 500 nm.

3.2.4 *Calculation of Results*

Cytochrome b_5 is determined as the absorption difference between the peak at around 424 nm and the trough at around 409 nm in the redox spectrum. Using the $\Delta A(424-409)$nm value cytochrome b_5 content is calculated by the same procedure as for cytochrome P450 (see Section 3.1.4) except that an extinction coefficient of 185 cm^2/mmol (29) is employed.

3.3 **Determination of Total Haem**

In the absence of either contaminating haemoglobin or the presence of cytochrome P420 the total haem content of the microsomal preparation should equal the sum of the contents of cytochromes P450 and b_5 (29). Microsomal total haem content may be easily estimated as the redox spectrum of dithionite-reduced against fully oxidised pyridine-haemochromogen in the presence of 0.1 M NaOH (29, 33).

3.3.1 *Equipment*
(i)   Dual beam recording spectrophotometer (see Section 3.1.1).
(ii)   Matched glass or quartz 1 cm pathlength stoppered cuvettes.

(iii)   Various pipettes.
(iv)   Parafilm and tissues.
(v)   Glass or plastic disposable test tubes.

3.3.2 *Media*

(i)   Homogenising medium.
(ii)   0.5 m NaOH. Dissolve 20 g of solid NaOH in 1 l of distilled water and store at room temperature.
(iii)   0.154 M KCl. Dissolve 11.5 g of KCl in·1 l of distilled water. This is stable for at least 12 weeks at +4°C.
(iv)   12.5 mM Potassium ferricyanide. Prepare fresh reagent for each set of assays by dissolving 205.8 mg $K_3Fe(CN)_6$ in 50 ml of distilled water.
(v)   Pyridine (analar grade) which should be kept and dispensed in an efficient fume cupboard.
(vi)   Sodium dithionite (general laboratory reagent grade).

3.3.3 *Procedure*

(i)   Pipette a suitable volume of the microsomal suspension (e.g., for rat liver microsomes use $0.25 - 1.5$ ml of a 0.25 g fresh tissue/ml suspension) into each of two matched 1 cm stoppered spectrophotometer cuvettes and then add 0.154 M KCl to each cuvette to a total volume of 1.5 ml.
(ii)   Add 0.5 ml of 0.5 M NaOH to each cuvette.
(iii)   Using an efficient fume cupboard add 0.5 ml of pyridine to each cuvette.
(iv)   To the test cuvette add a small quantity of solid sodium dithionite.
(v)   To the reference cuvette add 10 μl of 12.5 mM freshly prepared $K_3Fe(CN)_6$ solution.
(vi)   Stopper both the test and reference cuvettes and gently mix the contents by inversion.
(vii)   Place the cuvettes in a spectrophotometer and record the difference spectrum between 530 and 600 nm.

3.3.4 *Calculation of Results*

(i)   From the spectrophotometer chart determine $\Delta A(557-575)$nm, that is the difference between the absorption maximum at around 557 nm and 575 nm.
(ii)   The extinction coefficient of pyridine haemochromagen has been determined to be 34.7 cm²/mmol (29). Hence the total haem concentration (nmol/ml) of the sample is calculated for the formula:

$$\frac{\Delta A(557-575)\text{nm}}{1} \times \frac{1000}{34.7} \times \frac{2.5}{\text{volume of microsomal sample (ml)}}$$

(iii)   The total haem content of the sample may be expressed as either nmol/mg microsomal protein or nmol/g liver by simply dividing by either the sample protein (mg/ml) or tissue (g liver/ml) concentration, respectively.

3.3.5 *Additional Comments*

(i)   The literature contains various extinction coefficients for the determination of pyridine haemochromagen. Apart from extinction coefficients of 32.4 and 34.7 cm²/mmol (29) for $\Delta A(557-575)$nm, an extinction coefficient of 19 cm²/mmol (34) for $\Delta A(557-541)$nm may also be used.

(ii)   When comparing microsomal haem content to the amounts of cytochrome P450 and b_5, the amount of cytochrome P420 present can also be estimated (8) if required.

3.4 Determination of Reduced Cytochrome P450 — Ethyl Isocyanide Interaction Spectrum

Apart from carbon monoxide (see Section 3.1), ethyl isocyanide may also be used as a ligand for reduced cytochrome P450 (29). The resulting difference spectrum of dithionite reduced cytochrome P450 plus ethyl isocyanide against dithionite reduced cytochrome P450 has two absorption maxima at around 430 and 455 nm. Whilst these Soret bands are not utilised for the determination of cytochrome P450 content, the ratio of the absorbance maxima (expressed as 455/430 nm peaks) may be diagnostic of changes in the content of certain isoenzymes of cytochrome P450 (see Section 3.5). However, it should be noted that the absorbance ratio of the two peaks is also dependent on both the pH and ionic strength of the media used in the assay (35).

3.4.1 *Equipment*

See Section 3.3.1

3.4.2 *Media*

(i)   Homogenising medium.

(ii)   1.0 M Tris-maleate buffer, pH 7.4. Dissolve 23.72 g of Tris-maleate in approximately 80 ml of distilled water, adjust pH to 7.4 with concentrated NaOH and make up to 100 ml. This is stable for at least 12 weeks at +4°C.

(iii)   Ethyl isocyanide. Warning: this chemical has an exceptionally foul, irritating odour and must be used in an efficient fume cupboard. The compound may be destroyed by pouring into 4 M HCl and this is also recommended for initial decontamination of spectrophotometer cuvettes. Should ethyl isocyanide not be available commercially it may be synthesised by the procedure of Jackson and McKusick (36). The compound should be stored at −20°C in well-sealed containers and periodically redistilled.

(iv)   Sodium dithionite (general laboratory reagent grade).

3.4.3 *Procedure*

(i)   Add 1.0 ml of a suitable dilution of the microsomal fraction (e.g., for rat liver microsomes use a 0.25 g fresh tissue/ml suspension) to 4 ml of 1.0 M Tris-maleate buffer, pH 7.4.

(ii)   Add a small quantity of solid sodium dithionite and, using parafilm, mix gently by inversion.

(iii)   Divide the mixture between two matched 1 cm stoppered spectrophotometer cuvettes.

(iv)   To the test cuvette, using an efficient fume cupboard, carefully add 1 μl of ethyl isocyanide, stopper the cuvette and mix gently by inversion.

(v)   Place the cuvettes in a spectrophotometer and record the difference spectrum between 410 and 500 nm.

(vi)   Dispose of the contents of the test cuvette by pouring into acid.

3.4.4 *Calculation of Results*

(i)   From the spectrophotometer chart determine the absorbance of the peaks at around 430 and 455 nm relative to the baseline at around 490 nm.

(ii)   Calculate the absorbance ratio of the two Soret peaks by the formula:

$$\frac{\Delta A(455-490)nm}{\Delta A(430-490)nm}$$

3.4.5 *Additional Comments*

If required the pH of the isobestic point (i.e., where the absorbances of the 430 and 455 nm maxima are identical) of the ethyl isocyanide interaction spectrum may be determined by employing the above procedure and a suitable range of buffers.

3.5 Determination of the Spectral Interaction of Xenobiotics with Cytochrome P450

A large number of compounds are known to produce spectrally apparent interactions with microsomal cytochrome P450 (37). Such compounds are normally, but not exclusively, substrates for cytochrome P450-dependent mixed function oxidase reactions. The observed spectra result from characteristic perturbations of the haem iron of cytochrome P450 and are normally determined as difference spectra of oxidised microsomes plus xenobiotic against oxidised microsomes alone (37–39). Generally three major types of difference spectra are observed. Type I spectra, which are produced by compounds such as benzphetamine, biphenyl, aminopyrine and SKF 525A, are characterised by a peak at around 385 nm and a trough (i.e., a negative absorption with respect to the baseline) at around 420 nm. In contrast, Type II spectra, which are produced by compounds such as aniline, *n*-octylamine and nicotinamide, are characterised by a peak around 430 nm and a broad trough between 390 and 410 nm. Finally, Reverse Type I spectra, which are produced by compounds such as ethanol, *n*-butanol and phenacetin, are the mirror image of Type I spectra.

By applying Michaelis–Menten kinetics to spectral interaction studies a quantitative estimate of the affinity of the xenobiotic for cytochrome P450, determined as the apparent spectral dissociation constant K_s, and the maximal spectral change (ΔAmax) may be obtained. Particularly with Type I substrates, the similarities in K_s and K_m (the Michaelis constant) values suggest that these interaction spectra represent the formation of an enzyme–substrate complex (37,38).

In order to describe the technique of determining xenobiotic–cytochrome P450 interaction spectra, the spectral interaction of benzphetamine (Type I) and aniline (Type II) will be described as examples.

3.5.1 *Equipment*

(i) Dual beam spectrophotometer equipped with provision to handle turbid samples and preferably an automatic baseline correction system (see Section 3.1.1).

(ii) Matched glass or quartz 1 cm pathlength stoppered cuvettes. In order to prevent loss of liquid from the cuvettes when adding compounds it is preferable to replace the normal Teflon stoppers with silicone rubber stoppers (e.g. Suba-Seal stoppers size SYJ-350-150Y available from A.Gallenkamp and Co. Ltd., London, UK) which have thin centres, thus permitting the introduction of test compounds and solvents by injection through the stopper.

(iii) 5 and 10 μl glass syringes.

(iv) Stop-clock or timer.

(v) Various pipettes and tissues.

(vi) Domestic hair drier.

3.5.2 *Media*

(i) Homogenising medium.

(ii) 10 mM Benzphetamine. Prepare this freshly by dissolving 69 mg of benzphetamine hydrochloride in 25 ml of homogenising medium. Dilute a portion with homogenising medium to provide a 2 mM solution.

(iii) 250 mM Aniline. Prepare this freshly by dissolving 324 mg of aniline hydrochloride in 10 ml of homogenising medium. Dilute a portion with homogenising medium to provide a 62.5 mM solution.

3.5.3 *Procedure*

(i) Pipette 2.5 ml of microsomal suspension (2 mg protein/ml homogenising medium) into each of two clean, dry, matched 1 cm stoppered spectrophotometer cuvettes.

(ii) Stopper the cells and place them in the spectrophotometer.

(iii) Allow the cuvettes to equilibrate to instrument temperature for approximately 3 – 5 min.

(iv) Using an absorbance scale of ± 0.05 A units or less record the baseline between 350 and 500 nm. If the spectrophotometer is fitted with an automatic baseline correction facility, use this system to minimise any absorption differences between the cuvettes.

(v) When a satisfactory baseline has been obtained add the first concentration of the xenobiotic in solvent to the test cuvette and an equal volume of solvent to the reference cuvette. For example, if studying the Type I spectral interaction of benzphetamine inject 4 μl of 2 mM benzphetamine solution (in homogenising medium) through the stopper of the test cuvette and inject 4 μl of homogenising medium into the reference cell. Note that it is preferable to use separate syringes for test compound and solvent in order to avoid contamination.

(vi) Mix the cuvette contents gently by inversion and replace them in the spectrophotometer. Note that it is preferable to maintain a constant cell alignment by keeping the same optical face (e.g., the face with matching or other engraved markings) of both test and reference cells facing the photomultiplier.

(vii) When the absorbances of the cells have stablised (e.g., when any air bubbles have been removed from the liquid) record the difference spectrum between 350 and 500 nm. A typical Type I difference spectrum (37,38) should be obtained.

(viii) Add a further 1 μl of 2 mM benzphetamine to the test cuvette (i.e., so that the cell now contains a total of 5 μl of 2 mM benzphetamine) and 1 μl of homogenising medium to the reference cuvette. Repeat steps (vi) and (vii). The resultant spectrum should show a larger spectral change (i.e., the sum of the absorbance of peak to baseline plus trough to baseline) than the first difference spectrum recorded.

(ix) Continue to add increasing amounts of 2 mM benzphetamine and record difference spectra after the *total* additions of 4 [i.e. (v) above], 5 [i.e. (viii) above], 6, 8 and 10 μl. Then, using 10 mM benzphetamine, record difference spectra after the total additions of 1, 2 and 6 μl. In each instance add an appropriate amount of homogenising medium to the reference cuvette and gently mix cuvette contents by inversion before recording each difference spectrum. Note: when recording a number of difference spectra it is advisable to identify the peak and trough of each spectrum on the spectrophotometer chart in order to facilitate subsequent calculation of the results.

(x) After recording the final difference spectrum discard the cuvette contents, wash the cuvettes and stoppers well with distilled water and then with several washes of ethanol. Dry the cuvettes and stoppers between determinations with a domestic hair drier.

(xi) The spectral interaction of aniline can be similarly studied by employing the above procedure. The *total* additions to the test cuvette are 4, 5, 6, 8 and 10 μl of 62.5 mM aniline, followed by *total* additions of 1, 4 and 8 μl of 250 mM aniline.

3.5.4 *Calculation of Results*

(i) From the spectrophotometer charts determine the absorbance change of each difference spectrum as the sum of the peak-to-baseline and trough-to-baseline absorbances.

(ii) Determine the reciprocals of these values (i.e., calculate $1/\Delta A$).

(iii) From the additions of 2 and 10 mM benzphetamine it can be calculated that the cuvette concentrations (volume 2.5 ml) employed are 0.0032, 0.0040, 0.0048, 0.0064, 0.0080, 0.012, 0.016 and 0.032 mM. The reciprocals of these concentrations (i.e. $1/S$ vlaues) are 312.5, 250, 208.3, 156.3, 125, 83.3, 62.5 and 31.3.

(iv) Plot a graph of $1/\Delta A$ (y axis) against $1/S$ (x axis), determine the y ($1/\Delta A max$) and x ($-1/K_s$) axis intercepts and hence values of $\Delta A max$ (absorbance units/2 mg protein) and K_s (mM). These values may be more accurately determined by employing a suitable calculator equipped with a program for linear regression analysis.

(v) For aniline the cuvette concentrations employed are 0.10, 0.125, 0.15, 0.20, 0.25, 0.35, 0.65 and 1.05 mM. The reciprocals of these concentrations (i.e., $1/S$ values) are 10.0, 8.0, 6.67, 5.0, 4.0, 2.86, 1.54 and 0.95.

3.5.5 *Additional Comments*

(i) Whenever possible xenobiotics should be dissolved in aqueous media (homogenising medium, buffer or water). Insoluble compounds can be dissolved in solvents (e.g., dimethyl sulphoxide, ethanol, acetone) although these may also produce spectral changes. The total solvent volume added should be kept as low as possible (e.g., to around 20 μl) in order to minimise absorbance changes due to dilution of the microsomal suspension.

(ii)   It should be noted that the K_s values obtained are only accurate over the xenobiotic concentration range employed. As a general rule K_s values should be determined over at least a ten-fold range (e.g., from around one-third to three times the K_s value).

(iii)   The sensitivity of spectral interaction studies (i.e., the magnitude of the absorbance changes) can be improved by employing microsomal preparations from phenobarbitone treated rats (see Section 3.13).

(iv)   The spectral interaction of compounds which absorb in the 350−500 nm wavelength range can be studied by employing a tandem cuvette procedure (38).

3.6 Determination of NADPH−Cytochrome c (P450) Reductase

This microsomal flavoprotein enzyme (EC 1.6.2.4) is normally determined with cytochrome c as an artificial electron acceptor as the actual measurement of NADPH−cytochrome P450 reductase activity is relatively difficult and requires specialised apparatus (40). The principle of the assay is that oxidised (ferric) cytochrome c is converted to reduced (ferrous) cytochrome c which, unlike the oxidised form, has a characteristic absorption maximum at 550 nm.

3.6.1 *Equipment*

(i)   Dual beam recording spectrophotometer complete with a thermostatted cell compartment, heater and pump to maintain cuvette contents at 37°C.

(ii)   Matched glass or quartz 1 cm pathlength cuvettes.

(iii)   Stop-clock or timer.

(iv)   Various pipettes.

(v)   Parafilm and tissues.

3.6.2 *Media*

(i)   Homogenising medium.

(ii)   0.1 M Phosphate buffer, pH 7.6. This is stable for at least 12 weeks at +4°C.

(iii)   15 mM KCN. Carefully weigh out 97.7 mg of KCN and dissolve in 100 ml of distilled water. This is stable for at least 12 weeks at +4°C.

(iv)   0.125 mM Cytochrome c. This is freshly prepared by dissolving 1.55 mg/ml cytochrome c in buffer.

(v)   10 mM NADPH. Prepare this freshly by dissolving 8.3 mg/ml of NADPH in buffer. (Note: the exact weight of NADPH depends on the water content and purity of the batch used.)

3.6.3 *Procedure*

(i)   Into each of two matched 1 cm spectrophotometer cuvettes pipette 1.0 ml of cytochrome c solution and 0.2 ml of 15 mM KCN (warning: do not pipette this reagent by mouth).

(ii)   Pipette a suitable volume of the microsomal suspension (e.g., for rat liver microsomes use 0.1 ml of a 50 mg fresh tissue/ml suspension) into each cuvette.

(iii)   Add phosphate buffer so that the total volumes of the test and reference cuvette contents are 2.4 and 2.5 ml, respectively.

(iv)   Using parafilm gently mix the cuvette contents by inversion and transfer the cuvettes to the 37°C thermostatted cell compartment of the spectrophotometer.

(v)   After 3 min initiate the reaction by adding 0.1 ml of 10 mM NADPH to the test cuvette only, mix the cuvette contents and record the increase in absorbance with time at 550 nm.

3.6.4 *Calculation of Results*

(i)   From the spectrophotometer chart determine the initial rate of cytochrome c reduction in units of $\Delta A550$ nm/min.

(ii)   The extinction coefficient of reduced cytochrome c at 550 nm has been determined to be 21 cm^2/mmol (41). Hence NADPH−cytochrome c reductase activity (nmol/min/ml microsomal suspension) is calculated by the formula:

$$\frac{\Delta 550 \text{ nm/min}}{1} \times \frac{1000}{21} \times \frac{2.5}{\text{volume of microsomal suspension (ml)}}$$

added to cuvette

(iii)   Enzyme activity is expressed as either nmol/min/mg microsomal protein or nmol/min/g liver by dividing by the sample protein (mg/ml) or tissue (g liver/ml) concentration respectively.

3.6.5 *Additional Comments*

(i)   The literature contains many variations on the basic assay procedure for determining this enzyme activity. For example, enzyme activity is stimulated by increasing ionic strength (42) or by the addition of magnesium ions (43).

(ii)   Some workers assay enzyme activity at lower temperatures, for example at 30°C (8).

(iii)   KCN is added to prevent errors due to possible mitochondrial contamination of the microsomal suspension.

(iv)   A number of alternative extinction coefficients are used including 19.1 (43) and 19.6 (44) cm^2/mmol.

3.7 **Determination of Mixed Function Oxidase Enzyme Activities**

3.7.1 *General Considerations*

Many compounds have been used as model substrates for the measurement of cytochrome P450 dependent mixed function oxidase enzyme activities. Generally, assays should be selected where product formation and not substrate consumption (much less sensitive) is monitored. The choice of a particular substrate is dependent on the purpose of the experiment. For example, the induction of certain mixed function oxidase enzyme activities is diagnostic of particular forms of cytochrome P450. In addition, assays of certain enzyme activities (e.g., using radiometric and fluorimetric procedures) are more sensitive than others and hence are of greater use when either relatively inactive (e.g., extrahepatic) or small amounts of tissue are to be assayed.

For any given substrate the literature often contains many different sets of experimental conditions for assay of a given enzyme activity. Enzyme activity is obviously dependent on the concentration of tissue, substrate and cofactors, and pH but other factors can also affect the measurement of mixed function oxidase activities *in vitro*. For example, Fouts (45) reported variations in enzyme activity due to other factors including choice of buffer, gas phase (air or oxygen) and incubator shaking rate (i.e., cycles/min). Unless very extensive research is undertaken for each enzyme assay, it has to be assumed that the conditions chosen are valid for comparing the effects of experimental treatment with control or for comparing sex, strain and species differences in xenobiotic metabolism etc. However, under no circumstances should any work be performed using experimental conditions where linearity with respect to both time and tissue concentration have not been established. When assaying microsomal or other tissue preparations from animals pre-treated with a compound whose effects are unknown it may be prudent to assay several different combinations of tissue/incubation time; in order to ensure that linearity of product formation will be maintained under conditions of marked enzyme induction/inhibition.

The four methods for determination of mixed function oxidase enzyme activities given below were primarily developed for rat and mouse hepatic microsomal fractions. However, as they are relatively 'robust' methods they may be applied, with appropriate modifications as necessary, to other tissues and species.

3.7.2 *Apparatus*

The general apparatus required has been described in earlier sections of this chapter. A few additional items are described below.

(i)   Shaking water bath. Almost any type of incubator capable of producing 100 cycles/min is suitable. For certain assays where either the substrate or products are light sensitive a covered incubator is preferable.

(ii)   Fluorescence spectrophotometer. Almost any commercial instrument (e.g., Perkin-Elmer, Beaconsfield, Bucks, UK) that has diffraction grating rather than filter monochromators will be suitable. For certain assays the provision of a heated cell block is desirable.

(iii)   Low speed refrigerated centrifuge. Almost any type of instrument capable of centrifuging different sizes of incubation tubes at speeds of up to 2000 *g* will be suitable.

(iv)   Incubation tubes. Many types of plastic or glass tubes are available. Generally incubations can be adequately conducted either in 10 ml nominal capacity plastic disposable tubes or, for extractions, screw-threaded glass tubes (e.g., from Fisons PLC, Loughborough, Leics., UK) equipped with plastic caps and Teflon seals.

3.7.3 *Enzyme assays*

Table 1 shows the composition of the incubation mixtures (all 2 ml final volume) for

B.G.Lake

Table 1. Typical Incubation Conditions for Assay of some Mixed Function Oxidase Enzymes in Rat and Mouse Hepatic Microsomal Fractions.

| Enzyme activity | Species[a] | Substrate (mM) | Cofactors[b] | | | | Buffer pH (50 mM Tris-HCl[c]) | Tissue volume[d] (ml) |
			NADP+ (mM)	DL-Isocitric acid (mM)	Isocitric dehydrogenase (Units)	MgSO₄ (mM)		
Ethylmorphine N-demethylase	Rat	5.0	0.5	3.75	1	5	8.2	0.1
	Mouse	5.0	0.5	3.75	1	5	8.8	0.3
Aniline 4-hydroxylase	Rat	5.0	0.5	7.5	1	5	7.8	0.3
	Mouse	5.0	0.5	3.75	1	5	7.6	0.3
7-Ethoxycoumarin O-deethylase	Rat	0.5	0.75	7.5	1	5	7.8	0.05
	Mouse	0.5	0.5	7.5	1	5	7.6	0.05
7-Ethoxyresorufin O-deethylase	Rat	0.0005	0.5	7.5	1	5	8.4	0.05
	Mouse	0.0005	0.5	7.5	1	5	8.2	0.05

[a]The assays were developed for Sprague–Dawley and Wistar strain rats and C3H/He and C57BL/6 strain mice.
[b]The final volume of all incubations was 2 ml.
[c]The Tris-HCl buffers were prepared at room temperature.
[d]The tissue volumes recommended are for microsomal preparations from untreated rats (0.25 g fresh tissue/ml) and mice (0.1 g fresh tissue/ml).

assay of the four mixed function oxidase enzymes described below. For each tissue preparation it is recommended that a duplicate test and a single blank incubation be performed. In the assays described below the blank is a combined tissue/reagent blank i.e., the substrate is added at the end of the incubation period. For each animal treatment group a triplicate standard should be assayed.

The NADPH generating system shown in *Table 1* is based on isocitric acid/isocitric dehydrogenase. Some workers, however, employ glucose-6-phosphate/glucose-6-phosphate dehydrogenase. It is recommended that the cofactors (namely NADP$^+$, DL-isocitric acid, isocitric dehydrogenase and magnesium ions) are prepared in buffer and dispensed together in order to reduce pipetting and hence avoid errors. To do this the total number of tubes to be assayed is calculated and with an allowance for a few extra tubes the total number of μmol (and hence mg) of NADP$^+$ and DL-isocitric acid are calculated. For example, for ethylmorphine N-demethylase each tube requires 1.0 and 7.5 μmol of NADP$^+$ and DL-isocitric acid, respectively (see *Table 1*). Weigh these two cofactors in a glass beaker or other suitable container and add an appropriate quantity (to give 1 unit/tube) of isocitric dehydrogenase (usually supplied as a solution in glycerol). Add an appropriate quantity of Tris-HCl buffer and magnesium sulphate, as a 100 mM solution of MgSO$_4$.7H$_2$O (24.65 mg/ml), and then add the cofactor solution (kept on ice until use). The volume of magnesium sulphate solution to add is calculated to be 0.1 ml (i.e., 10 μmol)/tube, and the volume of buffer depends on the volume of substrate and tissue (total incubation volume 2 ml). Hence the maximum amount of cofactor solution per assay is calculated and multiplied by the number of tubes. This volume less that of the magnesium sulphate solution and isocitric dehydrogenase is then added to the cofactors.

3.8 Determination of Ethylmorphine N-Demethylase

The N-demethylation of this substrate results in the formation of formaldehyde which is trapped in the incubation medium by the presence of semicarbazide and subsequently estimated colorimetrically by means of the Hantzsch reaction (46).

3.8.1 *Media*

(i)   50 mM Ethylmorphine hydrochloride substrate (19.3 mg/ml) in distilled water. This reagent is prepared immediately before use.

(ii)   Formaldehyde standard solution. A stock standard solution is prepared by diluting commercial formaldehyde solution ($\sim 37\%$, w/w) 1 in 10 000 with distilled water. The exact formaldehyde concentration of the working standard is determined as described below.

(iii)   50 mM Tris-HCl buffer (see *Table 1*).

(iv)   50 mM Semicarbazide hydrochloride. Dissolve 557.7 mg in approximately 60 ml distilled water, adjust pH to 7.0 with NaOH and make up to 100 ml with distilled water.

(v)   Double strength Nash reagent. Dissolve 308.3 g of ammonium acetate in 1 l of distilled water containing 0.6% (v/v) glacial acetic acid. Prepare the working reagent immediately before use by dissolving 0.4 ml of redistilled acetylacetone per 100 ml of the ammonium acetate/acetic acid solution.

(vi)   5% (w/v) Zinc sulphate. Dissolve 50 g of $ZnSO_4.7H_2O$ in 1 l of distilled water.

(vii) Saturated barium hydroxide. Prepare this reagent by first boiling distilled water to remove dissolved carbon dioxide and then add excess solid barium hydroxide. Boil the mixture gently for a few minutes, allow to cool, filter and store in a tightly stoppered container.

Reagents ii, iii, iv and v are stable for at least 12 weeks at 4°C and reagents vi and vii are stable for at least 12 weeks at room temperature.

3.8.2 *Procedure*

(i)   Set up plastic disposable incubation tubes in a test tube rack on ice and add 0.2 ml of 50 mM neutralised semicarbazide solution, cofactor solution (see Section 3.7.3), tissue (see *Table 1*) and additional buffer as required to give a final volume of 1.8 ml in the test and blank incubation tubes and 1.5 ml in the standards.

(ii)   Mix the tube contents with a vortex mixer and pre-incubate in a shaking water bath at 37°C for 5 min.

(iii) Remove the rack of tubes from the water bath and rapidly add 0.2 ml of the substrate solution to the test incubation tubes only.

(iv)   Shake the rack to agitate the tube contents, place the rack back in the shaking water bath and incubate for 10 min.

(v)   Remove the rack of tubes from the water bath, plunge into ice and add 1 ml of 5% $ZnSO_4$ and then 1 ml of saturated barium hydroxide to all the tubes.

(vi)   Add 0.2 ml substrate solution to the blank and standard tubes and 0.3 ml of the formaldehyde standard solution to the standard tubes only.

(vii) Mix the contents of all the tubes with a vortex mixer.

(viii) Centrifuge the tubes at 2000 *g* for 15 min at 4°C.

(ix)   Remove 2 ml of the deproteinised supernatant from each tube into a clean glass or plastic disposable tube and add 2 ml of freshly prepared double-strength Nash reagent.

(x)   To calibrate the formaldehyde standard, prepare triplicate tubes each of 2.0 ml distilled water and 0.2 ml standard plus 1.8 ml of distilled water. Add 2 ml of Nash reagent to all these tubes.

(xi)   Mix all tube contents with a vortex mixer and place in a water bath at 37°C for 60 min.

(xii) Remove the tubes, allow to cool, mix the tube contents with a vortex mixer and measure and record the absorbance at 412 nm of all the tubes with a suitable spectrophotometer.

3.8.3 *Calculation of Results*

(i)   The formaldehyde concentration of the working standard solution is calculated by use of an extinction coefficient of 8 cm²/mmol for the formaldehyde colour complex formed with the Nash reagent (46). Subtract the mean of the triplicate calibration tubes (see Section 3.8.2). Calculate the formaldehyde concentration (mM) of the working standard by the formula:

$$\frac{A412 \text{ nm (calibration standard } - \text{ reagent blank)}}{1} \times \frac{1}{8} \times \frac{4}{0.2}$$

The concentration of the formaldehyde working standard should be in the range 1.0−1.5 mM.

(ii) For each sample determine the mean value of the absorbance of the test incubations and subtract the blank absorbance value.

(iii) Determine the mean value of the absorbance of each set of triplicate standard tubes and subtract the value of an appropriate blank.

(iv) Calculate enzyme activity (pmol/min/ml tissue) by the formula:

$$\frac{A412 \text{ nm}}{1}\text{(test}-\text{blank)} \times \frac{1}{10} \times \frac{1}{\text{Volume of tissue added (ml)}} \times \frac{\text{Concentration of standard (mM)} \times 0.3}{A412 \text{ nm (standard}-\text{blank)}}$$

(v) Enzyme activity in the sample may be expressed either as pmol/min/mg microsomal protein or as pmol/min/g tissue by dividing by either the sample protein (mg/ml) or tissue (g tissue/ml) concentration, respectively.

3.8.4 *Additional Comments*

(i) The colorimetric assay for formaldehyde described above can be applied to study the metabolism of other xenobiotic substrates which are either N-demethylated (e.g., benzphetamine and aminopyrine) or O-demethylated (e.g., codeine). In addition, a radiometric assay for benzphetamine N-demethylase has been reported (8).

3.9 **Determination of Aniline 4-Hydroxylase**

The 4-hydroxylation of this substrate results in the formation of 4-aminophenol which is estimated colorimetrically by the indophenol reaction (47).

3.9.1 *Media*

(i) 50 mM Aniline hydrochloride. Dissolve 648.0 mg in approximately 60 ml of distilled water, adjust pH to 7.0 with NaOH and make up to 100 ml with distilled water. Prepare this reagent immediately before use.

(ii) 0.50 mM 4-Aminophenol standard. Dissolve 54.6 mg in 1 l of distilled water.

(iii) 50 mM Tris-HCl buffer (see *Table 1*).

(iv) 0.5 M Tripotassium orthophosphate containing 1% (w/v) phenol. Dissolve 28.8 g tripotassium orthophosphate and 2.5 g phenol (analar grade) in 250 ml distilled water. Prepare this reagent immediately before use.

(v) Diethyl ether. If desired 1.5% (v/v) *iso*-amyl alcohol can be added to help prevent the formation of emulsions during extraction.

(vi) Sodium chloride.

Reagents ii and iii are stable for at least 12 weeks at 4°C.

3.9.2 *Procedure*

(i) Set up 15 ml glass stoppered tubes in a test tube rack on ice and add cofactor solution (see Section 3.7.3), tissue (see *Table 1*) and additional buffer as required to give a final volume of 1.8 ml in the test and blank incubation tubes and 1.7 ml in the standards.

(ii)   Mix the tube contents with a vortex mixer and preincubate in a shaking water bath at 37°C for 5 min.

(iii)   Remove the rack of tubes from the water bath and rapidly add 0.2 ml of the substrate solution to the test incubation tubes only.

(iv)   Shake the rack to agitate the tube contents, place the rack in the shaking water bath and incubate for 10 min.

(v)   Remove the rack of tubes from the water bath, plunge into ice and add 10 ml of diethyl ether and then a small amount of solid sodium chloride to all the tubes.

(vi)   Add 0.2 ml of substrate solution to the blank and standard tubes and 0.1 ml (50 nmol) of the 4-aminophenol standard solution to the standard tubes only.

(vii)   Carefully stopper all the tubes ensuring that each tube cap contains a correctly positioned Teflon seal.

(viii) Extract the tube contents for 10 min using a flatbed shaker or other suitable mixing device.

(ix)   If necessary centrifuge the tubes at 1000 g for 5 min at 4°C to break up any emulsions formed during extraction.

(x)   Remove 5 ml of the upper ether layer into clean 10 ml glass stoppered tubes and add 4 ml of the freshly prepared phenol reagent.

(xi)   Carefully stopper the tubes and again extract for 10 min.

(xii)   Stand the tubes for 60 min at room temperature to allow the completion of colour development.

(xiii) Aspirate the upper ether phase and measure and record the absorbance at 630 nm of the aqueous phase of all the tubes with a suitable spectrophotometer.

3.9.3 *Calculation of Results*

(i)   For each tissue sample determine the mean value of the absorbance of the duplicate test incubations and subtract the blank absorbance value.

(ii)   Determine the mean value of the absorbance of each set of triplicate standard tubes and subtract the value of an appropriate blank.

(iii)   Calculate enzyme activity (nmol/min/ml tissue) by the formula:

$$\frac{\text{A630 nm (test} - \text{blank)}}{1} \times \frac{1}{10} \times \frac{1}{\substack{\text{Volume of} \\ \text{tissue added} \\ \text{(ml)}}} \times \frac{50}{\substack{\text{A630 nm} \\ \text{(standard} - \text{blank)}}}$$

(iv)   Enzyme activity in the sample may be expressed either as nmol/min/mg microsomal protein or as nmol/min/g tissue by dividing by either the sample protein (mg/ml) or tissue (g tissue/ml) concentration, respectively.

3.10 Determination of 7-Ethoxycoumarin O-Deethylase

The O-deethylation of this substrate results in the production of 7-hydroxycoumarin which is readily estimated fluorimetrically. In the original procedure for determination of this mixed function oxidase enzyme activity tissue samples were assayed individually by performing each incubation in a fluorimeter cuvette and monitoring the increase in fluorescence intensity with time (48). However, the procedure described below was develped to permit the rapid measurement of activity of a number of samples by means of a single endpoint measure of fluorescence intensity (49).

3.10.1 *Media*

(i)   2 mM 7-Ethoxycoumarin (0.380 mg/ml) in 50 mM Tris-HCl buffer (see *Table 1*). Warm the substrate solution to dissolve and keep at around 37°C. Prepare this reagent immediately before use.

(ii)   0.1 mM 7-Hydroxycoumarin. Dissolve 40.5 mg of 7-hydroxycoumarin in 50 ml ethanol and dilute 5 ml to 250 ml with distilled water. Prepare this reagent immediately before use.

(iii)   50 mM Tris-HCl buffer (see *Table 1*).

(iv)   0.5 M Glycine-NaOH, pH 10.5. Dissolve 37.54 g glycine in approximately 600 ml distilled water, adjust the pH to 10.5 with NaOH and make up to 1 l with distilled water.

(v)   5% (w/v) Zinc sulphate (see Section 3.8.1).

(vi)   Saturated barium hydroxide (see Section 3.8.2).

Reagents iii and iv are stable for at least 12 weeks at 4°C and reagents v and vi are stable for at least 12 weeks at room temperature.

3.10.2 *Procedure*

(i)   Set up 10 ml plastic disposable test tubes in a test tube rack on ice and add cofactor solution (see Section 3.7.2), tissue (see *Table 1*), and additional buffer as required to give a final volume of 1.5 ml in the test and blank incubation tubes and 1.4 ml in the standards.

(ii)   Mix the tube contents with a vortex mixer and pre-incubate in a shaking water bath for 5 min.

(iii)   Remove the rack of tubes from the water bath and rapidly add 0.5 ml of the substrate solution to the test tubes only.

(iv)   Shake the rack to agitate the tube contents, place the rack in the shaking water bath and incubate for 10 min.

(v)   Remove the rack of tubes from the water bath, plunge into ice and add 1 ml of 5% $ZnSO_4$ and then 1 ml of saturated barium hydroxide to all the tubes.

(vi)   Add 0.5 ml of substrate solution to the blank and standard tubes and 0.1 ml (10 nmol) of the 7-hydroxycoumarin standard solution to the standard tubes only.

(vii)   Mix the contents of all the tubes with a vortex mixer.

(viii)   Centrifuge the tubes at 2000 g for 15 min at 4°C.

(ix)   Remove 1 ml of the deproteinised supernatant from each tube into a clean glass or plastic disposable tube and add 2 ml of 0.5 M glycine-NaOH buffer, pH 10.5.

(x)   Mix the contents of all the tubes with a vortex mixer.

(xi)   Set the excitation and emission monochromators of a suitable spectrofluorimeter to 380 and 452 nm, respectively.

(xii)   Measure and record the fluorescence of all the tubes.

3.10.3 *Calculation of Results*

(i)   For each tissue sample determine the mean value of the fluorescence of the duplication test incubations and subtract the blank fluorescence value.

(ii)   Determine the mean value of the fluorescence of each set of triplicate standard tubes and subtract the value of an appropriate blank.

(iii)   Calculate enzyme activity (nmol/min/ml tissue) by the formula:

$$\frac{\text{Fluorescence (test}-\text{blank})}{1} \times \frac{1}{10} \times \frac{1}{\substack{\text{Volume of} \\ \text{tissue added} \\ \text{(ml)}}} \times \frac{10}{\text{Fluorescence (standard}-\text{blank})}$$

(iv)   Enzyme activity in the sample may be expressed either as nmol/min/mg microsomal protein or as nmol/min/g tissue by dividing by either the sample protein (mg/ml) or tissue (g tissue/ml) concentration, respectively.

3.10.4 *Additional Comments*

(i)   The high sensitivity of this assay makes it suitable for studies with either small tissue samples or samples with low enzyme activity (e.g., extrahepatic tissues).

(ii)   As 7-hydroxycoumarin is rapidly conjugated with D-glucuronic acid and sulphate, care has to be taken when working with whole cell or possibly whole homogenate preparations. If necessary a crude β-glucuronidase or aryl sulphatase preparation can be employed after the incubation to liberate any conjugated 7-hydroxycoumarin.

(iii)   Should high blank values be encountered (e.g., with preparations of small intestine or liver cell homogenates) an extraction step is necessary. This is simply achieved by stopping the incubation with 0.5 ml of 4 M HCl and extracting into 6 ml of chloroform. The 7-hydroxycoumarin formed is then determined by extracting 4 ml of the chloroform layer with 4 ml of the glycine-NaOH buffer.

3.11 **Determination of 7-Ethoxyresorufin O-Deethylase**

The O-deethylation of this substrate results in the production of resorufin which is readily estimated fluorimetrically. As with the assay for 7-ethoxycoumarin O-deethylase (see Section 3.10) the original procedure (50) was developed for single sample estimation, whereas the method described below was developed for rapid measurement of several samples (51).

3.11.1 *Media*

(i)   5 μM 7-Ethoxyresorufin. The stock substrate is stored at $-20°C$. Add a small portion of 7-ethoxyresorufin to 5 ml of 50 mM Tris-HCl buffer (see *Table 1*) and dissolve by sonication for $10-15$ min in an ultrasonic water bath (e.g., Sonicon Instrument Corp., Copiague, NY, USA). Scan the substrate solution between 400 and 800 nm in a dual beam recording spectrophotometer and record the absorbance of the maximum at 482 nm (relative to the baseline). As the extinction coefficient for 7-ethoxycoumarin at 482 nm has been reported to be 22.5 cm^2/mmol (52) the concentration of the substrate solution is determined by the formula:

$$\text{substrate concentration (nmol/ml)} = \frac{\text{A482 nm}}{1} \times \frac{1000}{22.5}$$

Then dilute the substrate solution with 50 mM Tris-HCl buffer to give a final concentration of 5 nmol/ml. Prepare this reagent immediately before use. Note that as the substrate is light sensitive it should be protected by wrapping the container containing the working substrate solution in aluminium foil. The entire assay is best performed in subdued light and a covered incubator is recommended.

(ii)   Resorufin standard. The solid resorufin standard is kept at room temperature and a stock solution in ethanol may be kept at 4°C. Dilute a small portion of the stock resorufin solution in ethanol to 5 ml with 50 mM Tris-HCl buffer (see *Table 1*). Scan the standard solution between 500 and 800 nm in a dual beam recording spectrophotometer and record the absorbance of the maximum at 572 nm (relative to the baseline). As the extinction coefficient for resorufin has been reported to be 40.0 cm^2/mmol (52) the concentration of the standard solution is determined by the formula:

$$\text{standard concentration (nmol/ml)} = \frac{\text{A572 nm}}{1} \times \frac{1000}{40}$$

Then dilute the standard solution with 50 mM Tris-HCl buffer to give a final concentration in the range 2−5 nmol/ml. Prepare this reagent immediately before use.

(iii)   50 mM Tris-HCl buffer (see *Table 1*).

(iv)   0.5 M Glycine-NaOH buffer, pH 8.5. Dissolve 37.54 g glycine in approximately 600 ml of distilled water, adjust pH to 8.5 with NaOH and make up to 1 l with distilled water.

(v)   5% (w/v) Zinc sulphate (see Section 3.8.1).

(vi)   Saturated barium hydroxide (see Section 3.8.1).

Reagents iii and iv are stable for at least 12 weeks at 4°C and reagents v and vi are stable for at least 12 weeks at room temperature.

3.11.2 *Procedure*

(i)   Set up 10 ml plastic disposable test tubes in a test tube rack on ice and add co-factor solution (see Section 3.7.2), tissue (see *Table 1*) and additional buffer as required to give a final volume of 1.8 ml in the test and blank incubation tubes and 1.6 ml in the standards.

(ii)   Mix the tube contents with a vortex mixer and pre-incubate in a shaking water bath for 5 min.

(iii)   Remove the rack of tubes from the water bath and rapidly add 0.1 ml of the substrate solution to the test incubation tubes only.

(iv)   Shake the rack to agitate the tube contents, place the rack in the shaking water bath and incubate for 10 min.

(v)   Remove the rack of tubes from the water bath, plunge into ice and add 1 ml of 5% $ZnSO_4$ and then 1 ml of saturated barium hydroxide to all the tubes.

(vi)   Add 0.1 ml substrate solution to the blank and standard tubes and 0.2 ml of the 7-ethoxyresorufin standard solution to the standard tubes only.

(vii)   Mix the contents of all the tubes with a vortex mixer.

(viii) Centrifuge the tubes at 2000 *g* for 15 min at 4°C.

(ix)   Remove 1 ml of the deproteinised supernatant from each tube into a clean glass or plastic disposable tube and add 2 ml of 0.5 M glycine-NaOH buffer, pH 8.5.

(x)   Mix the contents of all the tubes with a vortex mixer.

(xi)   Set the excitation and emission monochromators of a suitable spectrofluorimeter to 535 and 582 nm, respectively.

(xii)   Measure and record the fluorescence of all the tubes.

3.11.3 *Calculation of Results*

(i)   For each tissue sample determine the mean value of the fluorescence of the duplicate test incubations and subtract the blank fluorescence value.

(ii)   Determine the mean value of the fluorescence of each set of triplicate standard tubes and subtract the value of an appropriate blank.

(iii)   Enzyme activity (nmol/min/ml tissue) is determined by the formula:

$$\frac{\text{Fluorescence (test}-\text{blank)}}{1} \times \frac{1}{10} \times \frac{1}{\text{tissue volume assayed (ml)}} \times \frac{\text{concentration of standard solution } (\mu M) \times 0.2}{\text{Fluorescence (standard}-\text{blank)}}$$

(iv)   Enzyme activity is expressed as either nmol/min/mg microsomal protein or as nmol/min/g liver by dividing by the sample protein (mg/ml) or tissue (g tissue/ml) concentrations, respectively.

3.11.4 *Additional Comments*

(i)   As with 7-ethoxycoumarin O-deethylase this fluorimetric enzyme assay is highly sensitive and hence useful for studies with small amounts of tissue and tissues with low levels of mixed function oxidase enzyme activity.

(ii)   Note that in the method described above, unlike some other procedures, the substrate is not added to the incubation mixture in an organic solvent which could result in an inhibition of enzyme activity.

(iii)   When using either hepatocyte whole homogenates or hepatic post-mitochondrial supernatant fractions as the enzyme source, errors can arise due to the further metabolism of the resorufin product. This problem can be overcome by including 10 μM dicumarol in the reaction mixture (53).

(iv)   The literature contains several extinction coefficients for both 7-ethoxyresorufin, namely 22.2, 22.3 and 22.5 cm²/mmol, and resorufin, namely 20.1, 40.0, 47.0 and 73.2 cm²/mmol (52,54). However, it should be stressed that these various extinction coefficients were not all derived under the same conditions.

3.12 **Determination of Other Xenobiotic Metabolising Enzyme Activities**

The four mixed function oxidase activities described above were chosen as examples of relatively rapid procedures which require a minimum of equipment. Other recommended mixed function oxidase assays include acetanilide 4-hydroxylase (55), aldrin epoxidase (10), aryl hydrocarbon (benzo[a]pyrene) hydroxylase (8,56,57) and lauric acid 11- and 12-hydroxylase (58). In addition, testosterone is oxidised by cytochrome P450 to give a variety of site- and spectro-specific metabolites and the particular profile observed may be useful in studies of induction of various forms of cytochrome P450 (10,59). Similarly, the use of a range of alkoxy resorufins (including 7-ethoxyresorufin) as mixed function oxidase substrates may also be diagnostic of the induction of particular forms of cytocrome P450 (60).

Assays are also available for several non-cytochrome P450 dependent enzyme activities including the mixed function amine oxidase (4), epoxide hydratase (57,61) and UDP-glucuronyltransferase (27).

3.13 **Treatment of Experimental Animals to Induce Hepatic Xenobiotic Metabolising Enzymes**

Many compounds are known to induce xenobiotic metabolising enzyme activities in experimental animals. As the literature contains many different treatment schedules for a single inducer the examples given below are presented as a guide only. In order to achieve maximum induction of any particular enzyme in any particular species or strain of animal the worker is advised to conduct studies using a range of inducer dose levels and pre-treatment times. All of the compounds listed below are administered by intraperitoneal injection employing either 0.9% (w/v) saline (sodium phenobarbitone) or corn oil (all other compounds) as the injection vehicle. The compounds are dissolved in the vehicle to give dosing volumes of 5 or 10 ml/kg body weight, depending on compound solubility. For comparative studies control animals should be given corresponding quantities of the injection vehicle.

3.13.1 *Sodium Phenobarbitone*

This compound is normally given at doses of either 80 (mice) or 100 (rats) mg/kg/day for periods of 3−5 days, animals being killed 24 h after the last injection. Some workers administer the compound in the drinking water (1 mg/ml) for periods of around 7 days.

3.13.2 *β-Naphthoflavone*

This compound is now preferred as a polycyclic hydrocarbon-type enzyme inducer (62) to the use of carcinogens such as 20-methylcholanthrene. The compound is administered at doses of either 80 (rats) or 40 (mice) mg/kg/day for periods of 3−4 days, animals being killed 24 h after the last injection.

3.13.3 *Aroclor 1254*

This compound is a complex mixture of polychlorinated biphenyls and is a 'mixed-type' inducer as it produces a pattern of enzyme induction with characteristics of the simultaneous administration of 'drug-type' (e.g., phenobarbitone) and 'polycyclic hydrocarbon-type' (e.g., 20-methylcholanthrene) enzyme inducers (63).

4. APPLICATIONS OF MICROSOMAL FRACTIONS IN TOXICOLOGY

Microsomal fractions have many uses in biochemical toxicology some of which are briefly reviewed below. Such fractions can be used as a tissue source to produce metabolites for other studies and of course to confirm that particular metabolites are produced in the endoplasmic reticulum. These fractions are frequently used for studying the activation of xenobiotics. Thus incubation with radiolabelled substrates (8) and subsequent exhaustive solvent extraction permits the detection of covalent binding of reactive metabolites to protein, lipid and nucleic acids (DNA can be added to the reaction mixtures if necessary). For further discussion of such studies see Chapter 5. Microsomal fractions can also be used as activating systems for mutagenicity tests and, by comparative studies with other fractions (e.g. post-mitochondrial supernatant and cytosol), addition of cofactors, inhibitors etc., can be used to elucidate activation and deactivation pathways. Finally, they can also be used in studies of other toxic processes such as lipid peroxidation (see Chapter 6).

Many xenobiotics are known to induce, or even inhibit, xenobiotic metabolising enzymes. For example, phenobarbitone and polycyclic hydrocarbons have markedly different effects on xenobiotic metabolising enzyme activities (51,63). The induction of multiple forms of cytochrome P450 can be studied with the various enzyme assays described in this chapter and also by other techniques including immunochemical procedures and sodium dodecyl sulphate-polyacrylamide gel electrophoresis.

Studies with microsomal or indeed with other subcellular fractions should be considered a valuable tool in biochemical toxicology. When studying the interaction of a xenobiotic with biological systems information can be generated at various levels of cellular organisation ranging from subcellular fractions, isolated cells, tissue slices, and perfused organs through to the intact animal. Each of these systems has their place in research and the choice of any particular system depends on the questions that are being asked. For example, if a xenobiotic is known to inhibit hepatic mixed function oxidase enzyme activities *in vivo* then studies with microsomal fractions can help identify the mechanism(s) involved, such as the kinetics of inhibition of cytochrome P-450, whether the parent compound or a metabolite is responsible for the effect, and whether reversible or irreversible binding to microsomal enzymes occurs. As with any *in vitro* system care needs to be taken in extrapolating the data obtained to the *in vivo* situation. However, data generated with microsomal fractions can provide valuable information on likely effects in the intact animal. For example, compounds which induce xenobiotic metabolising enzymes *in vitro* often also induce xenobiotic metabolism *in vivo*.

5. ACKNOWLEDGEMENTS

The author is indebted to Dr Mike Tredger (Liver Unit, King's College Hospital Medical School, London) and Dr Mike Collins (Central Toxicology Laboratory, ICI plc, Alderley Park, Cheshire) for helpful discussions.

6. REFERENCES

1. Siekevitz,P. (1965) *Fed. Proc.*, **24**, 1153.
2. DePierre,J.W. and Dallner,G. (1975) *Biochim. Biophys. Acta*, **415**, 411.
3. Mason,H.S., North,J.C. and Vanneste,M. (1965) *Fed. Proc.*, **24**, 1172.

4. Ziegler,D.M. and Poulsen,L.L. (1978) in *Methods in Enzymology*, Vol. **52**, Fleischer,S. and Packer,L. (eds.), Academic Pres, New York, p. 142.
5. Lu,A.Y.H. and Miwa,G.T. (1980) *Annu. Rev. Pharmacol. Toxicol.*, **20**, 513.
6. Boutin,J.A., Antoine,B., Batt,A.-M. and Siest,G. (1984) *Chem.-Biol. Interactions*, **52**, 173.
7. Lu,A.Y.H. and West,S.B. (1979) *Pharmacol. Rev.*, **31**, 277.
8. Guengerich,F.P. (1982) in *Principles and Methods of Toxicology*, Hayes,A.W. (ed.), Raven Press, New York, p. 609.
9. Dallner,G. (1974) in *Methods in Enzymology*, Vol. **31**, Fleischer,S. and Packer,L. (eds.), Academic Press, New York, p. 191.
10. Tredger,J.M., Smith,H.M., Davis,M. and Williams,R. (1984) *Biochem. Pharmacol.*, **33**, 1729.
11. Tredger,J.M. and Chhabra,R.S. (1976) *Drug. Metab. Dispos.*, **4**, 451.
12. Gram,T.E. (1974) in *Methods in Enzymology*, Vol. **31**, Fleischer,S. and Packer,L. (eds.), Academic Press, New York, p. 225.
13. Dallner,G. (1978) in *Methods in Enzymology*, Vol. **52**, Fleischer,S. and Packer,L. (eds.), Academic Press, New York, p. 71.
14. Gram,T.E., Schroeder,D.H., Davis,D.C., Reagan,R.L. and Guarino,A.M. (1971) *Biochem. Pharmacol.*, **20**, 1371.
15. Burke,M.D. and Orrenius,S. (1979) *Pharmacol. Therap.*, **7**, 549.
16. Shirkey,R.J., Chakraborty,J. and Bridges,J.W. (1979) *Anal. Biochem.*, **93**, 73.
17. Stohs,S.J., Grafström,R.C., Burke,M.D., Moldéus,P.W. and Orrenius,S.G. (1976) *Arch. Biochem. Biophys.*, **177**, 105.
18. Fang,W.-F. and Strobel,H.W. (1978) *Arch. Biochem. Biophys.*, **186**, 128.
19. Johannesen,K.A.M. and DePierre,J.W. (1978) *Anal. Biochem.*, **86**, 725.
20. Tangen,O., Jonsson,J. and Orrenius,S. (1973) *Anal. Biochem.*, **54**, 597.
21. Schenkman,J.B. and Cinti,D.L. (1978) in *Methods in Enzymology*, Vol. **52**, Fleischer,S. and Packer,L. (eds.), Academic Press, New York, p. 83.
22. Fry,J.R. and Bridges,J.W. (1975) *Anal. Biochem.*, **67**, 309.
23. Litterst,C.L., Mimnaugh,E.G., Reagan,R.L. and Gram,T.E. (1975) *Life Sci.*, **17**, 813.
24. Schenkman,J.B. and Cinti,D.L. (1972) *Life Sci.*, **11**, Part II, 247.
25. Lowry,O.H., Rosebrough,N.J., Farr,A.L. and Randall,R.J. (1951) *J. Biol. Chem.*, **193**, 265.
26. Heymann,E. (1982) in *Metabolic Basis of Detoxication, Metabolism of Functional Groups*, Jacoby,W.B., Bend,J.R. and Caldwell,J. (eds.), Academic Press, New York, p. 229.
27. Bock.K.W., v.Clausbruch,U.C., Kaufmann,R., Lilienblum,W., Oesch,F., Pfeil,H. and Platt,K.L. (1980) *Biochem. Pharmacol.*, **29**, 495.
28. Morgenstern,R., Lundquist,G., Andersson,G., Balk,L. and DePierre,J.W. (1984) *Biochem. Pharmacol.*, **33**, 3609.
29. Omura,T. and Sato,R. (1964) *J. Biol. Chem.*, **239**, 2370.
30. Mailman,R.B., Tate,L.G., Coons,L.B., Muse,K. and Hodgson,E. (1975) *Chem.-Biol. Interactions*, **10**, 215.
31. Matsubara,T., Koike,M., Touchi,A., Tochino,Y. and Sugeno,K. (1976) *Anal. Biochem.*, **75**, 596.
32. Paine,A.J. and Legg,R.F. (1978) *Biochem. Biophys. Res. Commun.*, **81**, 672.
33. Gilbert,D. (1972) *Biochem. Pharmacol.*, **21**, 2933.
34. Bond,E.J. and DeMatteis,F. (1969) *Biochem. Pharmacol.*, **18**, 2531.
35. Imai,Y. and Sato,R. (1958) *J. Biochem.*, **63**, 270.
36. Jackson,H.L. and McKusick,B.C. (1955) *Organic Syntheses*, **35**, 62.
37. Schenkman,J.B., Remmer,H. and Estabrook,R.W. (1967) *Mol. Pharmacol.*, **3**, 113.
38. Jefcoate,C.R. (1978) in *Methods in Enzymology*, Vol. **52**, Fleischer,S. and Packer, L. (eds.), Academic Press, New York, p. 258.
39. Schenkman,J.B., Sligar,S.G. and Cinti,D.L. (1981) *Pharmacol. Ther.*, **12**, 43.
40. Peterson,J.A., Ebel,R.E. and O'Keeffe,D.H. (1978) in *Methods in Enzymology*, Vol. **52**, Fleischer,S. and Packer,L. (eds.), Academic Press, New York, p. 221.
41. Williams,C.H.,Jr. and Kamin,H. (1962) *J. Biol. Chem.*, **237**, 587.
42. Phillips,A.H. and Langdon,R.G. (1962) *J. Biol. Chem.*, **237**, 2653.
43. Peters,M.A. and Fouts,J.R. (1970) *J. Pharmacol. Exp. Ther.*, **173**, 233.
44. Jansson,I. and Schenkman,J.B. (1977) *Arch. Biochem. Biophys.*, **178**, 89.
45. Fouts,J.R. (1970) *Toxicol. Appl. Pharmacol.*, **16**, 48.
46. Nash,T. (1953) *Biochem. J.*, **55**, 416.
47. Nakanishi,S., Masamura,E., Tsukada,M. and Matsumura,R. (1971) *Jap. J. Pharmacol.*, **21**, 303.
48. Ullrich,V. and Weber,P. (1972) *Hoppe-Seyler's Z. Physiol. Chem.*, **353**, 1171.
49. Lake,B.G., Foster,J.R., Collins,M.A., Stubberfield,C.R., Gangolli,S.D. and Srivastava,S.P. (1982) *Acta Pharmacol. Toxicol.*, **51**, 217.

50. Burke,M.D. and Mayer,R.T. (1974) *Drug Metab. Dispos.*, **2**, 583.
51. Lake,B.G. and Paine,A.J. (1983) *Xenobiotica*, **13**, 725.
52. Prough,R.A., Burke,M.D. and Mayer,R.T. (1978) in *Methods in Enzymology*, Vol. **52**, Fleischer,S. and Packer,L. (eds.), Academic Press, New York, p. 372.
53. Lubet,R.A., Nims,R.W., Mayer,R.T., Cameron,J.W. and Schechtman,L.M. (1985) *Mutat. Res.*, **142**, 127.
54. Smith,A.G., Francis,J.E. and Bird,I. (1986) *J. Biochem. Toxicol.*, **1**, 105.
55. Shimazu,T. (1965) *Biochim. Biophys. Acta*, **105**, 377.
56. Lake,B.G., Collins,M.A., Harris,R.A. and Gangolli,S.D. (1979) *Xenobiotica*, **9**, 723.
57. DePierre,J.W., Johannsen,K.A.M., Moroń,M.S. and Seidegård,J. (1978) in *Methods in Enzymology*, Vol. **52**, Fleischer,S. and Packer,L. (eds.), Academic Press, New York, p. 412.
58. Orton,T.C. and Parker,G.L. (1982) *Drug Metab. Dispos.*, **10**, 110.
59. Tredger,J.M., Smith,H.M. and Williams,R. (1984) *J. Pharmacol. Exp. Ther.*, **229**, 292.
60. Burke,M.D. and Mayer,R.T. (1983) *Chem.-Biol. Interactions*, **45**, 243.
61. Lu,A.Y.H. and Levin,W. (1978) in *Methods in Enzymology*, Vol. **52**, Fleischer,S. and Packer,L. (eds.), Academic Press, New York, Vol. 52, p. 193.
62. Boobis,A.R., Nebert,D.W. and Felton,J.S. (1977) *Mol. Pharmacol.*, **13**, 259.
63. Alvares,A.P. and Kappas,A. (1977) *J. Biol. Chem.*, **252**, 6373.

CHAPTER 9

Preparation and Use of Mitochondria in Toxicological Research

KELVIN CAIN and DAVID N.SKILLETER

1. INTRODUCTION

Mitochondria carry out a variety of biochemical processes, the most important of which is oxidative phosphorylation and in this respect they can be regarded as the 'power houses' of the cell. Thus, in the mitochondria, energy derived from the oxidation of fatty acids, carbohydrates and amino acids is conserved as ATP which is transported to the energy-requiring reactions of the cell. Mitochondrial oxidative phosphorylation in this respect is much more efficient than substrate level phosphorylation; for example, anaerobic glycolysis yields 2 mol of ATP per mol of glucose, whereas aerobic glycolysis produces 30 mol of ATP. Consequently mitochondria tend to be concentrated in those tissues in which the energy demand is high. Oxygen consumption by the kidney, for example, on a per gram tissue basis, is second only to that of the heart and 95% of the ATP produced aerobically is used to drive tubular reabsorption processes (1). In the manifestation of a toxic response a compound which inhibits mitochondrial oxidative phosphorylation can therefore have a profound effect on the metabolism of important organs like the heart, kidney, liver and brain. In this chapter, methods for the preparation of intact, functioning mitochondria from these organs are described. In the main, the methods refer to rat tissues but are equally applicable to other laboratory animals. Techniques for measuring mitochondrial respiration, phosphorylation, metabolite transport and other partial reactions related to mitochondrial metabolism are presented. Using these methods it is possible to determine the site of action of an inhibitor and examples of representative compounds are given. In this context, detailed mechanistic explanations are beyond the scope of this chapter and more comprehensive interpretations are given in the reference list (2 − 6). For similar reasons a detailed account of oxidative phosphorylation has not been given and the reader is referred to the appropriate literature for a detailed discussion (7 − 9). Finally, we have tried to illustrate with known examples the significance of 'in vitro' mitochondrial effects in relation to the 'in vivo' toxicity.

2. BASIC TECHNIQUES

2.1 Equipment/Apparatus

2.1.1 Homogenisers

The first step in the preparation of mitochondria is the disruption of the tissue to yield

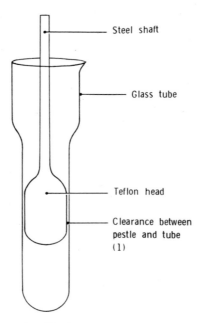

Steel shaft

Glass tube

Teflon head

Clearance between
pestle and tube
(1)

Figure 1. Potter-Elvehjem homogeniser. The clearance (1) between pestle and tube can be varied by skimming the Teflon head on a lathe. For recommended clearances see text.

a 'cell soup' or homogenate from which the mitochondria can be isolated by differential centrifugation. There are a variety of methods available to homogenise tissues, however, in the case of mitochondria it is important to optimise the conditions to produce intact, 'tightly coupled' mitochondria. Thus a vigorous homogenisation method will produce a good yield of mitochondria but the respiratory control ratio (Section 3.1.2) which is a measure of mitochondrial integrity will be poor. The type of tissue being homogenised also determines the method chosen. In the case of the liver, kidney and other soft tissues the preferred apparatus is the Potter-Elvehjem or Dounce Homogeniser which is shown in *Figure 1*. This homogeniser is easily obtainable from any good supplier of laboratory equipment (e.g., Jencons Scientific Ltd., Cherrycourt Way Industrial Estate, Leighton Buzzard, UK). The apparatus consists of a glass body with a glass or Teflon pestle. We recommend the Teflon pestle for two reasons: (i) the Teflon head is easily machineable to give different clearances; and (ii) the steel handle can be removed so the pestle can be motor driven. Typically, the homogenisers are supplied with a total clearance between body and pestle of $0.15 - 0.25$ mm. In our experience this is too tight a fit and we recommend that the pestle is machined to give a minimum and maximum clearance of 0.25 mm and 0.5 mm, respectively. It should be stressed that these values are guidelines only and it may be necessary to determine empirically the optimum clearance value. When the pestle is motor driven any overhead mounted electrical motor can be used, provided it has a high torque and variable speeds up to approximately 2000 r.p.m. In this laboratory we use a motor, normally geared to a constant 1100 r.p.m. and fitted with a drill chuck to take the steel shank of the pestle. A foot-operated on/off switch allows the operator to control the homogeniser with both hands.

The glass and Teflon homogenisers can be obtained in sizes varying from 5 ml to 55 ml capacity.

For tough fibrous tissues such as heart, the Potter-Elvehjem homogeniser is unsuitable and alternative methods have to be used. In the case of small animal hearts an 'enzyme digestion' step is used, followed by conventional homogenisation (Section 2.6). For larger animals (e.g., ox heart) a more vigorous method can be employed. There is a variety of apparatus available but they all rely on the use of very fast rotating blades to disrupt the tissues and are basically very similar to the household liquidiser/blender and can be obtained from any good supplier of laboratory equipment (e.g., Scientific Furnishings Ltd., London Road South, Poynton, Cheshire, UK). In any event the blender system of homogenisation invariably produces many broken mitochondria. However, these can be separated from the intact mitochondria by differential centrifugation. Invariably the mitochondrial yields are lower but this is compensated for by the fact that larger amounts of tissue are available (usually from the slaughterhouse) and therefore large quantities of mitochondria can be obtained.

Finally, sonication is sometimes used to disrupt cell membranes. This method is not suitable for whole tissues, although it can be used to disrupt cells in culture. Its main use in mitochondrial work is the preparation of submitochondrial particles (Section 2.8). In this respect, a variety of companies manufacture sonifiers which are suitable (e.g., M.S.E. Scientific Instruments, Manor Royal, Crawley, Sussex, UK).

2.1.2 *Centrifuges*

Mitochondria are isolated from the homogenate by differential centrifugation and a basic requirement is the availability of suitable centrifuges. It is essential that the centrifuge(s) can be refrigerated. A centrifuge similar to the MSE Europa 24M or Beckman J21 series is ideal. Both these centrifuges have 8 × 50 ml and 6 × 250 ml rotors which are capable of g_{max} values of approximately 48 000 g and 30 000 g, respectively. The 8 × 50 ml rotor is particularly useful as it accelerates up to speed rapidly and in the main this is the rotor of choice for mitochondrial preparations. The 6 × 250 ml rotor is only used for large-scale preparation of mitochondria, and in this respect a large capacity centrifuge like the MSE 6L is also very useful. For the preparation of submitochondrial particles an ultracentrifuge and rotor capable of 105 000 g are also required.

2.1.3 *Oxygen Electrode*

The prime biochemical reaction catalysed by mitochondria is respiration and this classically can be measured by Warburg manometry or alternatively, and more conveniently, with an oxygen electrode. The latter is now the method of choice and a schematic diagram of the apparatus is shown in *Figure 2*. There are a variety of instruments available commercially, and for our own studies we have used an apparatus supplied by Hansatech Ltd., Hardwick Industrial Estate, Kings Lynn, Norfolk, UK, available at reasonable cost. Alternatively, the instrument is based on an original design by Delieu and Walker (10) who give very detailed diagrams and instructions for constructing the electrode, and any competent workshop should be able to reproduce the apparatus.

219

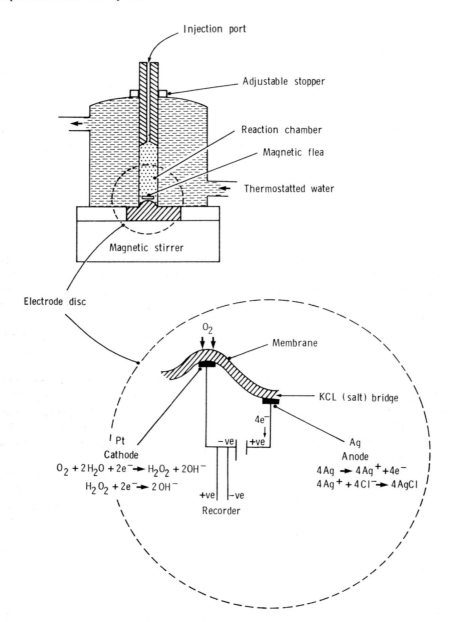

Figure 2. Schematic drawing of oxygen electrode. The apparatus shown is marketed by Hansatech Ltd. and is based upon the design of Delieu and Walker (10). The components of the instrument are made of Perspex, except for the reaction chamber which is borosilicate glass for efficient thermal transfer. The chamber is closed with an adjustable Perspex stopper which is pierced with a narrow injection port. The electrode is polarized with 0.7 V which causes the reduction of O_2 to OH^- at the $^-$ve Pt electrode. The rate of reduction is equal to the activity of the dissolved oxygen and is proportional to the current flowing, the latter is amplified and recorded on a potentiometric recorder using a full-scale deflection of 10 inches which is equivalent to 1 mV.

2.1.4 *General Equipment*

In addition to the above equipment it will be necessary to have the following.

(i)   Recording split or dual beam spectrophotometer with the facility for measuring turbid mitochondrial samples (usually this is a curvette holder located close to the photomultipliers to minimise the effects of light scattering).

(ii) β Scintillation counter.

(iii)   High speed bench centrifuge for routine assays — the Eppendorf 5412 centrifuge is ideal for this purpose and has the advantage that it can also be used for metabolite transport studies employing the silicone oil layer separation technique (Section 3.6.2).

(iv)   General laboratory equipment including shaking water bath, automatic pipettes and microlitre syringes.

2.2 Media/Buffers

Mitochondria are very easily damaged and often problems in obtaining good respiratory control can be avoided if some general precautions are taken. Detergents used in washing glassware and common cations like calcium will 'uncouple' mitochondria. Consequently, all buffers should be made up in double glass-distilled water and all glassware, etc., rinsed at least three times in double glass-distilled water. We recommend that a set of apparatus (scissors, homogenisers, centrifuge tubes, etc.) should be dedicated for mitochondrial use and washed separately from other laboratory glassware. Always use chemicals which are of the highest purity.

Many mitochondrial functions require mitochondrial integrity to be maintained and isotonicity of the media is usually achieved with either ionic (e.g., 100 mM KCl) or, commonly, with non-ionic agents like 0.25 M sucrose. Usually, the pH is buffered to around 7.5 with Tris [tris(hydroxymethyl)aminomethane] or other suitable buffering reagents. The composition of many buffers described in the mitochondrial literature often vary according to both tissue and laboratory. Although similarities exist, it is probably true to say that the composition of buffers owes more to historical usage and personal preference than systematic study. In this respect we have included only those buffers which we have used satisfactorily and these are detailed at the beginning of each method. This does not mean that the buffers described are essential for success and, if necessary, the buffers may be modified according to particular requirements. However, certain chemicals are essential for the normal functioning of the mitochondria and these are described in *Table 1*.

2.3 Isolation of Rat Liver Mitochondria

This method is based on the procedure originally described by Johnson and Lardy (12).

2.3.1 *Buffer*

0.25 M sucrose; 5 mM Tris-HCl pH 7.4; 1 mM EGTA or EDTA (optional).

2.3.2 *Method*

Ideally all operations should be performed in the cold room. In any event keep all

Table 1. Essential Ingredients for Mitochondrial Buffers.

Ingredient	Concentration	Rationale for use	Comments
Sucrose	0.25 – 0.3 M	Preservation of mitochondrial integrity	Most widely used osmotic agent
KCl	100 – 150 mM	Preservation of mitochondrial integrity	Not so widely used now – high ionic strength may leach out cytochrome c
K^+	10 – 20 mM	Preservation of membrane potential	Some inhibitors may have different effects if Cl is counter ion, e.g., triethyltin (11)
EDTA or EGTA	~1 mM	Chelation of Ca^{2+} and other divalent cations which uncouple mitochondria	Normally not used in assay buffers, but in some cases, e.g., kidney, it is necessary for good respiratory control
Mg^{2+}	~5 mM	Essential co-factor	Binds to ADP and ATP forming Mg-ADP and Mg-ATP complexes
PO_4^{2-}	~5 mM	Essential for ATP synthesis	Also needed for dicarboxylate transport
Bovine serum albumin	1 – 10 mg/ml	Sequestering fatty acid and acyl-CoAs which can inhibit mitochondria	Can also bind lipophilic inhibitors so caution is needed

apparatus (e.g., homogenisers, beakers and centrifuge tubes) and buffer at $0-4°C$ in an ice bucket. It is important that contamination with ice is avoided.

(i)   Stun and decapitate a $200-250$ g rat and bleed out the carcass under running cold water.

(ii)   Rapidly remove the liver (~ 10 g weight) and trim away all blood vessels and connective tissue before transferring to a 100-ml beaker containing about 70 ml of ice-cold buffer. Decant the liquid to remove any blood and add more buffer. Repeat the procedure $2-3$ times to wash out as much blood as possible.

(iii)   Transfer the liver to a 50-ml beaker combining $20-30$ ml of buffer and mince the tissue into small sections with scissors. This procedure releases more blood and the contaminated buffer must be discarded. Add more buffer and chop again, decant the liquid and then wash the minced liver into a 50-ml homogeniser with 30-ml of buffer.

(iv)   Homogenise the tissue with a motor-driven pestle (1100 r.p.m.). The best technique is to gradually advance the pestle to the bottom of the tube with successively longer strokes. When the pestle can be advanced to the bottom of the tube without excessive resistance (approximately six strokes) do a further four strokes to complete the homogenisation. Do not try to advance the pestle to the bottom of the tube on the first pass as this requires excessive force and may break the glass homogeniser tube. In addition, a vacuum effect may be producd when the pestle is withdrawn and this will cause mechanical damage to the mitochondria. Also try and keep the homogeniser at $0-4\,°C$ throughout the procedure, an ice jacket is usually sufficient.

(v)   Divide the homogenate into two 50-ml centrifuge tubes and dilute each aliquot to 45 ml with ice-cold buffer.

(vi)   Balance the centrifuge tubes against each other ('by eye' is surprisingly accurate and much quicker) and centrifuge at $4\,°C$ and 2000 r.p.m. (460 g_{max}) for 10 min (this includes the acceleration time). This spin pellets the nucleus, red blood cells and cell debris fractions.

(vii)   Decant the supernatants into two centrifuge tubes, balance and centrifuge at 10 000 r.p.m. (12 500 g_{max}) for 7 min. In decanting the supernatant be careful not to disturb the nuclei/cell debris pellet. If in doubt, aim for a reduced but purer yield of mitochondria.

(viii)   The pellets from the 12 500 g_{max} spin contain the mitochondrial fraction. In appearance the pellet has three distinct regions:
(a) lower layer (HL) which is white and red, this is composed of residual cell debris and red blood cells;
(b) middle layer (ML) which is medium to dark brown in colour and contains largely intact mitochondria and some lysosomes;
(c) an upper fluffy, light, pinkish brown layer (LM) containing broken mitochondria and microsomes.

(ix)   Decant all but 10 ml of the supernatant and gently swirl the remaining liquid around the pellet to dislodge the LM layer, which is discarded.

(x)   Add a few millilitres of fresh buffer to the centrifuge tube and with a chilled smooth glass or plastic rod very gently dislodge the ML fraction and transfer to a 50 ml homogeniser tube. Any mitochondria sticking to the sides of the centrifuge tube can be washed out with further aliquots of buffer. With care the mitochondrial fraction can be removed without disturbing the HL fraction at the bottom of the tube.

(xi)   Add buffer to the homogeniser tube to a final volume of 40 ml and then very 'gently' resuspend the mitochondria by hand with a few strokes of a loose-fitting pestle. Centrifuge the mitochondrial suspension at 12 500 g_{max} for 7 min, balancing against a water-filled centrifuge tube.

(xii)   Resuspend the mitochondrial layer (ML) as described above, except this time add a maximum of $2-3$ ml of buffer to transfer the mitochondria to a 5-ml homogeniser. Resuspend very gently with a loose-fitting pestle. The final yield is approximately $2-3$ ml of mitochondria with a protein concentration of 30 mg/ml.

2.2.3 *Troubleshooting*

The above procedure should produce a good yield of functionally and morphologically intact mitochondria. The key factor is speed and the preparation should take no more than 1 h; the axiom, 'practise makes perfect' is very true. Other possible causes of 'poor mitochondria' are described in *Table 11*.

2.4 **Isolation of Rat Kidney Mitochondria**

The method for this is basically as described for rat liver mitochondria with some modifications.

2.4.1 *Buffer*

0.25 M sucrose; 5 mM Tris-HCl pH 7.4; 1 mM EDTA or EGTA essential for good respiratory control.

2.4.2 *Method*

(i)   Kill and bleed out $3-4$ rats (250 g) and remove the kidneys rapidly (combined weight $7-10$ g).

(ii)   Remove the capsule by gently pinching the kidney between thumb and forefinger, causing the tissue to be extruded from the membrane. At this stage one of two different procedures can be carried out. With the first procedure the kidneys are minced in a similar manner to that described for rat liver. Alternatively, if only kidney cortex mitochondria are required then cut the kidneys sagitally and dissect away the medullary region before mincing. This latter procedure is not easy to do with the small rat kidney, although it is certainly feasible with larger animals (e.g., rabbit). It should also be pointed out that 70% of the kidney is cortex and the majority of the aerobic metabolism takes place in these mitochondria (1). Consequently, the contribution of medullary mitochondria to a mitochondrial preparation from the whole kidney must be small. In any event the idea that the cortex mitochondria are homogeneous must be treated with some scepticism as the cortex is composed of glomeruli, proximal, distal and collecting tubes. At best the cortex preparation will be heterogeneous and in the case of the rat the 'extra purity' must be balanced against the time needed to dissect out the cortex before homogenisation. In the preparation described here the whole kidney was homogenised.

(iii) Transfer the mince to a 50-ml homogeniser tube and make up to 30 ml with buffer as described for rat liver mitochondria. Homogenise and centrifuge in exactly the same way as described for rat liver mitochondria.

(iv)   The pellet, after the high-speed (i.e., 12 500 g_{max} for 7 min) spin, has a slightly different appearance from that described for the rat liver preparation and consists of four bands. The bottom layer (HL) is cell debris and residual blood cells, the next layer (L) is dark brown, almost black in colour and is largely made up of lysosomal granules (13,14). The mitochondrial layer (ML) immediately follows and is covered with the light mitochondrial layer (LM).

(v)   Remove the LM fraction and resuspend the mitochondrial layer as previously described (Section 2.3.3).

(vi)   Re-centrifuge at 12 500 g_{max} for 7 min and resuspend the mitochondria in $2-3$ ml. The yield and protein concentration is similar to the rat liver preparation.

2.4.3 *Troubleshooting*

The problems encountered are similar to those described for rat liver mitochondria in Section 2.3.3 and *Table 11*.

2.5 Isolation of Rat Brain Mitochondria

This procedure is based on the original method described by Clark and Nicklas (15) which relies on a Ficoll discontinuous gradient to separate the mitochondria from synaptosomes.

2.5.1 *Buffers*

(i)   Homogenisation buffer; 0.25 M sucrose, 10 mM Tris-HCl, 0.5 mM potassium EDTA, pH 7.4.

(ii)   Gradient buffer A; 3% Ficoll, 0.12 M mannitol, 0.03 M sucrose, 25 μM potassium EDTA, pH 7.4.

(iii) Gradient buffer B; 6% Ficoll, 0.24 M mannitol, 0.06 M sucrose, 50 μM potassium EDTA, pH 7.4.

2.5.2 *Method*

(i)   Kill eight rats (body weight ~200 g) by decapitation and quickly trim the skin from the top of the skull. Cut along the lateral sutures of the cranium with strong sharp scissors and expose the brain. Remove the cerebral hemispheres and transfer the tissue (~1.5 g) to a beaker containing chilled homogenisation buffer.

(ii)   Wash the hemispheres three times with buffer and chop finely with scissors. Again discard contaminated buffer and then add 40 ml of buffer to transfer the chopped brain to a 50-ml homogeniser tube.

(iii) Homogenise the tissue in two stages. Initially use a loose-fitting pestle (minimum clearance 0.5 mm) and homogenise until the pestle can be passed smoothly to the bottom of the tube. The final homogenisation is carried out with a pestle with a 0.25 mm clearance. Homogenise slowly with three strokes of the pestle. In this preparation the motor-driven homogeniser should be set at 1950 r.p.m. instead of the 1100 r.p.m. speed used for the rat liver and kidney methods.

(iv) Divide the homogenate between two centrifuge tubes and make up each aliquot to a final volume of approximately 40 ml with chilled homogenisation buffer and centrifuge at 4000 r.p.m. (2000 g_{max}) for 3 min.

(v)   Transfer the supernatants to another two centrifuge tubes and re-spin at 10 000 r.p.m. (12 500 g_{max}) for 8 min. Discard the supernatants and resuspend the 'crude mitochondrial' pellet in 10 ml of gradient buffer A.

(vi) Carefully layer the mitochondrial suspension onto 20 ml of the gradient buffer B solution in a centrifuge tube and balance (by weighing) against a water blank. Centrifuge at 9500 r.p.m. (11 500 g_{max}) for 30 min.

(vii) Decant the supernatant and slough away the slightly fluffy fraction overlaying the mitochondrial pellet. Resuspend the latter in homogenisation media (~40 ml volume) and centrifuge at 12 500 g_{max} for 10 min.

(viii) Discard the supernatant and gently resuspend the mitochondria in 2 ml of homogenisation buffer. The final protein concentration should be about 20 mg/ml.

2.5.3 *Troubleshooting*

Speed of operation is very critical as the brain is particularly susceptible to anoxia. The operator should try to keep the time for killing and removal of the brain to an absolute minimum. This will inevitably mean that the first few preparations attempted will not be up to standard. Potassium salts must be used as they enhance respiration in brain mitochondria.

2.6 Isolation of Rat Heart Mitochondria

This method is derived from the procedure described by Chappell and Hansford (16). The technique involves the use of a proteolytic digestion step to aid homogenisation of the tissue. The enzyme used is Nagarse (Subitilisin BNP$^-$) and can be obtained from Sigma Chemicals.

2.6.1 *Buffers*

0.21 M mannitol, 0.07 M sucrose, 1 mM EGTA, 5 mM Tris-HCl, pH 7.4.

2.6.2 *Method*

(i)   Kill 4 − 5 (250 g) rats and rapidly remove the hearts (∼8 g). As each heart is removed, immediately transfer to a beaker containing 30 ml of buffer (4°C). Lightly chop the hearts to release any blood and decant the liquid. Add fresh buffer to wash out the blood. Repeat this procedure three times.

(ii)   Resuspend the chopped hearts in 20 ml of buffer containing 1 − 2 mg of Nagarse and transfer to a 50-ml homogeniser and add a further 20 ml of buffer before carrying out an initial homogenisation with 3 − 4 strokes of the motor-driven pestle. Incubate for 1 min (4°C) and then re-homogenise with three strokes of the pestle.

(iii) Divide the homogenate into two 50-ml centrifuge tubes and make each aliquot up to 40 ml with buffer and centrifuge at 10 000 r.p.m. for 7 min.

(iv)   Discard the supernatants and gently resuspend each pellet to its original volume with a loose-fitting homogeniser.

(v)   Isolate the mitochondria from the resuspended homogenate with the same centrifugation procedures described for rat liver. Resuspend the heart mitochondria in a final volume of about 2.5 ml and protein concentration of approximately 20 mg/ml.

2.6.3 *Troubleshooting*

(i)   In this preparation the time taken to remove the heart and wash it with buffer is very important. With practise the animal can be killed and the heart removed and chopped in ice-cold buffer within 30 sec.

(ii)   The use of Nagarse allows the heart to be homogenised with minimum force. It is essential, however, that the enzyme is removed after homogenisation and the initial high-speed spin is carried out so that the majority of the Nagarse is removed with the supernatant. However, residual enzyme can be carried over into the final mitochondrial suspension and cause damage to the mitochondria which will be 'uncoupled'. This problem can usually be averted by progressively decreasing the amount of Nagarse used until good 'coupled mitochondria' are produced.

2.7 Isolation of Ox Heart Mitochondria

Mitochondria prepared from ox heart have the advantage that slaughterhouse material is very easily obtainable and the preparation produces large quantities of functionally intact mitochondria which can be stored frozen (see Section 2.7.2). The procedure described is a modification of the method of Low and Vallin (17).

2.7.1 Buffers

(i)   Homogenisation buffer; 0.25 M sucrose, pH maintained at 7.5 with 1 M Tris-base (see Section 2.7.2).

(ii)   Resuspension buffer; 0.25 M sucrose, 1 mM EDTA, 20 mM Tris-HCl, pH 7.5.

2.7.2 Method

(i)   Obtain $2-3$ ox hearts ($3-4$ kg) from freshly-killed animals and transport, packed in ice, back to the laboratory. Most slaughterhouses will come to an agreement by which they will inform you at what time they are killing the animals, thereby allowing the hearts to be chilled and transported to the laboratory soon after killing.

(ii)   All operations should be carried out in a cold room at $4°C$; pre-cooling of all apparatus is essential.

(iii) Trim away any connective tissue and cut the hearts into small (~ 3 cm square) cubes. Discard any papillary muscle or blood clots (the best instruments for this purpose are ordinary sharp butchers' knives).

(iv)   Feed the heart cubes through an electric meat grinder and collect the mince in 500 g aliquots in 2 l plastic beakers. To each beaker add 1 l of 0.25 M sucrose and stir with a glass rod, check the pH and adjust to approximately 7.5 with cold 1 M Tris-base. For the sake of speed, pH test papers can be used for this purpose but do not dip the test paper into the suspension.

(v)   Homogenise aliquots of the suspension for 1 min in a large Waring blender using the maximum speed. Check and adjust the pH with Tris-base to pH 7.5. Repeat the homogenisation until all the minced suspension has been processed. Distribute the homogenate between 6×1 l centrifuge pots and spin at 2000 r.p.m. (1250 g_{max}) for 15 min.

(vi)   Decant the dark red supernatant (in a good preparation this will account for 50% of the contents of the centrifuge pot) through four layers of cheesecloth. The pellet from this spin yields largely cell debris, red blood cells and myelin. It has a solid gelatinous appearance and must be disposed of as soon as possible, otherwise it will be very difficult to clean the centrifuge pots. Do not dispose of the debris down the sink as it is very easy to block the drains.

(vii) Centrifuge the strained supernatant in 6×250 ml containers at 10 000 r.p.m. (15 000 g_{max}) for 20 min (this scale of preparation will normally require two runs to process all the supernatant).

(viii) Discard the supernatant from each pot, the pellet has two layers; a lower dark, reddish brown layer (heavy mitochondria) and an upper lighter brown layer (light mitochondria). To each pot add a small volume of resuspension buffer and by gently swirling, resuspend the light mitochondrial layer. Pour off the latter and add further volumes of buffer before resuspending the heavy mitochondria with a 'cooled' glass rod. Use a 50-ml homogeniser with a loose-fitting pestle to gently fully resuspend (by hand) the mitochondria. Transfer the mitochondria to a beaker kept in ice whilst repeating the same procedure for the remaining mitochondrial pellets.

(ix) Distribute the 'pooled' mitochondrial suspension from both 10 000 r.p.m. spins into 6 × 250 ml centrifuge pots and make up to volume with resuspension buffer. Re-spin at 10 000 r.p.m. for 20 min.

(x) Resuspend the heavy mitochondria as before to a protein concentration of approximately 30−40 mg/ml. From 3−4 kg of ox heart the yield should be approximately 1.5 g of mitochondrial protein.

(xi) Ox heart mitochondria can be stored in the deep freeze and for this purpose divide the preparation into aliquots (glass scintillation vials are ideal) and freeze at −20°C.

2.7.3 Troubleshooting

Ox heart mitochondria are extremely robust and survive the rigours of preparation very well. The main problem is in maintaining the pH at 7.5 and this should be checked frequently throughout the procedure. It is particularly critical during homogenisation because an acid pH seems to promote the polymerisation of actin and myosin. If this happens, the first spin (2000 r.p.m.) does not yield a 50/50 split between supernatant and pellet and up to 80% of the volume of the centrifuge pot can be taken up by the debris pellet. The yield and quality of mitochondria from the supernatant is then very much reduced.

2.8 Preparation of Submitochondrial Particles

Submitochondrial particles are formed by sonication of mitochondria which produces inverted mitochondrial vesicles. These particles contain all the enzymes for oxidative phosphorylation without the matrix or outer membrane enzymes. They are useful for looking at respiratory chain oxidation, ATPase activity and other partial reactions. The preparation described below can be applied to any mitochondrial preparation.

2.8.1 Buffers

0.25 M sucrose, 1 mM EDTA, 10 mM Tris-HCl, pH 7.5.

2.8.2 Method

(i) Pellet the mitochondria with a 10 000 g_{max} × 10 min spin.

(ii) Resuspend the pellet to approximately 20 mg/ml in ice-cold preparation buffer.

(iii) Sonicate the mitochondria for 1 min at 15 sec intervals (allowing time for cooling between each sonication) with an MSE 60W sonicator (or equivalent) at full power. Depending on the sonicator, the exact amplitude will have to be determined empirically. Usually this is not hard to determine as the right gain setting causes the mitochondrial suspension to 'sizzle'.

(iv)   Dilute the sonicated mitochondria with two volumes of buffer and centrifuge for 15 min at 14 000 r.p.m. (29 000 g_{max}) to sediment the unbroken mitochondria and larger mitochondrial fragments.

(v)   Transfer the supernatant to ultracentrifuge tubes and spin at 100 000 g_{max} for 30 min.

(vi)   The submitochondrial particles pellet from this spin is red in colour with a resin-like appearance. Resuspend the pellet in buffer to the same volume and centrifuge again at 100 000 g_{max} for 30 min.

(vii)   Resuspend the pellet to a final concentration of 20 mg/ml.

2.8.3 *Troubleshooting*

It is essential to keep the mitochondria cool throughout sonication. Otherwise, the preparation is straightforward.

2.9 **Protein Assays**

All mitochondrial functions are usually expressed on a per mg of protein basis, which can be determined by either the Folin and Ciocalteau (18) method or a modification of the Biuret method (19).

2.9.1 *Buffers*

(i)   Biuret; 0.15% w/v cupric sulphate $CuSO_4. 5H_2O$), 0.6% w/v sodium potassium tartrate, 0.75 M NaOH; 5% w/v sodium deoxycholate.

(ii)   Folin and Ciocalteau; 1% w/v cupric sulphate (solution A); 2% w/v sodium potassium tartrate (solution B); 2% w/v sodium carbonate, 0.1 M NaOH (solution C).

2.9.2 *Methods*

(i)   *Biuret.*

(a)   Mix the protein sample with 0.1 ml of 5% w/v deoxycholate and water to 1.5 ml.

(b)   Add 1.5 ml of Biuret reagent and whirlmix, leave for 30 min and then read the absorbance at 540 nm using a reagent blank.

(c)   Approximately 1 mg of bovine serum albumin/ml gives an absorbance of 0.1. Above this level, however, the colour is not linear and a standard curve should be used.

(d)   With some mitochondria a white cloudy precipitate is often produced and this should be removed by centrifugation.

(ii)   *Folin and Ciocalteau.*

(a)   Mix 0.5 ml of solutions A and B with 49 ml of solution C to give solution D. This should be made fresh each day.

(b)   Add 3.0 ml of D, protein sample and water to a final volume of 3.6 ml and whirlmix.

(c)   After 15 min add 0.3 ml of a 1:1 dilution of Folin and Ciocalteau's reagent and whirlmix.

(d)   Read the absorbance at 750 nm after 20 min against a reagent blank.

(e)   Use a standard curve of $0-200$ μg of bovine serum albumin to determine the unknown protein concentration.

(f)   For mitochondrial suspension (protein $10-30$ mg/ml) dilute 4- to 5-fold with water and use 10 and 20 μl aliquots in the assay.

2.9.3 *Troubleshooting*

Biuret is a quick and easy assay but because of its relatively poor sensitivity requires a lot of protein and lipids can cause interferences. The Folin and Ciocalteau reagent method is more sensitive, but beware of interferences from common buffer constituents, e.g., sucrose and Tris. In the method above, the dilution of the mitochondria with water also has the advantage of diluting the sucrose and Tris. This, coupled with the use of 10 and 20 μl aliquots, does not result in any interference problems. It is, however, always advisable to check that the buffers being used are not giving any reaction with the protein assay.

3. CHARACTERISATION OF MITOCHONDRIAL FUNCTIONS

The quality of the mitochondrial preparation is affected by possible contamination with other subcellular organelles (e.g., lysosomes and peroxisomes) and/or the functional integrity of the mitochondria. The latter is the major factor and can only be assessed by measuring the various metabolic functions.

Mitochondria carry out a variety of functions which are dependent on oxidative phosphorylation. The ability of the mitochondrion to carry out this process can be affected by the quality and purity of the preparation or the presence of a toxin which is causing inhibition. Obviously, the decision that the latter is happening is based upon the belief that the method of preparing the mitochondria is satisfactory. In this respect, the best methods available to assess the purity and quality of the mitochondrial preparation rely on assaying the various metabolic functions. This section therefore details the more important mitochondrial reactions which can be assayed easily and routinely. These methods are also used to determine the site of inhibition of a putative mitochondrial inhibitor.

With regard to contamination with other subcellular organelles, it should be remembered that the range of buoyant densities of mitochondria and lysosomes is very close and in the differential centrifugation methods described herein some cross-contamination is inevitable. The degree of contamination will be different from tissue to tissue since the mitochondrial mass as a proportion of the cellular mass varies. Thus in the heart the cells are 'packed' with mitochondria but contain few lysosomes and consequently isolated mitochondrial preparations are virtually free of contamination. In the hepatocyte there are proportionally more lysosomes and the mitochondrial preparations will inevitably exhibit some contamination. Practically, for assessing the effect of a toxin on mitochondrial function *in vitro*, this is not a problem as it is more important to preserve functional integrity, and in this respect speed of preparation is the key factor. If it is known that the toxin interacts with lysosomes or peroxisomes, then further purification can be achieved by isopycnic centrifugation using a discontinuous sucrose density gradient (15, 30, 50 and 70% w/v sucrose, 20 mM Tris-HCl, 1 mM EDTA, pH 7.5) centrifuged for 3 h at 90 000 g_{max} (20). The success of the purification procedure can then be monitored using lysosomal and peroxisomal markers (see Chapter 10). However, a major drawback to this type of purification is that the functional integrity of the mitochondria will suffer. Thus, although enzymic specific activities such as succinate dehydrogenase may increase, the coupled functions such as respiratory control ratio will decrease. Furthermore, the purified fraction must be regarded as a subfraction of the total mitochondrial population of the tissue. Also, in dealing with mitochondria isolated from toxin-treated animals, other problems arise. A toxic substance may cause structural damage which is restricted to a proportion of the mitochondrial population, the purification procedures would discard damaged mitochondria and the toxic effect might be overlooked. Alternatively, redistribution of the toxin may occur as a result of cell fractionation and, in addition, substantial losses may be produced by the purification methods which involve extensive washing procedures. Consequently an effect on mitochondrial function 'in vivo' may remain undetected.

3.1 Respiration Studies with Mitochondria

The mitochondrial oxidative phosphorylation system consists of two distinct but closely linked multi-enzyme complexes, the respiratory chain and the ATPase/synthetase enzyme. A full description of these enzyme complexes is given in the appropriate references (7 – 9). However, for the purpose of this section, the process can be described schematically (*Figure 3*). In simple terms the respiratory chain accepts reducing equivalents

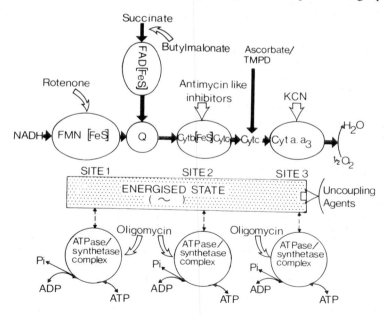

Figure 3. Mitochondrial oxidative phosphorylation. The figure shows schematically the process of oxidative phosphorylation. The elipses represent the five respiratory complexes which make up the electron transfer chain (for review, see 7, 8 and 9). The solid arrows represent the transfer of reducing equivalents within or into the respiratory chain. The dotted area (energised state) is the coupling mechanism (high energy state; ~ in conventional terms) and is equivalent to the protonmotive force in the chemiosmotic theory. Sites 1, 2 and 3 refer to the sites of ATP synthesis (i.e., P/O ratios). Inhibitors are represented by open arrows and note that uncoupling agents dissipate the energised state which can also be used to drive solute transport.

231

Table 2. Composition of Respiration Buffers.

Buffer Composition	Suitable for

A.   0.25 M Sucrose
    5 mm KH_2PO_4
    10 mM KCl
    5 mM $MgCl_2$
    10 mM Tris-HCl
    pH 7.4
	Rat heart mitochondria
	Ox heart mitochondria
	Rat liver mitochondria

B. As A + 1 mM EGTA	Rat kidney mitochondria

C.   58 mM KCl
    26 mM NaCl
    6 mM $MgCl_2$
    10 mM Tris-phosphate
    pH 7.4
	Rat brain mitochondria

Respiration buffer should be thoroughly aerated and kept at the correct temperature (usually 30°C or 37°C) before use in the oxygen electrode.
Store buffers at 5°C or, better still, in frozen aliquots. Discard buffer at the end of the day, i.e., buffers which have been thawed and aerated.

Table 3. Substrates Used in Mitochondrial Respiration Studies.

Mitochondria	Substrates known to be metabolised	Recommended	Comments
Rat liver	All TCA cycle intermediates, 3-hydroxybutyrate, glutamate, proline and fatty acids (21,22)	pyruvate + malate[a] glutamate + malate[b] 3-hydroxybutyrate and succinate	
Rat kidney	Malate, pyruvate, citrate, glutamate, α-ketoglutarate, 3-hydroxybutyrate, fatty acids and succinate (1,23)	pyruvate + malate glutamate + malate and succinate	EGTA is needed for good coupling (see text)
Rat heart and ox heart	Pyruvate, malate, glutamate, α-ketoglutarate, fatty acids, 3-hydroxybutyrate, succinate (24,22)	pyruvate + malate glutamate + malate fatty acids	Do not oxidise endogenous citrate or isocitrate (22)
Rat brain	Malate, pyruvate, citrate, glutamate, oxaloacetate, α-ketoglutarate, 3-hydroxybutyrate and succinate (15,25,26)	pyruvate + malate 3-hydroxybutyrate + oxaloacetate and succinate	Very susceptible to uncoupling with fatty acids

The concentrations of stock substrates are determined on the basis that $10-20$ μl aliquots are injected into the oxygen electrode chamber giving final concentrations of $5-10$ mM. This holds for all substrates except fatty acids which are given as the fatty acyl carnitine derivative at a final concentration of approximately 25 μM. In all cases stock solutions should be adjusted to pH 7.4 with buffer or NaOH and stored frozen in small aliquots (1 ml is ideal). Use each aliquot for one set of experiments and discard the remainder.
[a]Usually added in 1:1 ratio but a 10:1 ratio also works well.
[b]Usually 1:1 ratio.

from the various substrates and their respective dehydrogenases. H^+ and e^- are then transferred down the respiratory chain to reduce oxygen to water. The energy made available from this series of redox reactions is used to drive ATP synthesis at the indicated energy conservation sites. With intact mitochondria, oxidation is tightly coupled

Table 4. Other Reagents Used in Respiration Studies.

Reagent	Stock concentration	Volume added to electrode chamber	Purpose of compound added	Comments
ADP	0.2 M pH 6.8	10 − 20 μl of 20 mM solution	Phosphate acceptor i.e., ADP + Pi → ATP	Stimulates O_2 consumption and used to determine RCR
Rotenone	1 mg/ml in dimethyl formamide	1 − 2 μl	Inhibits between NADH dehydrogenase and respiratory chain	Use when succinate respiration is measured as it prevents oxaloacetate inhibition
CCCP (carbonyl cyanide m-chloro-phenyl-hydrazone)	100 μg/ml in 95% EtOH	1 − 2 μl	Potent uncoupling agent − used to demonstrate respiratory control in mitochondria	Too much CCCP will inhibit rather than stimulate respiration − useful tool in inhibitor studies (see text)
Antimycin A	1 mg/ml in 95% EtOH	1 − 2 μl	Respiratory chain inhibitor	Useful tool in inhibitor studies (see text)
Oligomycin	1 mg/ml in 95% EtOH	1 − 2 μl	ATPase/synthetase inhibitor	Inhibits ADP-stimulated respiration. Useful tool in inhibitor studies
KCN	0.1 M − make up fresh	5 μl	Cytochrome oxidase inhibitor	Useful tool in inhibitor studies
Na dithionite	Solid	Few crystals	Used to demonstrate oxygen electrode is working	Strong reducing agent also used in cytochrome studies (see text)

to ATP synthesis. The measurement of oxygen uptake with various substrates and co-factors is therefore a powerful analytical method for studying mitochondrial metabolism.

3.1.1 Buffers

Table 2 describes the standard respiratory buffers which are used in oxygen electrode experiments. Appropriate aliquots of mitochondria are added to the reaction chamber (*Figure 2*) and the substrates, ADP and inhibitors added *via* the injection port. *Table 3* details the substrates, and *Table 4* other reagents which are often used.

3.1.2 Methods

(i) *Setting up the electrode.*

(a)   Switch on the electrode, magnetic stirrer, chart recorder and thermostatted water circulator and allow to equilibrate for at least 30 min. The thermostat should be adjusted to give a temperature of 30°C as measured in the electrode chamber.

(b)   The oxygen electrode should be assembled according to the manufacturer's instructions and it is important to check that the silver and platinum electrodes are untarnished and the Teflon membrane unbroken. The performance and calibration of the electrode is easily checked (see below).

(c)   Pipette 2 ml of respiration buffer (equilibrated at 30°C and fully saturated with air) into the chamber and start the stirrer and recorder (1 cm/min).

(d)   Allow $1-2$ min for everything to stabilise and then adjust the recorder to the 95% position. This represents the 'air line', that is buffer fully saturated with air, and over the next 5 min there should be no upward or downward deflection of the pen. If there is movement of the pen then the problem is usually to do with the electrode assembly and the operator should refer to the manufacturer's instructions for troubleshooting.

(e)   If there is no significant movement of the pen, switch off the magnetic stirrer. This should produce an almost instantaneous downward deflection. Switching on the stirrer again should cause the pen to return to the 'air line'. This response is due to the electrode consuming oxygen. With the stirrer switched off this oxygen uptake, although small, is greater than the normal diffusion processes and a drop in pO_2 occurs at the membrane surface. If the pen responds sluggishly to switching off the stirrer, then it is likely that the electrodes are tarnished and they should be cleaned (scouring powder and cotton buds are very good for this).

(f)   Switching on the stirrer again mixes the chamber contents and the recorder pen should go back rapidly to its previous position.

(g)   Having ensured that the electrode response is adequate, add a few crystals of sodium dithionite which will react with the dissolved O_2 as follows:

$$Na_2S_2O_4 + O_2 + H_2O \rightarrow NaHSO_4 + NaHSO_3$$

This will cause the pen to drop rapidly and after a few minutes it will stabilise at a much lower position on the chart paper. At this stage the buffer is anaerobic and the pen position is referred to as the 'N_2 line'. The difference between the latter and the 'air line' represents the oxygen concentration of the buffer. Usually the N_2 line is the position the recorder pen assumes when the recorder zero is set (zero voltage from the electrode).

(ii) *Recording mitochondrial respiration.*

(a)   Pipette 1.9 ml of respiration buffer and 100 μl of mitochondria into the electrode chamber. Replace the stopper, adjusting its position so that the buffer rises 1 cm up the injection port. Make sure that there are no air bubbles. The electrode chamber is now sealed and the mitochondria's only source of oxygen is the dissolved gas in the respiration buffer.

(b)   Additions to the electrode chamber can be made without appreciably altering the oxygen concentration of the buffer. Microlitre syringes are ideal for this purpose and no more than 20 μl should be added at a time. With care, the injections can be made without causing the recorder pen to 'jump'.

(iii) *Measuring respiratory control.* The exact additions made to the electrode chamber will depend on the requirements of the experiment. However, it is important to determine whether or not the mitochondria are intact, with respiration tightly coupled to ATP synthesis. In this respect 'coupled' mitochondria exhibit a phenomenon known as 'respiratory control'. The respiratory control ratio (RCR), originally systematically described by Chance and Williams (27), is simple to determine, although in practice good RCRs are not always easy to obtain. The RCR is defined as the ratio of oxygen consumption in the presence and absence of ADP (when the substrate concentration

Table 5. Metabolic States of Mitochondria.

State	Metabolite concentrations			Respiration rate	Rate-limiting factor
	O_2	ADP	Substrate		
1	High	Low	Low	Slow	ADP level
2	High	High	~0	Slow	Substrate level
3[a]	High	High	High	Fast	Respiratory chain
4[a]	High	Low	High	Slow	ADP level
5	0	High	High	0	Oxygen level

Adapted from Chance and Williams (27).
[a]Respiratory control ratio (RCR) = $\dfrac{\text{state 3}}{\text{state 4}}$
See *Figure 4* for experimental examples.

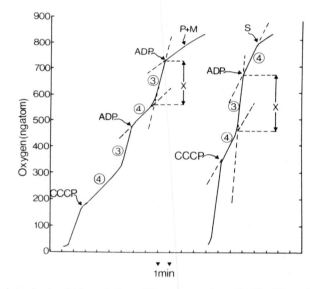

Figure 4. Measuring mitochondrial respiration with an oxygen electrode. The figure shows two traces of mitochondria respiring on succinate (S) and pyruvate plus malate (P + M). With succinate approximately 4 mg of rat liver mitochondria were added to the chamber with respiration buffer to a final volume of 2 ml. The respiration rate was recorded for $1-2$ min and the following additions made: S = 10 μl of succinate (1 M), ADP = 20 μl of ADP (20 mM), CCCP = 2 μl of CCCP (100 μg/ml). With pyruvate plus malate similar conditions were used except 10 μl of 1 M pyruvate plus 0.1 M malate were added as substrate. The numbers 3 and 4 refer to the state of respiration and the dotted lines are drawn to show how the RCR and ADP values are calculated [see Section 3.1 (vi) for details].

is not limiting). Experimentally, this is determined by measuring the state 3 and 4 rates of respiration. The various states of respiration are described in *Table 5* and *Figure 4* shows a typical experimental determination of RCR values for rat liver mitochondria using pyruvate/ malate and succinate as substrates.

The RCR is determined for each substrate as follows:

(a)   Record endogenous respiration (i.e., state 1) for 2 min and then add substrate. This stimulates the respiration and the mitochondria are effectively in state 4.

(b)   Record state 4 for 2 — 3 min and then add a known amount of ADP (400 nmol is ideal), which stimulates the respiration to give state 3.

(c)   After the ADP is used up the respiration will return to state 4 (*Figure 4*).

(d)   Determine the state 3 and state 4 rates a shown in the figure and calculate the RCR.

(e)   Note that RCR value should be determined using the state 4 rates obtained after addition of ADP. This is because the initial state 4 rate is always slower than that seen after ADP is added.

In *Figure 4* the RCR values for rat liver mitochondria respiring on pyruvate/ malate and succinate are 4 — 5 and 3 — 4, respectively. These are acceptable ratios and the mitochondria would be expected to maintain this level of respiratory control for 2 — 3 h. Similar values can be obtained for mitochondria from other tissues. Ox heart is particularly robust and high RCRs are usual. Rat heart and rat brain are a little more temperamental as the presence of residual Nagarse (heart) and fatty acids (brain) can lead to 'poor coupling'. In these cases, the addition of 1 mg/ml of bovine serum albumin usually improves the RCR. Rat kidney mitochondria require 1 mM EDTA or EGTA to produce good respiratory control. There is no hard and fast rule as to what constitutes an acceptable RCR. Values of well over 10 have been recorded with the vibrating platinum oxygen electrode (28) and it may be even higher in the intact cell. In our experience RCR values around 4 — 6 and 3 — 4 for NADH- and succinate-linked respiration, respectively, are indicative of intact and well coupled mitochondria (see refs 8 and 9 for further discussion on this topic). RCR values below 3 for both substrates should be considered to be indicative of a poor preparation.

Respiratory control can also be demonstrated by the addition of an uncoupling agent. This is shown in *Figure 4* and 'coupled' mitochondria should always undergo a marked stimulation in respiration when an uncoupler is added. This respiration rate is only limited by the respiratory chain (comparable with state 3) and respiration will continue until all the oxygen is used up. Be careful not to add excess uncoupler as this causes an inhibition of respiration.

(iv) *Measuring substrate oxidation and ADP/O ratios.* Measuring the rate of respiration with various substrates is simple (see *Figure 4* for example) and *Table 3* lists the recommended substrates. Exogenous NADH does not cross the inner mitochondrial membrane and is therefore not metabolised. It is therefore necessary to measure NADH-driven respiration by adding substrates which are fed into the respiratory chain *via* NADH-linked dehydrogenases. These substrates are transported into the mitochondrion by a complex system of carriers. These are described in detail by Nicholls (9) and are summarised in *Table 6*. It is important to realise from this table that the tricarboxylic acid carrier is absent in heart mitochondria and that malate acts as a counterion for a number of substrates. Malate is used as a 'sparker' substrate for pyruvate and fatty acid respiration as it generates oxaloacetate which condenses with acetyl-CoA to give citrate. The use of succinate as a substrate does not present any problems, except that oxaloacetate is a potent feedback inhibitor of succinate oxidation. This inhibition can be overcome by adding rotenone.

The P/O ratio describes the stoichiometric relationship betwen phosphorylation and respiration and can be measured very accurately by Warburg manometry. Alternatively, the P/O ratio is equal to the ADP/O ratio which can be calculated from the experi-

Table 6. Metabolite Uptake in Mitochondria.

Carrier	Metabolite imported into mitochondria	Metabolite exported by mitochondria
Adenine nucleotide translocator	ADP	ATP
Phosphate	Pi	OH$^-$
Dicarboxylate	malate	Pi
Tricarboxylate[a]	citrate + H$^+$	malate
α-Ketoglutarate	α-ketoglutarate	malate
Glutamate[b]	glutamate	OH$^-$
Pyruvate	pyruvate	OH$^-$
Carnitine	acylcarnitine	carnitine
Glutamine[c]	glutamine	glutamate

[a]Virtually absent in heart mitochondria.
[b]Exclusive for liver.
[c]Probably restricted to kidney.

Table 7. Classes of Mitochondrial Inhibitor.

Class	Example
Respiratory chain inhibitor	
(a) Site I	rotenone
(b) Site II	antimycin A
(c) Site III	KCN
Uncoupling agent	CCCP or DNP
ATPase/synthetase inhibitor	Oligomycin
Substrate transport	Butylmalonate (dicarboxylate transport)
Ion transport	e.g., Ruthenium red (Ca^{2+} transport)

ments used to determine the RCR. In *Figure 4* the value X is the amount of oxygen consumed with the addition of a known amount of ADP. For this particular experiment the ADP/O ratios for pyruvate/malate and succinate were 2.6 and 1.7, values which are close to the theoretical values of 3 and 2. The value for the ADP/O ratio will depend on the condition of the mitochondria and the accuracy of measurement of ADP and O_2 concentrations. In this respect, calibration of the electrode is important and this is discussed below.

(v) *The effect of inhibition on mitochondrial respiration.* There are many compounds which inhibit mitochondrial oxidative phosphorylation and the reader is referred to references 1 − 9 for a detailed discussion and bibliography. The importance of inhibition in mitochondrial studies cannot be overstated as their use has contributed greatly to the understanding of oxidative phosphorylation. Any study on a putative mitochondrial inhibitor can be greatly aided by comparing the effects with known inhibitors. In this respect mitochondrial inhibitors can be put into different classes according to their effects on mitochondrial respiration. *Table 7* lists these classes and in *Figure 5* the effects on coupled mitochondrial respiration are shown. Traces A to E can be explained as follows.

(A)   Mitochondrial respiration on an NADH-linked substrate inhibited by rotenone. Adding succinate stimulates respiration by feeding reducing equivalents into the

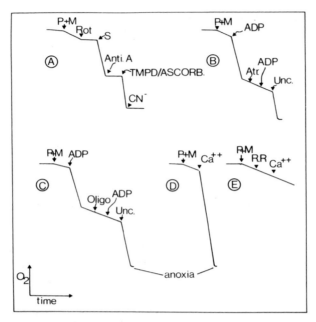

Figure 5. Schematic representation of the effects of inhibition on mitochondrial respiration. The five traces (A – E) show the effects of various inhibitors on mitochondrial respiration. The additions are: P + M = pyruvate plus malate, S = succinate, Rot = rotenone, Anti-A = antimycin A, TMPD/Ascorb = tetramethyl-*p*-phenaline diamine/ascorbate; CN⁻ = potassium cyanide; ATr = atractyloside, Unc = uncoupler, oligo = oligomycin, R.R = ruthenium red. An explanation of each trace is given in the text.

     respiratory chain beyond the inhibitor block (see *Figure 3*). Antimycin A, however, blocks the respiratory chain after succinate. Tetramethyl-*p*-phenaline diamine (TMPD)/ascorbate is an artificial electron donor feeding electrons in after antimycin A and respiration is again stimulated before being blocked by potassium cyanide.

(B)   In this trace a typical state 3 to state 4 transition is initiated, then atractyloside is added which does not affect state 4. ADP has no effect as the inhibitor blocks ADP/ATP translocation. An uncoupling agent, however, stimulates the respiration as normal. It should be noted that if an inhibitor was acting on the respiratory chain, then respiration would not be stimulated by an uncoupling agent.

(C)   A similar trace is produced with oligomycin, except that this inhibitor acts on the ATPase/synthase complex. Thus the same effect on respiration is produced by two different inhibitors and, as we shall show later, other tests are required to distinguish the site of action.

(D/E) These traces show how Ca^{2+} stimulates respiration which is blocked by ruthenium red, a specific inhibitor of Ca^{2+} transport. There are other compounds which effect ion transport, e.g., valinomycin and nigericin, and these are dealt with in the literature (2 – 9).

    The examples in *Figure 5* show how simple studies with the oxygen electrode can yield a great deal of data about a compound's effect on mitochondria. It is important to realise that many classical inhibitors are active at micromolar concentrations and

Table 8. Oxygen Content of Water and Buffered Solutions.

Temperature (°C)	Oxygen content water[a]	(ng atom O/ml) in buffer[b]
0	884	
5	772	
10	682	
15	610	575
20	582	510
25	506	474
30	460	445
35	438	410
40		380

[a]Taken from Delieu and Walker (10).
[b]Taken from Chappell (21) in which the buffer composition was 80 mM KCl, 15 mM Pi, 20 mM triethanolamine hydrochloride, 1 μm cytochrome c, pH 7.2.

care should be taken to ensure that the electrode chamber is washed thoroughly between runs. With water-insoluble compounds it is important to test the effect of the solvent itself and it may be necessary to use ethanol to wash out any traces of the compound after each run.

(vi) *Calculation and expression of data.* Qualitatively, the results from respiration studies can be presented by reproducing the oxygen electrode traces (see *Figure 4*). This method of presentation is both informative and easy to understand. When comparative studies are being carried out it is necessary to quantify the oxygen uptake of the mitochondria. The oxygen electrode measures the activity and not the concentration of oxygen in water. The activity coefficient of oxygen in solution can be calculated if the ionic strength is known. It is however difficult in complex media to determine the ionic strength and also oxygen solubility can vary with temperature and buffer composition (*Table 8*). It is therefore necessary to determine the oxygen content directly and a variety of methods are available. A convenient and reliable method has been used by Estabrook (29) and Chappell (21). Submitochondrial particles or lysed mitochondria are incubated in buffer in the oxygen electrode. As these preparations are freely permeable to NADH and have no endogenous respiration (unlike mitochondria) a straight horizontal trace is recorded. Limiting amounts of NADH added to the electrode produce bursts of respiration which are seen as a step-like recorder trace. Each step is proportional to the amount of oxygen used which is stoichiometrically (1:1) related to the NADH added. The latter is standardised spectrophotometrically (i.e., ϵ_{340nm} 6.22 × 10^3 litre/mol/cm).

In the example shown in *Figure 4* the oxygen concentration of the respiration buffer is 420 ngatom O/ml and the total oxygen content of the solution is 840 ngatom O. The pen recorder was adjusted to give 95 divisions from the N_2 to air line. Thus one vertical division is equivalent to 8.8 ngatom O. From this it follows that the respiration rate (ngatom O/mg of protein/min) is given by the deflection of the pen (i.e., number of chart divisions) in 1 min multiplied by 8.8 and divided by the protein concentration. For example, in *Figure 4* the state 4 respiration rate with succinate is:

$$\text{respiration rate} = \frac{10.5 \times 8.8}{4}$$

$$= 23 \text{ ngatom O/mg/min}$$

For pyruvate and malate the rate was 10 ngatom O/mg/min and the state 3 rates for succinate and pyruvate and malate were 94 and 51 ngatom O/mg/min, respectively. The main error in these calculations is determining the number of divisions moved in 1 min and the best method is to draw a line tangential to the trace (see *Figure 4*) and take the reading over several minutes (i.e., 2 or 3 cm of chart paper).

For ADP/O ratio determinations the amount of oxygen consumed (X) is determined from the trace as shown. The ADP concentration can be standardised spectro-photometrically (ϵ_{259nm} 15.4 litre/mmol/cm).

3.2 Respiration Studies with Submitochondrial Particles

Submitochondrial particles do not have coupled respiration, soluble tricarboxylic acid (TCA) cycle enzymes or endogenous substrates, and are freely permeable to NADH. Consequently, respiration in submitochondrial particles is only limited by the dehydrogenase activity of the respiration chain. Respiration studies with submitochondrial particles are useful for studying the effects of respiratory chain inhibitors.

3.2.1 *Method*

The methods for measuring respiration in submitochondrial particles are identical to those described for mitochondria except for the following.

(i) NADH and succinate are used as substrates.

(ii) Respiration is not stimulated by uncoupling agents or ADP and is not inhibited by ATP synthetase inhibitors as they possess little or no respiratory control.

(iii) As the respiration is only limited by the respiratory chain the rate of oxygen uptake is faster and much less protein is needed in the electrode (i.e., ~1 mg).

3.3 Measuring ATP Synthesis

An alternative method to the oxygen electrode for assaying oxidative phosphorylation is to measure the amount of ATP synthesised. This can be done with a glucose/hexo-kinase trap system (30) in which ADP is continually regenerated from ATP and thus the mitochondria are maintained in state 3. ATP formation is determined by the disappearance of Pi (31). This method is simple and does not require specialised equipment. In addition, it is possible to carry out many assays whilst the mitochondria are still well coupled (i.e., 2 − 3 h after preparation). This is particularly useful for carrying out inhibition studies, as the effect of a wide range of inhibitor concentrations can be assayed simultaneously. One other advantage is that ATP synthesis, unlike coupled oxygen consumption, is less affected by the quality of mitochondria. It is therefore possible to get normal rates of ATP synthesis from preparations which show poor respiratory control as measured by the oxygen electrode. For example, kidney mitochondria need EDTA or EGTA to exhibit good RCR values. However, oxidative phosphorylation as measured by ATP synthesis is much less dependent on the presence of chelating agents (K.Cain, unpublished results).

3.3.1 *Buffers and Reagents*

(i) ATP synthesis buffer; 0.25 M sucrose, 22 mM glucose, 5 mM K_2PO_4, 2 mM

MgCl$_2$, 2 mM ADP, 1 mM EDTA (optional), 10 mM Tris-HCl, pH 7.5. Store frozen in 15 ml aliquots.

(ii)   Hexokinase (Sigma Type V) made up in 0.25 M sucrose, 10 mM Tris-HCl, pH 7.5 to give 20 units (i.e., EC units − 1 μmol substrate/min) per 20 μl.

(iii)   Phosphate assay; 10% TCA; stock Pi solution (3 mM KH$_2$PO$_4$ in 0.1 N H$_2$SO$_4$); ANSA reagent 10.5 g aminonaphtholsulphonic acid, 30 g NaHSO$_3$.7H$_2$O in 250 ml. Store cold and dark.

3.3.2 Method

In this assay it is essential to aerate the reaction mixture throughout the procedure and this is achieved with a shaking water bath (90 cycles/min). Small conical flasks or scintillation vials can be used as reaction vessels, or alternatively we recommend disposable, flat-bottomed plastic tubes. Sarstedt Ltd. (68 Boston Road, Beaumont Leys, Leicester LE41 1AW, UK) market a particularly suitable product (Cat. No. 58487) which has a 12 ml volume, 140 mm × 23 mm. With assay volumes of 1−2 ml there is optimal aeration. The method is as follows.

(i)   Set up the assay tubes as shown in *Table 9* and incubate with shaking for 5 min at 30°C. Add the substrate at 15 sec intervals (use the same substrates described for respiration studies) and continue the incubation.

(ii)   Stop the reaction after 10 min by transferring 200 μl of each assay tube to an Eppendorf 1.5 ml conical centrifuge tube containing 200 μl of 10% TCA solution. Whirlmix and centrifuge at maximum speed for 5 min in a bench top centrifuge.

(iii)   Transfer 100 μl of the supernatant to a test tube, add 3 ml of water, 0.3 ml ammonium molybdate and whirlmix. Add 0.2 ml of ANSA reagent and whirlmix again. Read the blue colour at 750 nm after 10 min. Determine the Pi content from a standard curve (0−0.8 μmol Pi).

The difference in Pi content between the control or test assays and the blank tubes represents the amount of Pi transformed into ATP and the results are normally given as nmol ATP/mg of protein/min. Typical values for pyruvate plus malate and succinate for liver and kidney mitochondria are 100 and 300 nmol/mg of protein/min, respectively. Under these conditions 1−2 mg of mitochondria will give a linear rate of syn-

Table 9. Protocol for ATP Synthesis Assay.

Assay description	Buffer	Hexokinase	Mitochondria	Inhibitors, etc.	H$_2$O	Substrate
Blank	1 ml	20 μl	50 μl	−	30 μl	−
Blank	1 ml	20 μl	50 μl	−	30 μl	−
Control	1 ml	20 μl	50 μl	−	20 μl	10 μl
Control	1 ml	20 μl	50 μl	−	20 μl	10 μl
Test	1 ml	20 μl	50 μl	10 μl[a]	10 μl	10 μl
1−8 assays	1 ml					

[a]Make up to 10 μl with water.

thesis over a 10-min period. The linearity of the reaction should always be verified by doing a time course experiment. ATP synthesis is blocked by uncouplers, ADP/ATP translocase inhibitors, ATPase/synthetase and respiratory chain inhibitors.

3.4 Measuring ATP Hydrolysis (ATPase)

The end product of oxidative phosphorylation is the condensation of ADP and Pi to ATP which is catalysed by the ATPase/ATP synthetase complex. This is a multi-subunit enzyme which is integral to the inner mitochondrial membrane. The functioning of this enzyme and its importance in the mechanism of oxidative phosphorylation have been studied in great detail (see references 9 and 32 for reviews). In 'coupled' mitochondria the energy made available from respiratory chain oxidation is used to drive ATP synthesis. The ATPase/synthetase complex is thus acting predominantly as an ATP synthetase enzyme. In preparations in which there is poor respiratory control the ATPase reaction is energetically more favourable. The ATPase activity of a mitochondrial preparation is therefore a very useful index of respiratory control. A variety of inhibitors interact with the ATPase complex. Thus, uncoupling agents stimulate ATPase activity, whereas oligomycin is a potent inhibitor. Atractyloside has no effect on ATPase activity (compare this with the similar effect that this compound and oligomycin have on state 3 respiration, *Figure 5*).

3.4.1 *Buffers*

(i)   Buffer A (mitochondrial ATPase buffer); 0.125 M sucrose, 2 mM $MgCl_2$, 60 mM Tris-HCl, pH 7.4.
(ii)   Buffer B (submitochondrial particles ATPase buffer); 2 mM $MgCl_2$, 50 mM Tris-HCl, pH 8.5.
(iii)   ATP solution; 0.1 M, pH to neutrality with NaOH and store frozen in aliquots.
(iv)   Pi reagents are as described in Section 3.3.1.

3.4.2 *Method*

The method used for both submitochondrial particles and mitochondria is as essentially described by Cain *et al.* (30). It should be remembered that Pi contamination can be a problem (many detergents used in glass washing contain Pi) and we therefore recommend the use of disposable plastic tubes for all assays. The method for both mitochondria and submitochondrial particles is the same, apart from the choice of buffer.

(i)   Prepare the assay tubes as described in *Table 10*. A convenient protocol is to do 12 assays at a time, eight of which can be test assays (an inhibitor dose-response, for example). Pre-incubate at 30°C for 10 min, then add 50 μl of ATP solution and whirlmix.
(ii)   Incubate for a further 5 min, then transfer 500 μl to a conical centrifuge tube containing an equal volume of 10% TCA. Mix and centrifuge at maximum speed for 5 min in a bench centrifuge to pellet the protein.
(iii)   Remove 500 μl of the supernatant to a test tube containing 2.5 ml of water, 0.3 ml of ammonium molybdate, mix and add 0.2 ml of ANSA reagent. Read at 750 nm after 10 min and determine the Pi content from the standard curve.

Table 10. Protocol for ATPase Assay.

Assay description	Buffer[a]	Protein[b]	Additions[c] (inhibitors)	H_2O	ATP
Blank	0.9 ml	–	–	50 μl	50 μl
Blank	0.9 ml	–	–	50 μl	50 μl
Control	0.9 ml	10 μl	–	40 μl	50 μl
Control	0.9 ml	10 μl	–	40 μl	50 μl
Test assay (1 – 8)	0.9 ml	10 μl	10 μl	30 μl	50 μl

[a]For mitochondria it is important to use buffer A which is isotonic.
[b]For both mitochondria and submitochondrial particles the protein concentration is adjusted to 10 mg/ml, giving 100 μg/assay.
[c]For inhibitors and uncouplers see references 2, 3 and 5.

(iv)   Results are corrected for the blank value and expressed as μmol Pi/mg of protein/min. The reaction is normally linear for 5 min, providing no more than 2 μmol Pi (total 5 μmol Pi/assay) are produced.

(v)   Typical values for coupled rat liver mitochondria are 5 – 10 nmol Pi/mg/min which can be stimulated 20- to 40-fold by the action of uncoupling agents (33,34). Submitochondrial particles are more active; those of ox heart, for example, have specific activities around 2 – 3 μmol Pi/mg/min (30) and are not stimulated by uncoupling agents. Inhibitors like oligomycin will inhibit the ATPase activity in both mitochondria and submitochondrial particles and this inhibition is not relieved by uncoupling agents (compare with the effects on respiration). Also ATP/ADP translocator inhibitors like atractyloside will inhibit the ATPase activity in intact mitochondria but not submitochondrial particles (in contrast with oligomycin) and this is an important difference between this type of inhibitor and an ATPase inhibitor.

3.5 Spectrophotometry of the Respiratory Chain

Many oxidative phosphorylation inhibitors act directly on the respiratory chain and respiration studies can be used to show this inhibition (see *Figure 3*). An alternative and complementary method for looking at these effects is to make use of the spectral properties of the cytochromes and flavoproteins of the respiratory chain which can be readily determined in a split beam spectrophotometer, the principles of which are shown in *Figure 6a*. This type of spectrophotometry effectively overcomes the difficulties of light scattering and non-specific absorption which are experienced with turbid samples. Usually, the mitochondrial suspension in the reference cuvette is oxidised and the mitochondria in the sample cuvette are reduced. A typical reduced-oxidised (difference) spectrum is shown in *Figure 6b* which shows how the a, b and c cytochromes can be visualised. Each cytochrome exhibits three absorption maxima on reduction, known as the α, β and γ (Soret) bands. On oxidation, only a reduced in size Soret band remains which, (depending on the cytochrome), has an absorption maximum between 400 and 460 nm. The X bands absorb between 550 and 610 nm and with difference spectra it is the latter region which is the most useful. Reducing the respiratory chain with substrate produces a spectrum similar to that of *Figure 6b*. When antimycin A

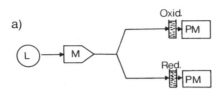

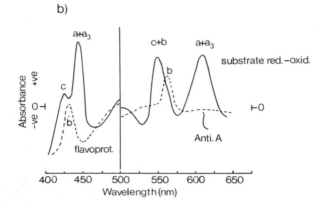

Figure 6. Reduced-oxidised difference cytochrome spectra of the respiratory chain. In (**a**) the principles of a split beam spectrophotometer are shown where L is the light source, M the monochromator and P the photomultiplier tube. Oxid. refers to the air-oxidised mitochondrial sample and red. to the substrate- or dithionite-reduced sample. The red.-oxid. signal is plotted on a recorder and in (**b**) a typical spectrum [redrawn from Chance and Williams (27)] is shown. In this succinate red.-oxid. spectrum, the cytochromes (a+a$_3$, b and c) are clearly seen. The effect of antimycin A demonstrates the cross-over point technique (see text).

is added only cytochrome b is visualised. This inhibitor acts between cytochromes c and b and, as a result, the respiratory chain components after the inhibitor block are oxidised whereas that part of the chain prior to the point of inhibition remains reduced. This is an example of a 'cross-over point' which is a useful way of determining the sites of inhibition (27,8).

3.5.1 *Buffer*

0.25 M sucrose, 10 mM Tris-HCl, pH 7.4.

Substrates and inhibitors are as previously described in respiration studies.

3.5.2 *Method*

Mitochondria or submitochondrial particles can be used, the latter are perhaps better because there are no permeability problems. Any split beam spectrophotometer can be used provided it has a turbid sample holder (this is basically a pair of cuvette holders which are situated close to the photomultipliers to minimise the effects of light scattering).

(i)   Dilute the mitochondrial preparation with buffer to give an approximately 2 mg/ml suspension. Pipette 3 ml aliquots of this suspension into two cuvettes and record the difference spectrum from 400 to 650 nm with a full scale deflection of $0 - 0.2$ absorbance units and a scan speed of 0.5 or 1 nm/sec. This gives an essentially straight line and represents the oxidised-oxidised spectrum.

(ii) Rewind the chart paper and record a reduced-oxidised spectrum by adding dithionite (a few crystals) to the sample cuvette and oxygenating the reference suspension (shaking the stoppered cuvette will suffice).

(iii) Record a substrate reduced-oxidised spectrum in the same way by replacing the dithionite with an appropriate substrate.

(iv) The effects of inhibition are easily determined by running the substrate reduced spectrum and rewinding the chart paper before running a spectrum on a sample containing the inhibitor and substrate.

3.6 Methods of Measuring Metabolite Transport

The mitochondrial inner membrane acts as a barrier to most metabolites. Substrates, ADP/ATP and various ions have to be transported across the mitochondrial membrane. These transport mechanisms allow the mitochondria to maintain a pH and membrane potential gradient which is essential for the maintenance of oxidative phosphorylation and also to regulate the transfer of substrates and ADP/ATP to the cytosol. Various methods are available to measure transport; for example, osmotic swelling often results from ion transport and this can be measured by a decrease in light scattering. This technique can be used to look at substrate and ion transport [see Chappell (35), for example]. Ideally, however, metabolite transport is best measured directly and this normally requires a radioactive metabolite and a method for rapidly separating mitochondria from the incubating medium. Two of the more widely used methods are filtration and silicone oil layer centrifugation, the procedures for which are outlined below.

3.6.1 *Millipore Filtration*

(i) The method described herein is based on the methodology of Winkler *et al.* (36) and was used for measuring ADP/ATP translocation in yeast mitochondria (20), although the technique is applicable to all types of mitochondria. A Millipore filter (pore size 0.45 μm, 25 mm diameter) is used to separate the mitochondria from the medium. The key to the success of this method is the speed of filtration which should separate mitochondria from medium in under 2 sec. The usual commercial filter holder is inserted into a small Buchner flask which is evacuated on a water pump. In our experience this is normally too slow to filter mitochondria rapidly. However, a simple multi-sample system can be constructed very cheaply. It consists of 10 filter holders and Buchner flasks which are connected to a glass manifold. Each arm of the manifold has a separate gas tap and the manifold itself is connected to a 2-l Buchner flask which is evacuated by an oil-driven high vacuum pump. This effectively produces a large negative pressure reservoir which can be instantly applied to the appropriate filter holder when the top is opened.

(ii) *Buffers and reagents* (ADP/ATP translocation assays). Incubation buffer; 0.25 M sucrose, 10 mM Hepes-KOH, pH 7.0, 1 mM EGTA.

ADP solution; 15 μM ADP made up in incubation buffer and spiked with ^{14}C-labelled ADP to give a specific activity of 6000 d.p.m./nmol ADP.

(iii)   *Method.* ADP transport at physiological temperatures is so rapid that it is very difficult to measure initial rates of transport. Therefore, ADP translocase assays are normally carried out at temperatures below 10°C. The method is as follows.
(1) Set up tubes containing 1 ml of buffer and 0.5 − 1 mg of mitochondria and pre-incubate at 5°C for 10 min.
(2) Initiate the reaction by adding 0.5 ml of [^{14}C]ADP solution (5°C) and rapidly mixing.
(3) Terminate the assay after 30 sec by filtering a 1-ml aliquot. Wash the filter and mitochondria with 5 ml of ice-cold buffer. Remove and dry the filter in a scintillation vial before adding 10 ml of scintillant and counting in the usual manner.

Good rates of ADP uptake require intact coupled mitochondria and all assays should be carried out within 3 h of the preparation of the mitochondria. In rat liver mitochondria the rate of ADP transport at low temperatures should be around 3 − 6 nmol/mg/min (37) and be linear over the 30-sec assay period. ADP transport is potently inhibited by atractyloside, carboxyatractyloside and bongkrekic acid [see Stubbs in ref. 2 for review and also Vignais (37)]. Note that the latter inhibitor is a penetrant inhibitor and sufficient incubation time with the compound (20 min at 5°C) is necessary to give complete inhibition. Actractyloside competes for the ADP binding site on the outside of the inner membrane and acts very rapidly, a feature which has been exploited in the inhibitor stop method for translocase assays (38). Uncoupling agents and oligomycin do not affect ADP uptake (38).

3.6.2 *Silicone Oil Centrifugation Technique*

This technique involves the centrifugation of mitochondria through a silicone oil layer which removes water, into a quenching layer of $HClO_4$. For optimal results a centrifuge and rotor with a low inertia are used. The Eppendorf bench centrifuge is ideal for this purpose. The standard rotor has 12 positions and can accelerate to 10 000 g in 5 sec. Adaptor tubes are used which take 0.4-ml microcentrifuge tubes (approximate size 46 mm × 6 mm). The method is based on the Harris and Van Dam technique (39) as described by Skilleter (40). The sedimentation of the mitochondria through the silicone invariably drags through some water which, together with fluid contained in the intermembrane space (i.e., the space between the inner and outer mitochondrial membranes), requires a correction factor to be applied. This is achieved by the use of parallel experiments with U-^{14}C-labelled sucrose which measures the sucrose-permeable space (i.e., carry over + intermembrane space radioactivity) and ^3H_2O measurements from which the matrix volume (i.e., sucrose impermeable space) can be measured (see calculation below).

(i) *Buffers and chemicals.* Incubation buffer; this will vary according to the type of uptake experiment. In this example [1-^{14}C]pyruvate uptake is measured (40) and the buffer is as follows: 100 mM KCl, 14 mM $Mg(NO_3)_2$, 1 mM EDTA, 3 mM ATP, 16.7 mM glycylglycine, 5 mM [1-^{14}C]pyruvate (0.0667 μCi/μmol), 1 mM fumarate, pH 6.8.

Other reagents, 1.5 M $HClO_4$, silicone oil [General Electric 1017 B158, density 1.058 for sucrose media and Versilube (R) F50, density 1.038 for KCl medium], [U-^{14}C]sucrose (25 μCi/ml) and ^3H_2O (0.5 mCi/ml).

(ii) *Method.*

(a) Prepare the centrifuge tubes with 40 μl of $HClO_4$ and two drops of silicone oil carefully pipetted on top to give a 3-mm thick layer. It is important to avoid air bubbles when preparing the tubes.

(b) For the assay of pyruvate uptake, add 0.15 ml (3 mg of protein) of mitochondria to 2.85 ml of buffer and incubate with shaking for 5 min at 37°C. Take 0.2 ml replicates and pipette carefully on top of the silicone layer and centrifuge at maximum speed for 2 min. Remove a 20-μl aliquot from the top layer before aspirating the remaining aqueous medium and silicone layer. A 20-μl aliquot is then taken from the $HClO_4$ layer.

(c) For the determination of the sucrose-permeable and impermeable spaces, the same conditions are employed except that unlabelled pyruvate is used and the 3-ml reaction medium is spiked with 30 μl of [U-^{14}C]sucrose and 15 μl ^3H_2O.

(4) the 20-μl aliquots are assayed for radioactivity in a scintillation counter.

(iii) *Calculation of results.*

(a) ^3H_2O *space (i.e., carry over d.p.m. + total mitochondrial d.p.m.)*

Let $x = {}^3H_2O$ space (μl), $b = $ d.p.m./μl of $HClO_4$ and $t = $ d.p.m./μl of upper suspending medium.

Then $x.t$ (total counts carried out) $= b(40 + x)$

$$\text{Rearranged } x = \frac{40b}{t-b}, \text{ or } x = \frac{40a}{1-a}$$

where $a = b/t$, i.e., the ratio of d.p.m. in the bottom layer/d.p.m. in top layer.

(b) [^{3}U-^{14}C]*sucrose space (i.e., carry over d.p.m. + intermembrane d.p.m.)*

Let $y = $ [U-^{14}C]sucrose space (μl), b and t as above (except that it is the ^{14}C d.p.m. values which are used).

Then yt (total d.p.m. carried into $HClO_4$) $= b(40 + x)$ where x is ^3H_2O space.

Therefore $y = b/t$ $(40 + x)$

$\qquad\qquad = a(40 + x)$

The sucrose-impermeable (matrix) space is given by $x-y$.

(c) [^{14}C]*Pyruvate uptake*

To calculate pyruvate uptake, let $y = $ sucrose space, $x = {}^3H_2O$, $t = $ d.p.m./μl in top layer and $b = $ d.p.m./μl in bottom layer.

Total d.p.m. in $HClO_4$ $= (40 + x)b$

Total sucrose-permeable d.p.m. $= yt$

$\therefore$ pyruvate uptake $= (40 + x)b - yt$

This value can then be expressed as nmol/mg/protein or alternatively it can be converted to a concentration value using the matrix space volume. Typically, in rat liver mitochondria the sucrose-permeable space is about 4 μl/mg of protein and the matrix value about 0.4 μl/mg of protein. These values can change with the conditions of assay (40) and it is important to determine the sucrose-permeable and sucrose-impermeable spaces for each mitochondrial preparation. It is also worth pointing out that the resolution of the method is such that with very fast rates of uptake, it is steady-state con-

Table 11. Guidelines for Assessing Mitochondrial Dysfunction.

Mitochondrial activity	Possible finding	Conclusion	Suggested course of action
Respiratory control	Low RCR (<3) and poor response to uncoupler	(a) Poor preparation or assay problems	Check preparation and assays with standard mitochondrial preparation (e.g., rat liver mitochondria)
		(b) If mitochondria from treated animals then toxin is probably an inhibitor	As above but check toxin in vitro assays, e.g., ATPase activity
	Respond to uncoupler but not ADP	(a) ADP solution suspect	Check out assay
		(b) If from treated animals suspect inhibitor is an oligomycin-like compound	Assay ATPase activity in mitochondria and submitochondrial particles
ATPase activity	High activity and not stimulated by uncouplers	(a) Poor preparation mitochondrial preparation	Check procedures with standard
		(b) If from treated animal, suspect toxin is an uncoupler	Do appropriate in vitro experiments
NADH oxidase	Respire rapidly when NADH is added	(a) Poor preparation	Check procedures with standard mitochondrial preparation
		(b) If from treated animal, it is possible that toxin is making the mitochondria more fragile	As above and if possible obtain electron microscopic evidence in vivo
Respiration on TCA cycle intermediates	Mitochondria do not respire on specific substrates	(a) Mitochondria do not have carrier transporter or incorrect combination of substrates	Check assays with standard mitochondrial preparation
		(b) If from treated mitochondria, suspect compound is an inhibitor of substrate transport or blocks the respiratory chain at an appropriate point	Do appropriate in vitro experiments
	Mitochondria do not respire	(a) Poor preparation or assay problems	Check procedures with standard mitochondrial preparation
		(b) If from treated animals, suspect toxin is a respiratory chain inhibitor	Do appropriate in vitro experiments

ditions which are being measured rather than initial velocities. This, however, does not preclude using the method for looking at the possible effects of an inhibitor on substrate transport.

3.7 Guidelines for Assessing Mitochondrial Dysfunction

The conclusion that a toxin is interacting with mitochondria to produce the observed pathological effects is usually based on a combination of '*in vivo*' and '*in vitro* ' experiments. The latter experiments examine the effects of the compound when it is added to isolated mitochondria whilst '*in vivo*' experiments involve pre-treating the animals with the toxin and then isolating mitochondria from an appropriate tissue and comparing its metabolic reactions with those of control mitochondria. In both types of experiment it is necessary to confirm that any differences from normal are not artefactual (e.g., brought about by poor preparation technique). *Table 11* describes some possible problems and how they can be overcome.

4. MITOCHONDRIA AS POTENTIAL TARGETS OF TOXICITY

As stated in the Introduction, the main objectives of this chapter are to describe methods of preparing and assaying mitochondria. Implicit in the decision to use such techniques is the belief that a particular compound exerts its toxicological effect by a primary perturbation of mitochondrial oxidative phosphorylation. It is therefore necessary to consider what sort of experimental evidence and/or approach would point to a mitochondrial involvement. This is a difficult problem and there is no easy answer. In this section we present data for some examples along with guidelines which should enable a productive approach to the problem. *Figure 7* illustrates how some of the *in vitro* tests described in the previous section may be used to characterise a toxin as a mitochondrial inhibitor and thus implicate mitochondria as a major target of toxicity.

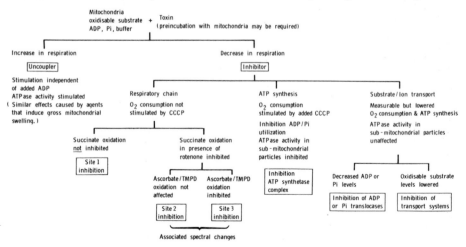

Figure 7. Possible effects of a toxin on *in vitro* mitochondrial reactions.

The end results of any toxic insult are the observed symptoms of toxicity. These may result from acute or chronic exposure and in both cases the route of entry and subsequent distribution in the body are very important as they determine which organs will receive a toxic dose. If these organs are large and rely heavily on oxidative phosphorylation for their energy supply then the overall effect on the whole animal's metabolism may be extreme and easily noticeable. Two good examples of this are uncoupling agents and the plant toxin atractyloside. In the example of uncoupling agents, the classical compound dinitrophenol was used as a slimming agent (41) and a related compound dinitrocresol (DNC) as a slimming agent and pesticide (42). In the use of both compounds, severe side effects including death were recorded and resulted in the withdrawal of use. Bidstrup and Payne (42), for example, described eight case histories in which all subjects died. The symptoms included hyperpyrexia, tachycardia, excessive sweating, increased heart rate, dehydration and increased metabolic rate and respiration. These are all consistent with the metabolic effects of uncoupling oxidative phosphorylation. Thus respiration (i.e., oxygen consumption) is stimulated and the energy of oxidation is liberated as heat rather than conserved as ATP. The body responds with mechanisms aimed at increasing heat loss (e.g., sweating) and is also less able to respond to increases in environmental temperature. Atractyloside and carboxyatractyloside are potent inhibitors of the ADP/ATP translocase and both compounds are natural toxic ingredients in the thistle *Atractylis gummifera*. This plant grows in the countries bounding the Mediterranean and has been known for centuries as an extremely poisonous plant (see 43 for historical background). The accidental poisoning in 1955 of a class of Italian schoolchildren stimulated research into its toxicity and eventually led to the discovery that the compound was a translocase inhibitor [see Stubbs for review (2)]. In whole animals, atractyloside is extremely toxic and Lucianni *et al.* (43) have described in detail the '*in vivo*' responses to these compounds. In mice, rats, rabbits and dogs atractyloside produces hyperglycaemia, followed by hypoglycaemia, convulsions, acidosis and decreased oxygen consumption. Significantly, atractyloside is more toxic at 0°C than at 25°C and this is in contrast with uncoupling agents which are more toxic at higher temperatures. The mitochondria are the major site of heat production in the cell (44) and the enhanced toxicity of atractyloside at low temperatures is not unexpected. Generally, marked changes in body temperature, respiration and blood sugar can be used as preliminary indications of an effect on mitochondria. However, it should be pointed out that such effects are likely to occur only in the acute, high dose situation and in the case of chronic low level exposure such effects may not be obvious. Furthermore, the control of body temperature and the process of thermogenesis are very complicated mechanisms (44,45) and an effect on body temperature may be misleading. For example, tri-butyl S,S,S-phosphorotrithioate (DEF), an organophosphorous defoliant, abolishes the thermogenic response to a drop in body temperature (46). As brown adipose mitochondria are a major source of heat production in cold-induced thermogenesis it is not inconceivable that DEF has a primary action on these mitochondria. However, this has not been substantiated (K.Cain and V.J.Cunningham, unpublished results) and it is more likely that the effect of DEF is centrally mediated rather than a direct action on mitochondria (47).

From the above discussion it is clear that data from whole animal studies can be used only as an indication of mitochondrial involvement in the toxic process. It is therefore necessary to determine the distribution and pathological effects of the toxin within the body and to obtain histological data on the affected tissues. Although light microscopy has not got the magnification to identify aberrant mitochondria, it can locate those cells that are affected. This information can be of considerable use as many organs (e.g., liver and kidney) are composed of heterogeneous populations of cells which have morphological and metabolic differences. In the liver, for example, gluconeogenesis is restricted to parenchymal cells, whilst the non-parenchymal cells have a low complement of mitochondria and are rich in glycolytic enzymes (48), indicating that glycolysis must play a major role in meeting the energy requirements of these cells. In this respect it is interesting that Cd^{2+} is a potent oxidative phosphorylation inhibitor (49) and a hepatotoxin which is preferentially accumulated in the parenchymal cells of the liver (50). This results in a characteristic mid-zonal necrosis. Significantly, oxidative phosphorylation in mitochondria isolated from the livers of Cd^{2+}-injected animals is depressed and this inhibition (at early times after injection) can be reversed by chelating agents (51). Another good example of the pathological symptoms being indicative of mitochondrial inhibition is the fatty liver syndrome produced by the anti-convulsant drug, valproate. As discussed by Turnbull (52) there is good evidence from *in vivo* and *in vitro* studies that this compound is a potent inhibitor of fatty acid oxidation and this explains the build up of fatty acids in the liver. A similar effect is produced by 2-mercaptoacetate which is an inhibitor of 3-hydroxybutyrate and long chain fatty acyl-CoA dehydrogenases (53).

The kidney also exhibits marked differences in morphology and metabolism (1). The cortex accounts for 70% of the kidney by weight and contains the proximal convoluted tubules in which many of the energy-requiring transport/absorption processes are located (1). As a result, this region of the kidney has the highest oxygen consumption. The heavy metals Cd^{2+}, Hg^{2+} and Pb^{2+} are potent nephrotoxins producing distinctive histological lesions in the proximal tubular cells. All three metals are also potent inhibitors of oxidative phosphorylation and there is strong evidence that these mitochondrial effects are important factors in producing the toxic symptoms (54).

Any damage which can be observed by light microscopy usually means that the toxic effect is severe. Under such conditions it may be difficult to decide whether or not the mitochondrial effects are the cause of toxicity or are secondary to the death of the cell. Ideally, mitochondrial damage should be observed prior to the onset of clinical symptoms. In this situation, electron microscopy can be very useful as mitochondria have distinctive conformations which alter according to their metabolic state (55). Innumerable studies have shown that mitochondrial changes are often very early indications of cell injury. However, some caution must be observed in interpreting the data as mitochondria can undergo reversible changes in conformation which may reflect changes in the osmolality of the cell rather than a direct mitochondrial inhibition.

Histopathological techniques can therefore provide tentative evidence that mitochondrial oxidative phosphorylation is being inhibited. Additional evidence could also be derived from isolated and perfused organ techniques. The tachycardia, for example, produced by uncouplers in whole animals can be demonstrated in the perfused heart preparation (56). This cardiac depressant activity has been shown to be correlated with uncoupler potency (57). Isolated cell cultures can also be used to look at the metabolic effects of a mitochondrial inhibitor. The effects of valproate on fatty acid oxidation, for example, have been shown in the whole animal and isolated rat hepatocytes (52).

Ultimately, however, it is necessary to investigate the effect of a compound on isolated mitochondria and demonstrate that this inhibition is produced *in vivo*. In order to prove the latter is is important to provide evidence that the toxin actually reaches the mitochondria *in vivo*. One simple approach is to administer a radiolabelled compound and then to carry out fractionation studies to determine the subcellular distribution of the toxin in target tissues taken from the treated animals. This methodology requires that the toxic compound is bound very tightly to mitochondria and there is no significant redistribution of the compound during the cell fractionation procedures. In addition, it is important to determine that the radiolabel bound to the mitochondria is the original compound, rather than a metabolite. If the latter is the case, then it may be necessary to extract and then identify the compound. The metabolite must then also be checked out as a possible mitochondrial inhibitor. The redistribution problem is harder to overcome, although simple binding studies to mitochondria *in vitro* can determine whether or not the compound will stay bound to the mitochondria in the presence of other subcellular organelles and/or during the various centrifugation procedures employed in fractionation studies. It is also worth remembering that the subcellular distribution of a compound shortly after an acute administration may not be the same as that produced by chronic administration. A good example of this is the subcellular distribution of Cd^{2+} which changes very dramatically when metallothionein synthesis is induced (58). Also, the fact that the mitochondria from treated animals contain appreciable quantities of the toxin does not mean that these mitochondria are inhibited. It is essential to test that the various oxidative phosphorylation reactions are effected and that the findings are analogous to results achieved with parallel *in vitro* experiments. For example, kidney mitochondria isolated from rats fed on 100 p.p.m. Cd^{2+} have normal oxidative phosphorylation with pyruvate plus malate and succinate (K.Cain, unpublished results). However, the mitochondria contain 2 ngion of Cd^{2+}/mg of protein which, on the basis of *in vitro* studies, should abolish pyruvate plus malate and markedly depress succinate-driven phosphorylation. Clearly, in the mitochondria from treated animals the Cd^{2+} is bound to a site which is not inhibitory.

In conclusion, the decision that an inhibition of mitochondrial oxidative phosphorylation is a primary event in a toxic syndrome can only be made after a careful assessment of all the available data. This is even more difficult in a chronic situation where the inhibition of mitochondrial function may not be total. In this respect the overall consequences of a reduced supply of energy on the normal functioning of the cell are unknown. It is worth remembering that not only may a toxic compound have several effects on mitochondria *per se* but also it is unlikely that a compound will be specific for mitochondria and thus the cell death produced may be the summation of a variety of biochemical effects.

5. ACKNOWLEDGEMENTS

We thank Dr. J.E.Cremer for helpful advice on the preparation of brain mitochondria.

6. REFERENCES

1. Cohen,J.J. and Kamm,D.E. (1981) in *The Kidney,* 2nd Edition, Brenner,B.M. and Rector,F.C. (eds.), Saunders,Philadelphia, p. 144.

2. Erecinska,M. and Wilson,D.F., (eds.) (1981) *Inhibitors of Mitochondrial Function, International Encyclopedia of Pharmacology and Therapeutics*, published by Pergamon Press, Oxford, UK.
3. Linnett,P.E. and Beechey,R.B. (1979) in *Methods in Enzymology*, Vol. **55**, Fleischer,S. and Packer,L. (eds.), Academic Press Inc., London and New York, p. 472.
4. Reed,P.W. (1979) in *Methods in Enzymology*, Vol. **55**, Fleischer,S. and Packer,L. (eds.), Academic Press Inc., London and New York, p. 435.
5. Heytler,P.G. (1979) in *Methods in Enzymology*, Vol. **55**, Fleischer,S. and Packer,L. (eds.), Academic Press Inc., London and New York, p. 463.
6. Singer,T.P. (1979) in *Methods in Enzymology*, Vol. **55**, Fleischer,S. and Packer,L. (eds.), Academic Press Inc., London and New York, p. 455.
7. Lehninger,A.L. (1977) *Biochemistry*, published by Worth Publishers Inc., New York.
8. Jones,C.W. (1981) *Biological Energy Conservation Oxidative Phosphorylation, Outline Studies in Biology*, published by Chapman and Hall, London and New York.
9. Nicholls,D.G. (1982) *Bioenergetics, An Introduction to the Chemiosmotic Theory*, published by Academic Press, London, New York.
10. Delieu,T. and Walker,D.A. (1972) *New Phytol.*, **71**, 201.
11. Dawson,A.P. and Selwyn,M.J. (1974) *Biochem. J.*, **138**, 349.
12. Johnson,D. and Lardy,H. (1967) in *Methods in Enzymology*, Vol. **10**, Estabrook,R.W. and Pullman,M.E. (eds.), Academic Press Inc., London and New York, p. 95.
13. Maunsbach,A.B. (1964) *Nature*, **202**, 1131.
14. Beaufay,H. (1969) in *Lysosomes in Biology and Pathology*, Vol. **2**, Dingle,J.T. and Fell,H.B. (eds.), North Holland Publishing Co., Amsterdam, p. 515.
15. Clark,J.B. and Nicklas,W.J. (1970) *J. Biol. Chem.*, **245**, 4725.
16. Chappell,J.B. and Hansford,R.G. (1972) in *Subcellular Components*, Bernie,G.D. (ed.), Butterworth, London, p. 77.
17. Low,H. and Vallin,I. (1963) *Biochim. Biophys. Acta*, **69**, 361.
18. Lowry,O.H., Rosebrough,N.J., Farr,A.L. and Randall,R.J. (1951) *J. Biol. Chem.*, **193**, 265.
19. Gornall,A.G., Bardawill,C.J. and David,M.M. (1949) *J. Biol. Chem.*, **177**, 751.
20. Cain,K., Lancashire,W.E. and Griffiths,D.E. (1974) *Biochem. Soc. Trans.*, **2**, 215.
21. Chappell,J.B. (1964) *Biochem. J.*, **90**, 225.
22. Nedergaard,J. and Cannon,B. (1979) in *Methods in Enzymology*, Vol. **55**, Fleischer,S. and Packer,L. (eds.), Academic Press Inc., London and New York, p. 3.
23. Weiner,M.W. and Lardy,H.A. (1973) *J. Biol. Chem.*, **248**, 7682.
24. Tyler,D.D. and Gonge,J. (1967) in *Methods in Enzymology*, Vol. **10**, Estabrook,R.W. and Pullman,M.E. (eds.), Academic Press Inc., London and New York, p. 75.
25. Lai,J.C.K. and Clark,J.B. (1979) in *Methods in Enzymology*, Vol. **55**, Fleischer,S. and Packer,L. (eds.), Academic Press Inc., London and New York, p. 51.
26. Cremer,J.E. and Somogyi,J. (1970) *FEBS Lett.*, **6**, 174.
27. Chance,B. and Williams,G.R. (1956) *Adv. Enzymol.*, **17**, 65.
28. Chance,B. (1959) *Regulation of Cell Metabolism*, Ciba Foundation Symposia, 1958, p. 91.
29. Estabrook,R.W. (1967) in *Methods in Enzymology*, Vol. **10**, Estabrook,R.W. and Pullman,M.E. (eds.), Academic Press Inc., London and New York, p. 41.
30. Cain,K., Hyams,R.L. and Griffiths,D.E. (1977) *FEBS Lett.*, **82**, 23.
31. King,E.J. (1932) *Biochem. J.*, **26**, 292.
32. Pedersen,P.L. (1975) *Bioenergetics*, **6**, 243.
33. Beechey,R.B. (1966) *Biochem. J.*, **98**, 284.
34. Aldridge,W.N. and Street,B.W. (1971) *Biochem. J.*, **124**, 221.
35. Chappel,J.B. (1968) *Br. Med. Bull.*, **24**, 150.
36. Winkler,H.H., Bygrave,F.L. and Lehninger,A.L. (1968) *J. Biol. Chem.*, **243**, 20.
37. Vignais,P.V. (1976) *Biochim. Biophys. Acta*, **456**, 1.
38. Pfaff,E. and Klingenberg,M. (1968) *Eur. J. Biochem.*, **6**, 66.
39. Harris,E.J. and Van Dam,K. (1968) *Biochem. J.*, **106**, 759.
40. Skilleter,D.N. (1975) *Biochem. J.*, **146**, 465.
41. Astwood,E.B. (1970) in *The Pharmacological Basis of Therapeutics*, 4th Edition, Goodman,L.S. and Gilman,A. (eds.), p. 1481.
42. Bidstrup,P.L. and Payne,D.J.H. (1951) *Br. Med. J.*, **2**, 16.
43. Santi,R. and Lucianni,S., (eds.) (1978) *Atractyloside, Chemistry, Biochemistry and Toxicology*, published by Piccin Medical Books.
44. Himms-Hagen,J. (1976) *Annu. Rev. Physiol.*, **38**, 315.
45. Nicholls,D.G. (1979) *Biochim. Biophys. Acta*, **549**, 1.
46. Ray,D.E. (1980) *Br. J. Pharmacol.*, **69**, 257.

47. Ray,D.E. and Cunningham,V.J. (1985) *Arch. Toxicol.*, **56**, 279.
48. van Berkel,Th.J.C. (1979) *Trends Biochem. Sci.*, **4**, 202.
49. Jacobs,E.E., Jacob,M., Sanadi,R.D. and Bradley,L.B. (1956) *J. Biol. Chem.*, **223**, 147.
50. Cain,K. and Skilleter,D.N. (1980) *Biochem. J.*, **188**, 285.
51. Southard,J., Nitisewojo,P. and Green,D.E. (1974) *Fed. Proc.*, **33**, 2147.
52. Turnbull,D.M. (1983) *Adv. Drug React. Ac. Pois. Rev.*, **2**, 191.
53. Bauche,F., Sabourault,D., Giudicelli,Y., Nordmann,J. and Nordmann,R. (1982) *Biochem. J.*, **206**, 53.
54. Cain,K. (1984) *Hum. Toxicol.*, **3**, 322.
55. Hackenbrock,C.R. (1968) *J. Cell. Biol.*, **37**, 345.
56. Wollenberger,A. and Karsh,M.L. (1952) *J. Pharmacol. Exp. Ther.*, **105**, 477.
57. Weetman,D.F., Cain,K. and Sweetman,A.J. (1973) *Arch. Int. Pharmacodynam.*, **203**, 342.
58. Webb,M. (1979) in *The Chemistry, Biochemistry and Biology of Cadmium*, Webb,M. (ed.), Elsevier, North Holland Biomedical Press, Amsterdam, p. 285.

CHAPTER 10

Preparation and Use of Lysosomes and Peroxisomes in Toxicological Research

MILOSLAV DOBROTA

1. INTRODUCTION

The term lysosome encompasses a wide spectrum of membranous organelles which are all involved in a dynamic process of catabolism and processing of exogenous and endogenous cellular material, macromolecules and some small molecules. Broadly, lysosomes may be classified into primary lysosomes, two kinds of secondary lysosomes (heterolysosomes and autophagosomes) and residual bodies (1). In most mammalian tissues they are observed morphologically as electron-dense structures varying in size between 0.15 and 0.8 μm, but this variation might be considerably greater (e.g., 0.04 μm for primary lysosome precursor vesicles to 3 μm for autophagosomes or kidney protein droplets). Subfractionation of lysosomes by rate sedimentation and equilibrium banding techniques has highlighted this variation in size and in physical properties. This heterogeneity of lysosomal populations (2) represents the major difficulty in achieving high purifications of lysosomal preparations. The approaches which may be employed to solve the purification problems are discussed briefly in Section 3.

Because of the great heterogeneity of lysosomes, in most mammalian tissues, two different methods will be recommended for preparing lysosomes: (i) a relatively simple procedure which gives a high degree of purification and is suitable for further *in vitro* studies; (ii) a method which provides a spectrum of lysosomal populations, in relation to other subcellular organelles. The latter method, whilst not necessarily yielding high purity, is extremely useful for following the association of exogenous compounds with cellular structures involved in endocytic uptake of lysosomes and transport *in vivo*.

Peroxisomes are defined as organelles which contain catalase and enzymes which produce hydrogen peroxide (3). Their main function in most mammalian tissues appears to be the oxidation of long chain fatty acids (4). Peroxisomes, of a single tissue such as the liver, are much more uniform in size and sedimentation properties than lysosomes, and are thus relatively easier to isolate. Much of our current knowledge of fractionating peroxisomes stems from the systematic characterisations of lysosomal preparations in which peroxisomes are common contaminants. Approaches to subfractionating peroxisomes have been critically examined by Connock and Temple (5).

It is important to stress that neither lysosomes nor peroxisomes can be obtained 100% pure by any known fractionation technique. Thus, in order to be aware of the degree and consistency of contamination it is essential that all preparations are biochemically characterised (with a range of marker enzymes) as discussed in Section 3.

2. BASIC TECHNIQUES

2.1 **Equipment**

2.1.1 *Homogenisers*

The Potter-Elvehjem type homogeniser (size C, A.H.Thomas, Philadelphia, PA, USA) consisting of a glass vessel and a Teflon pestle rotated at 2400 – 3000 r.p.m. in a standard electric drill is ideal for homogenising both liver and kidney. The methods described in Sections 2.7 and 2.8 for homogenising the kidney cortex recommend 1000 r.p.m.

2.1.2 *Centrifuges and Rotors*

These are available from the major centrifuge manufacturers:

> Beckman-Spinco Division, Palo Alto, California, USA,
> MSE Scientific Instruments Ltd., Crawley, Sussex, UK,
> Kontron, Zurich, Switzerland,
> Dupont-Sorvall,
> Damon-IEC,
> Heraeus-Christ.

The methods require the following apparatus.

(i)   A low-speed refrigerated (or unrefrigerated but used in a cold room) centrifuge, which can be accurately (and reproducibly) spun at speeds less than 1000 r.p.m. and which is fitted with a convenient size swing-out rotor (e.g., accepting 50-ml centrifuge tubes).

(ii)   A high-speed refrigerated centrifuge capable of up to 18 000 r.p.m. and fitted with a standard (e.g., 8 × 50 ml) angle rotor. For large-scale work (see Section 2.5) a larger rotor, such as a 6 × 250 ml, may be required.

        For subfractionation of liver and kidney 'ML' (mitochondrial-lysosomal) fractions by rate sedimentation (Sections 2.5 and 2.8) an HS zonal rotor (MSE Ltd.) is required. [Lack of availability of zonal rotors (or fear of their use) may be readily overcome by scaling down the zonal method and performing the same separation in the appropriate swing-out rotor (or possibly even a vertical or angle rotor)]. This rotor can only be used in the now obsolete HS18 and HS21 centrifuges. However, the ZR-11 zonal rotor (Heraeus-Christ) may be used in the appropriate centrifuge to achieve exactly the same separation as with the obsolete HS zonal rotor.

(iii)   An ultracentrifuge is needed for spinning a 6 × 38 ml swing-out rotor (e.g., Beckman SW28, Dupont-Sorvall AH 627, Kontron TST 28.38, etc.) and also, in the case of the method given in Section 2.5.2 a B-14 zonal rotor.

2.1.3 *Gradient-making Apparatus*

Continuous density gradients, linear with volume in the case of swing-out rotors and exponential with volume in the case of the zonal rotors need to be generated for the methods described in Sections 2.5, 2.7 and 2.8. Gradient-making systems for both types

Figure 1. A typical example of a well designed density gradient fractionation unit (MSE Scientific Instruments Ltd.) for displacing and collecting gradients from tubes. Similar devices are available from all the major centrifuge manufacturers.

of rotor are available from the centrifuge manufacturers. Practical aspects of making density gradients of various profiles have been described in detail by Hinton and Dobrota (6).

2.1.4 *Fractionation and Collection of Density Gradients*

Gradients from swing-out tubes may be collected as discrete fractions by a 'Tube-slicer' (Beckman) or by unloading the gradient downwards, with the aid of a tube 'Piercing Unit' (Kontron) or by upward displacement with the 'Gradient extraction unit' (MSE Ltd.), or 'Fraction Recovery System' (Beckman-Spinco Inc.).

An example of such a system is shown in *Figure 1*. Collection of fractions from zonal rotors is achieved as described in detail elsewhere (6) and in manufacturers' instructions for operating zonal rotors.

2.1.5 *Recording Absorbance Monitor*

For purposes of locating separated bands and retaining a permanent record of the profile of the separation it is extremely useful (but not obligatory) to pass the gradient during unloading and fractionation through a spectrophotometer, or a column-type monitor, fitted with a flow-through cell and a recorder.

Whilst dedicated instruments for this purpose are available (Density Gradient Fractionator, Model 640, ISCO Inc., Lincoln, Nebraska, USA), modern spectrophotometers fitted with a low volume flow-through cell are quite adequate. Since in all the cases quoted in this article the relatively turbid bands can be easily detected at visible wavelengths (e.g., 450 − 650 nm) it is unnecessary to employ u.v. wavelengths.

2.1.6 *Density Measurements*

It is convenient to check the density gradient profile by means of either a direct reading density meter (e.g., DMA 35, Anton Paar Scientific, Graz, Austria) or an Abbé type refractometer. Measurements from the latter can be readily converted to density with the aid of conversion tables, published in reference books (7), or in data sheets supplied by gradient media manufacturers.

2.2 **Media and Buffers**

2.2.1 *Homogenisation Media − Sucrose*

Isotonic sucrose, 0.25 M, is the recommended homogenisation medium for preparation of lysosomes and peroxisomes from liver and kidney. There is some debate about kidney lysosomes being better preserved in hypertonic sucrose, for example the method described in Section 2.7 uses 0.3 M sucrose. The important point is that hypotonic media must be avoided for the preparation of lysosomes and peroxisomes from any mammalian tissues.

Ideally the chosen sucrose should be buffered. Amongst a wide variety of buffers employed for this purpsoe, isotonic sucrose solutions are most commonly buffered with 5 mM Tris-HCl or 5 mM Hepes. It is also common to include 1 − 2 mM EDTA in the sucrose.

It is suggested that 0.25 M sucrose (85.4 g/l), containing 1 mM EDTA and 5 mM Tris-HCl pH 7.4 is suitable as the homogenisation medium for all the preparative methods quoted in this article. Any variations are given in Section 2.3 and in the individual methods (Sections 2.4 − 2.10).

2.2.2 *Density Gradient Materials*

(i) *Sucrose.* For the purpose of making density gradients with sucrose it is most convenient to keep a stock solution of 2 M sucrose, containing 684.6 g per litre. This solution has a density of 1.2572 g/ml and a refractive index of 1.4295 (at 20°C). The 2 M stock solution can be accurately diluted to the required concentration and may also be used for making the isotonic 0.25 M sucrose for homogenisation, provided that the concentration of this is checked (density 1.0311, refractive index 1.3452 at 20°C).

(ii) *Metrizamide.* (Nyegaard and Co. A/S, Oslo, Norway). The concentrations of different solutions of metrizamide, required for isolating liver lysosomes and peroxisomes, are specified in Sections 2.4 and 2.6.

2.2.3 *Displacement of Density Gradients*

The stock solution of 2 M sucrose is ideal for the upward displacement of sucrose gradients from swing-out tubes (Section 2.7) and for unloading of gradients from zonal rotors (Sections 2.5, 2.8, 2.9).

Gradients of metrizamide (Sections 2.4 and 2.6) in swing-out tubes may be collected by upward displacement with either 60% w/v metrizamide or with 'Maxidens'. Whilst metrizamide would appear to be expensive for this purpose it should be pointed out that this displacing solution may be carefully collected and re-used a number of times, provided its concentration is checked and readjusted if necessary. 'Maxidens' supplied by Nyegaard and Co., Oslo, is an inert fluorocarbon which is manufactured by 3Ms Co. as FC43. Since it is completely immiscible with aqueous solutions and is extremely stable it can be re-used for numerous experiments.

2.3 **Biological Starting Material**

2.3.1 *Liver Homogenate*

(i)   Put freshly excised livers, from male Wistar albino rats, into ice-cold buffered 0.25 M sucrose. It is convenient to place the livers into a pre-weighed vessel containing the sucrose and weigh again to obtain the liver weight by difference.

(ii)   Cut the liver into small pieces, then transfer them into the pre-cooled glass vessel of the size C homogeniser (A.H.Thomas Inc.).

(iii)   Having previously pre-cooled the Teflon pestle, assemble this into the drill and homogenise the liver with 3−4 complete strokes of the pestle rotated at 2400−3000 r.p.m.

(iv)   If a large amount of liver needs to be homogenised the process needs to be repeated a number of times since only 10−15 g of liver can safely be homogenised at one time with the size C Thomas homogeniser.

(v)   Dilute the homogenate with more 0.25 M sucrose to give a concentration of 7−10 ml/g of liver.

2.3.2 Preparation of Liver Light Mitochondrial ('L') Fraction

(i)   Homogenise the livers with the Potter-type homogeniser (see Section 2.1.1) in buffered 0.25 M sucrose (see Section 2.2.1).

(ii)   Centrifuge the filtered homogenate at 3000 $g._{av}$ (or 5000 r.p.m. in a typical 8 × 50 angle rotor) for 10 min in a refrigerated high-speed centrifuge (see Section 2.1.2).

(iii)   Remove and retain the supernatant. Resuspend the pellet by re-homogenisation, and spin again at 3000 $g._{av}$ for 10 min.

(iv)   Combine the supernatant with the first supernatant [as in (iii) above] and spin at 12 000 $g._{av}$ (10 000 r.p.m. in an 8 × 50 rotor) for 20 min.

(v)   Remove the supernatant together with the light fluffy layer on top of the pellet and wash the pellet by resuspension and centrifugation as above (iv).

(vi)   After careful removal of the supernatant and fluffy layer, resuspend the pellet, representing the light mitochondrial fraction 'L' in buffered 0.25 M sucrose with a hand-operated Potter-type homogeniser. For the methods given in Sections 2.4

and 2.6, resuspend the L fraction to give a concentration of 1 ml/3 g. For the method given in Section 2.5, resuspend the L fraction in 15 – 25 ml of buffered 0.25 M sucrose (containing 5 mM EDTA) irrespective of the initial weight of liver.

2.3.3 *Kidney Cortex Homogenate*

(i)   Remove the kidneys and place them in cold buffered isotonic sucrose (see Section 2.2.1). [**Note:** in some methods it is recommended that after anaesthetising the animal, the kidneys are perfused with ice-cold isotonic sucrose *via* the abdominal aorta (e.g., for the method described in Section 2.7). In our experience there is no obvious advantage in this perfusion. Indeed comparison of lysosomal yields from perfused and unperfused kidneys suggests that perfusion may reduce the yield of lysosomes.]

(ii)   Excise the cortices carefully. Cut the kidney longitudinally, starting at the point of entry of the blood vessels, into two equal halves. Each half reveals a greyish triangular papilla, a red medulla and a brownish cortex. Hold one half on a piece of filter paper, with the cut surface exposed. Insert a sharp point of a pair of small pointed scissors into the brown cortex (start near the papilla), just missing the red medulla. Carefully excise, following the red interface until the outer brown layer cortex is completely free.

(iii) Homogenise with about five strokes of the Potter-Elvehjem homogeniser rotating at 1000 r.p.m. in 0.44 M sucrose, containing 1 mM EDTA, buffered to pH 7.0 (for methods given in Sections 2.7 and 2.9), or 0.25 M sucrose, buffered with 5 mM Tris-HCl, pH 7.4 (for the method given in Section 2.8).

(iv)   Filter the homogenate through a coarse sieve (a plastic tea strainer is ideal) and make it up to 10 ml/g of cortex (10% w/v).

(v)   Since kidney homogenates are prone to aggregate and settle out on standing, it is recommended that any subsequent centrifugations are carried out without any delay. If an aliquot of the homogenate is retained for further analysis this problem of aggregation may be overcome by vigorously homogenising the aliquot with a Turrax- or Polytron-type homogeniser in the cold.

2.4 **Isolation of Liver Lysosomes**

Amongst the large number of methods for purifying liver lysosomes the procedure developed by Wattiaux *et al.* (8) stands out as undoubtedly the best compromise between purification, yield, convenience and reproducibility. This method yields the most highly purified lysosomes (from normal rat liver) of any centrifugation method quoted (~ 73-fold) and is clearly very reproducible as judged by the number of times it has been successfully employed and cited by other workers. Relative to the high yield and convenience, the high cost of metrizamide, the gradient material, is only a minor disadvantage.

The procedure, described in detail below, is summarised in *Figure 2*.

(i)   To 1 volume of the resuspended 'L' fraction (see Section 2.3.2), add 2 volumes of 85.6% (w/v) metrizamide solution.

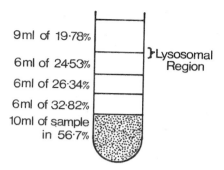

9ml of 19·78%

6ml of 24·53% }Lysosomal Region

6ml of 26·34%

6ml of 32·82%

10ml of sample in 56·7%

Figure 2. Discontinuous metrizamide gradient for isolating liver lysosomes. The dotted region represents the sample consisting of 10 ml of resuspended 'L' fraction in 56.7% metrizamide. After spinning the SW28 rotor (or equivalent) at 27 000 r.p.m. for 2 h the lysosomes band at the interface between 19.8% and 24.5% metrizamide, as indicated. Percent values refer to w/v of metrizamide.

(ii)   Place 10 ml of this mixture into a 35 ml centrifuge tube (for the SW28 or equivalent rotor, see Section 2.1.2).

(iii) Then layer over the mixture successively and with care:
6 ml 32.82% (w/v) metrizamide
6 ml 26.34% (w/v) metrizamide
6 ml 24.53% (w/v) metrizamide
9 ml 19.78% (w/v) metrizamide

(iv)   Spin the 6 × 38 ml swing-out rotor (SW28 or equivalent) at 95 000 $g_{.av}$ (27 000 r.p.m.) for 2 h at 4°C. [**Note:** It is important to realise that in metrizamide, lysosomes and mitochondria have exactly the same equilibrium banding densities and thus, under normal circumstances, cannot be separated. In this method however, the flotation of the sample from the bottom of the tube where a high hydrostatic pressure is generated, causes damage to mitochondria and increases their banding density thus leaving the lysosomal region free of mitochondria. Since the lysosomal purification is dependent on causing damage to mitochondria it is clear that a successful purification cannot be achieved with a swing-out rotor which generates lower hydrostatic pressure (e.g., shorter radial distance or centrifuged path) than the SW28.]

(v)   After centrifugation fractionate the gradient by means of a tube slicing device (see Section 2.1.6) ensuring that only the region found at the interface between 19.78% and 24.53% metrizamide and indicated in *Figure 2* is collected. Alternatively, the gradient may be fractionated by means of a gradient extraction device (see Section 2.1.6), which employs upward displacement of the gradient by a high density medium, through a tapered adaptor attached to the top of the tube. Practical aspects of fractionating gradients from tubes have been described in detail elsewhere (6).

(vi)   The pooled region of lysosomes as indicated in *Figure 2*, should contain lysosomes which have been purified some 73-fold, as judged by the average relative specific activity (relative to that of the homogenate) of a number of acid hydrolase marker enzymes. This preparation should contain about 11% of the total homogenate acid hydrolase activity.

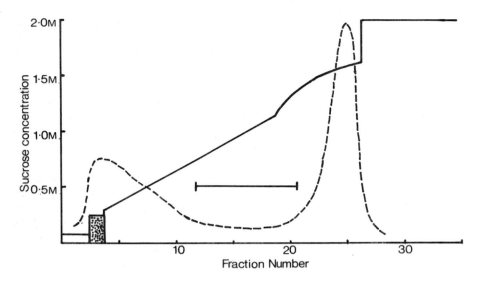

Figure 3. Pattern of 650 nm absorbance (---) after rate sedimentation of a liver 'L' fraction in a zonal HS rotor at 9000 r.p.m. for 40 min. The exact profile of the sucrose gradient (——) and the position of the sample (dotted) are also indicated. The horizontal bar represents the lysosome-rich region which is pooled for subsequent equilibrium spin.

(vii)   The pooled preparation contains about 22% metrizamide, whose osmolality is approximately 195 mOSm. At this concentration of metrizamide, lysosomes appear to be well preserved. Should it be necsssary to remove the metrizamide this is best achieved by diluting out the metrizamide, ensuring the diluted fraction is still isotonic (lysosomes readily rupture in hypotonic medium) and centrifuging at 25 000 $g._{av}$ for 30 min. Successive washes and spins may be used to remove all the metrizamide. However, it must be stressed that during such procedures a significant proportion of lysosomes will be ruptured.

2.5 Large-scale Preparation of Liver Lysosomes

This two-step procedure (9) involves the use of zonal rotors whose operation is described in detail elsewhere (6). Particular advantages of the method are its large scale (lysosomes from up to 200 g of liver may be prepared) and that it provides continuous profiles of lysosomal populations (in relation to other organelles) separated by rate sedimentation and equilibrium banding.

2.5.1 *Rate Sedimentation*

(i)   Fill the HS zonal rotor (or the Heraeus-Christ ZR-11) with a sucrose gradient, containing 5 mM Tris-HCl pH 7.4 and 2 mM EDTA, of exactly the profile indicated in *Figure 3*. This gradient consists of a 400 ml linear (with volume) portion rising in concentration from 0.3 M to 1.15 M, then a 150 ml exponential portion from 1.15 to 1.6 M and the remainder of 2 M sucrose.

(ii)   Load the resuspended 'L' fraction (see Section 2.3.2) into the centre of the rotor (equivalent to being layered at the top of the gradient in a tube). After pushing the sample away from the centre with some 20 – 30 ml of 'overlay' (0.1 M sucrose), in order to form a narrow sample band spin the rotor at 9000 r.p.m. for 40 min.

(ii)   After decelerating the rotor, displace the gradient and collect it in 20 ml fractions. It is most useful to monitor the absorbance (at 650 nm) of the outflowing gradient (see Section 2.1.5) to obtain a 'turbidity' profile (which approximates very closely to protein values determined on each fraction). A typical 650 nm profile of a rate zonal subfractionation of the liver 'L' fraction is presented in *Figure 3*.

(iv)   With the aid of the 650 nm profile, the lysosomal region, found in the centre of the gradient, can be easily located as the clear region between the very 'turbid' mitochondrial region (rapidly sedimenting) and 'microsomes' (slow sedimenting). These three regions are illustrated in *Figure 3*. By visually observing the turbidity of the fractions it is possible to locate the lysosome-rich region without the aid of a flow-through absorbance monitor.

(v)   Pool the fractions of the lysosome-rich region for the subsequent equilibrium spin (Section 2.5.2) but ensure that an aliquot of each fraction (e.g., 4 ml from 20) is retained for various assays.

(vi)   Analyse the fractions for a series of marker enzymes (see Section 3.2) in order to determine the distribution of lysosomes, mitochondria, peroxisomes, endoplasmic reticulum (ER) and plasma membrane.

(vii)   Measure the density (or refractive index) of the fractions to confirm the profile of the sucrose gradient.

(viii) The fractions may also be analysed for their content of an exogenous toxicologically or pharmacologically active compound (or metabolites) which might have been administered to the animals at an appropriate time before fractionation.

2.5.2 *Equilibrium Banding*

(i)   Fill a B14 zonal rotor with 400 ml of linear sucrose gradient (0.8 – 2 M) and 240 ml of 2 M sucrose. Again the sucrose solutions are buffered with 5 mM Tris-HCl pH 7.4 and contain 2 mM EDTA. The gradient profile is illustrated in *Figure 4*.

(ii)   Check that the density of the pooled lysosome-rich region (see above) is lower than (or at least equal to) that of 0.8 M sucrose.

(iii)   Load the lysosome-rich region (120 – 200 ml, typically) into the centre of the zonal rotor, followed by a small volume (e.g., 20 ml) of 'overlay' (0.1 M sucrose).

(iv)   Spin the rotor either for 2.5 h at 47 000 r.p.m., which is sufficient for equilibrium, or overnight (16 – 18 h) at 44 000 r.p.m. It is practically more convenient to spin overnight since this avoids the rush of having to complete (and analyse) two zonal spins in one day.

(v)   After completion of the spin, collect the gradient in 20-ml fractions. During unloading it is useful to monitor the absorbance of the gradient. However, since the sample is now unlikely to have much turbidity it is appropriate to monitor

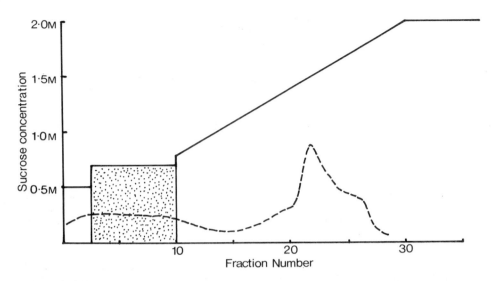

Figure 4. Pattern of 450 nm absorbance (---) after equilibrium sedimentation of the lysosome-rich region (see *Figure 3*) in a B-14 zonal rotor at 44 000 r.p.m. for 16 h. The profile of sucrose gradient concentration (——) is also shown. Lysosomes are found between fractions 24 and 27. The position of the sample is dotted.

at lower wavelengths, e.g., 450 nm. A typical 450 nm trace is shown in *Figure 4*, together with the approximate positions of lysosomes.

(vi)   Analyse the fractions for a series of marker enzymes (see Section 3.2) in order to locate the major subcellular components and to assess the purity of the lysosomal band.

(vii)   The lysosomes should be found as a narrow band (typically in three fractions) at a median density of 1.205 and should be about 60-fold purified, as judged by the relative specific activity of acid hydrolases. It may be advantageous to remove the rather large amount of sucrose present in the lysosomal preparation. Dilute with ice-cold distilled water to give a sucrose molarity of 0.25 and pellet the lysosomes by centrifuging at approximately 25 000 $g._{av}$ for 30 min. It should be noted that even from 200 g of liver the amount of material (protein) recovered in the final lysosomal preparation is very small, therefore very small pellets will be obtained.

2.6 Isolation of Liver Peroxisomes

Amongst a variety of methods available for purifying liver peroxisomes the method of Wattiaux and Wattiaux-De Coninck (10) which is rather similar to the method described for lysosomes (Section 2.4), appears to be the best compromise between purification and convenience. Methods based on fractionation in Percoll gradients (11,12), although rapid and convenient, do not yield preparations which are as pure as those of the recommended methods, described below. It is interesting to note that the purification of peroxisomes reported by Leighton *et al.* (13) in their comprehensive fractionation study using sucrose gradients was about 36-fold. This purification is very similar

264

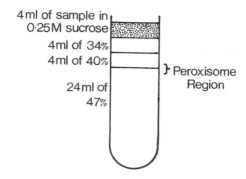

4 ml of sample in
0·25 M sucrose
4 ml of 34%
4 ml of 40%
24 ml of
47%
} Peroxisome
Region

Figure 5. Discontinuous metrizamide gradient used for isolating liver peroxisomes. The dotted area represents the sample (see text). After spinning the SW28 rotor (or equivalent) at 27 000 r.p.m. for 2 h the peroxisomes are found at the interface between 40% and 47% metrizamide. Concentrations of metrizamide are given as % w/v.

to that achieved by the method recommended here but it was achieved with the aid of reducing lysosomal contamination by modifying the density of lysosomes with Triton WR-1339.

(i)   Resuspend the 'L' fraction (see Section 2.3.2) in buffered 0.25 M sucrose to give a concentration of about 1 ml/g of liver.

(ii)   Prepare metrizamide gradients, in 38-ml swing-out tubes (for the Beckman SW28 rotor or equivalent — see Section 2.1.2) by successively layering:

    24 ml 47% (w/v) metrizamide
    4 ml 40% (w/v) metrizamide
    4 ml 34% (w/v) metrizamide
    4 ml of resuspended 'L' fraction [(i) above]

(iii)   Centrifuge the 6 × 38 ml swing-out rotor (or equivalent) at 95 000 $g_{\cdot av}$ (or 27 000 r.p.m.) for 2 h.

(iv)   Fractionate the gradient as described in step (v) of Section 2.4. The peroxisomal band, at the interface between the 40% and 47% metrizamide at a density of approximately 1.23 g/ml, is indicated in *Figure 5*. Analysis of peroxisomal marker enzymes (e.g., catalase) should reveal the peroxisomes to be about 36-fold purified with respect to the homogenate.

(v)   For further studies it may be necessary to remove the metrizamide from the preparation by diluting (ensuring that the mixture does not become hypotonic), spinning down the peroxisomes and resuspending in an appropriate isotonic medium.

Note. Peroxisomes are rather fragile and are easily ruptured by homogenisation and high hydrostatic pressure. The latter is indeed the reason for the very short gradient employed in the above method since this keeps the radial distance between the top of the liquid column and the interface with the 47% metrizamide at a minimum in this type of rotor. As suggested by Wattiaux and Wattiaux-De Coninck (10) a vertical tube rotor, with its very short centrifugal path, should be ideal for preparing peroxisomes by their method.

2.7 Isolation of Kidney Lysosomes

The method of choice for preparing the kidney lysosomes is that described by Mauns-bach (14) in meticulous detail. Although somewhat complex, the method is reliable and yields lysosomes of high purity. It is however, important to note that whilst the kidney cortex contains lysosomes of great heterogeneity the lysosomes prepared by this method represent a single population of unique kidney lysosomes which originate from proximal tubule cells. These lysosomes, representing secondary lysosomes in-volved in the catabolism of reabsorbed proteins and also called 'protein droplets' or 'kidney granules', are rather large and dense and are therefore relatively easy to isolate. The same population of lysosomes may also be purified by the methods described in references 15 and 16.

(i)   Spin the homogenate prepared as in Section 2.3.3, at 150 $g._{av}$ for 10 min and discard the pellet.

(ii)   Respin the supernatant at 150 $g._{av}$ for 10 min and again discard the pellet.

(iii)   Layer the supernatant (32 ml) over 8 ml of 0.44 M sucrose, containing 1% (w/v) glycogen and 1 mM EDTA, buffered to pH 7.0 in a standard 50-ml centrifuge tube (e.g., screw cap 'Oak Ridge Bottle').

(iv)   Centrifuge at 9000 $g._{max}$ for 3 min. The time of 3 min excludes acceleration and deceleration time. Pour off and discard the supernatant.

(v)   Carefully add a few millilitres of 0.44 M sucrose to the tube and gently swirl to resuspend the top, light-coloured fluffy layers only. Repeat the process until all the light upper layers of the pellet have been removed leaving only the dark bottom layer.

(vi)   Resuspend the dark part of the pellet in 0.44 M sucrose containing 1 mM EDTA, pH 7.0 by swirling and by jetting the sucrose with a Pasteur pipette (see Section 2.1.3).

(vii)   Prepare linear sucrose gradients (see Section 2.1.3), ranging from 1.1 M to 1.9 M sucrose, in 38-ml tubes of the 6 × 38 ml swing-out rotor (SW28 or equivalent). Each tube should contain about 30 ml of gradient.

(viii) Layer 2−4 ml of the resuspended dark pellet on top of the linear gradient and spin the 6 × 38 ml swing-out rotor at 90 000 $g._{max}$ (23 000 r.p.m.) for 2.5 h.

(ix)   After centrifugation the gradients are fractionated (see Sections 2.1.4 and 2.2.3) by upward displacement with 2 M sucrose and 1-ml fractions are collected. The gradient should contain three bands ('upper fraction', mitochondria, lysosomes or granule fraction) and a pellet. The lysosomes are found in the lowest band which is located about a third of the way up the tube.

(x)   The original method (14) assesses the purity of the lysosomal preparation as being 97.7% pure, as judged by quantitative morphological analysis. As judged by the relative specific activity of acid hydrolase marker enzymes, purifications of 10- to 15-fold have been reported for this method.

   For discussion of assessing purity refer to Section 3.3.

2.8 Isolation of Different Populations of Kidney Lysosomes

The method described below is based on that of Andersen *et al.* (17) and only differs from the original in the omission of the perfusion of the kidneys prior to cortical ex-

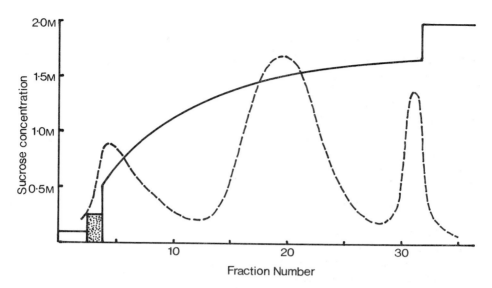

Figure 6. Pattern of 650 nm absorbance after rate sedimentation of a rat kidney cortex 'ML' fraction in a HS zonal rotor at 8000 r.p.m. for 1 h. The exact profile of the sucrose gradient (—) and the position of the sample (dotted) are also shown. Large lysosomes (protein droplets, hyaline droplets) are found in fractions 30 – 33 whilst small lysosomes are spread between fractions 12 and 25.

cision and homogenisation. It involves the subfractionation of the cortical 'ML' fraction by rate zonal sedimentation and in a single step gives a 'profile' of all the lysosomal populations present in the ML fraction. Although very simple in principle it is technically somewhat complicated by the use of a zonal rotor, but it does permit large-scale fractionation of kidney lysosomes.

(i)   Centrifuge the cortical homogenate (we have successfully fractionated >20 g of kidney cortex by this method) at 700 $g._{av}$ for 3 min at 4°C to sediment the nuclear fraction (N). Retain the supernatant.

(ii)   Resuspend the 'N' (nuclear) pellets in 0.25 M sucrose, buffered with 5 mM Tris-HCl pH 7.4 and spin again exactly as in (i).

(iii)   Combine the two supernatants and centrifuge at 9200 $g._{av}$ for 3 min (excluding acceleration and deceleration time) at 4°C to pellet the 'ML' fraction.

(iv)   Resuspend the 'ML' fraction in 15 – 20 ml of 0.25 M sucrose with a hand-operated Potter-Elvehjem type homogeniser and leave at 4°C.

(v)   Load the zonal HS rotor with 550 ml of exponential sucrose gradient (with 5 mM Tris-HCl pH 7.4) ranging from 0.5 M to 1.7 M then fill it up completely with 150 ml of 2 M sucrose. The profile of this gradient is shown in *Figure 6*. Practical details of operating zonal rotors are described elsewhere (6).

(vi)   Load the resuspended 'ML' fraction into the centre of the rotor (layer it over the gradient) and spin the rotor at 8000 r.p.m. for 1 h.

(vii)   Decelerate the rotor and collect the displaced gradient in 20-ml fractions.

During unloading and collection the gradient may be monitored with a flow-through spectrophotometer. A typical 650 nm trace of the separated components including two distinct populations of lysosomes, is shown in *Figure 6*.

The most rapidly sedimenting region of the gradient (consisting of 3 — 4 fractions), contains large lysosomes, which are purified about 35-fold with respect to the homogenate. This is somewhat greater purification than that of the preparation described in Section 2.7. These lysosomes may be harvested for further studies, by diluting out the sucrose and sedimenting down the lysosomes, exactly as described in Section 2.5.2. The central region of the gradient contains a broad peak of smaller lysosomes which are not separated or resolved from mitochondria, peroxisomes, brush border plasma membranes and consequently are not at all purified with respect to the homogenate. However, the distribution of acid hydrolases is quite distinct from the distributions of other marker enzymes, thus lysosomes can be distinguished from other organelles despite the contamination.

The regions of small lysosomes prepared in this manner may, if required, be further fractionated by equilibrium banding. The appropriate fractions should be pooled and loaded onto a linear sucrose gradient and spun to equilibrium. This may be achieved either with a B-14 zonal rotor, exactly as described for liver lysosome purification (see Section 2.5.2) or by putting smaller samples on to gradients in swing out tubes and spinning these, preferably in a fairly large capacity rotor such as the 6 × 38 ml. Since the large lysosomes obtained in the method described are extremely pure there is no particular advantage in subjecting these to an equilibrium spin, other than simply to confirm their rather high banding density of 1.23 g/ml. However, the small lysosome region may be most usefully resolved into a number of subcellular populations. The lysosomes fractionate into two distinct populations with mean densities of 1.20 and 1.23 g/ml. In addition, both populations are separated from brush border membranes, some of the mitochondria (undamaged) and some of the ER membranes. In this manner, some purification of the smaller lysosomes of the kidney cortex may be achieved.

2.9 Isolation of Kidney Peroxisomes

A survey of the methodology for separating kidney peroxisomes reveals that there is apparently no satisfactory published method for their isolation from the rat kidney. This seems to be due to the relative lack of interest in kidney peroxisomes and not to any fundamental properties which would make them particularly difficult to isolate. Peroxisomes have however, been successfully separated from mouse kidney (5) and from dog kidney (18). Peroxisomes from dog kidney were found at an equilibrium density of 1.23 g/ml where they were apparently free from lysosomes. Whilst the large dense lysosomes of rat kidney band at exactly this density, the distribution of lysosomal enzymes suggests that dog kidney lysosomes have much lower banding densities. Similarly, in the mouse kidney it appears that separation of peroxisomes by equilibrium banding is facilitated by the relatively low banding density of the lysosomes (5).

The problem of isolating rat kidney peroxisomes may be solved by employing metrizamide rather than sucrose gradients. As demonstrated by Wattiaux *et al.* (8), liver lysosomes have much lower banding densities in metrizamide than in sucrose, whilst the density of peroxisomes in either medium is the same. If the same is true of kidney lysosomes and peroxisomes, it follows that, since lysosomes will be less dense, the peroxisomes should be free of contaminating lysosomes. Indeed, this principle has been demonstrated with the separation of mouse kidney peroxisomes in metrizamide (19).

A brief description of the method of Kitano and Morimoto (18) for the preparation of dog kidney peroxisomes is presented below. This method is chosen in preference to the method for isolating peroxisomes from mouse kidney (5) only because it is rather simpler and only involves a single equilibrium spin in a zonal rotor.

(i)   Centrifuge the kidney cortex homogenate at 500 $g_{\cdot av}$ for 10 min, pour off the supernatant and discard the pellet.

(ii) Load 50 ml of the post-nuclear supernatant into a B-14 zonal rotor containing 460 ml of linear sucrose gradient (1.0−2.0 M) and 130 ml of 2 M cushion.

(iii) After loading a small volume of overlay to push the sample away from the very centre of the rotor, the rotor is spun at 23 000 r.p.m. for 2 h.

(iv) After the spin, fractions are collected, as described in Sections 2.5 and 2.8 and analysed for the appropriate marker enzymes.

(v)   The peroxisomes, which are found at a median density of 1.23 g/ml (1.73 M sucrose), should be about 15-fold purified with respect to the homogenate.

2.10 Lysosomes and Peroxisomes of Other Tissues

It is not possible within the scope of this chapter to cover the isolation of lysosomes and peroxisomes from the variety of tissues in which these organelles have been characterised. For the reader interested in these organelles in tissues other than the liver and kidney a partial literature survey of some key references is provided in *Table 1*.

3. CHARACTERISATION OF PREPARATIONS

3.1 Sedimentation Properties of Lysosomes

The great majority of methods for separating cellular organelles depend on differences

Table 1. Isolation of Lysosomes and Peroxisomes from Tissues Other than Liver and Kidney.

Tissue	References	
	Lysosomes	Peroxisomes
Adrenal medulla	20	—
Arterial wall	21	—
Bone	22	—
Brain	23	—
Brown fat	—	24
Fibroblasts	25	—
Enterocytes	—	26
Hepatoma	27	28,27
Intestine	—	24
Leucocytes	29	—
Lung	30	—
Lymphoid tissue	31	—
Macrophages	32	—
Muscle-skeletal	33	—
-cardiac	34	—
Platelets	35	—
Skin	36	—
Thyroid	37	—
Uterus, prostate	38	—

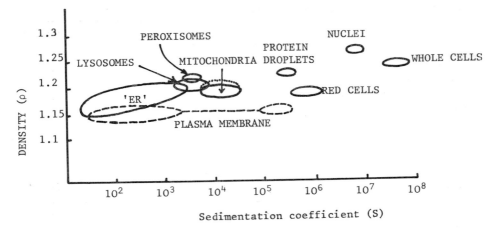

Figure 7. S-ϱ diagram defining the sedimentation properties of cells and organelles in sucrose gradients. The dotted line represents damaged mitochondria. Protein droplets are the large lysosomes of kidney proximal tubule cells.

in size and/or density. [Lysosomes have also been fractionated on the basis of their surface charge by free flow electrophoresis (39) and on the basis of sugar composition of the membrane glycoproteins by affinity chromatography (40).] This relationship can best be illustrated on a two-dimensional diagram (*Figure 7*), with one dimension representing size (S or sedimentation coefficient) and the other dimension density (ϱ). In such a diagram, membrane-bound organelles occupy a distinct area, but, as can be clearly seen in *Figure 7*, there is considerable overlap between different organelles. An awareness of this overlap of properties, which is so well illustrated in the S-ϱ diagram, is extremely useful in assessing the possible contaminants of a specific preparation.

It should also be emphasised that the density dimension of the S-ϱ diagram illustrated in *Figure 7* is unique to sucrose. As has already been mentioned in some of the methods (Sections 2.4 and 2.6) the banding densities of organelles in other gradient media may be quite different. Hence different S-ϱ diagrams for each of the other commonly used gradient materials, e.g., metrizamide/Nycodenz, percoll, ficoll, should be constructed.

The properties of size and density of subcellular organelles, which are most suitably expressed in the S-ϱ diagram, have both been applied in the isolation procedures described for liver and kidney lysosomes and peroxisomes. In the large scale purification of liver lysosomes, the hepatic 'L' fraction is first fractionated by rate sedimentation, i.e., according to size to a first approximation, and then by an equilibrium spin of the lysosome-enriched region, i.e., according to density.

3.2 Marker Enzymes

Marker enzymes (for definitions and criteria see ref. 41) are extremely important for locating the subcellular organelles being isolated (e.g., in a complex profile obtained after gradient centrifugation), determining their purity and also for estimating the degree of contamination by other subcellular components. It is fortunate that the majority of

Table 2. Marker Enzymes for Subcellular Membranous Structures.

Enzymes	Reference
Lysosome enzymes	
Acid phosphatase (β-glycerophosphatase)	9,46
β-galactosidase	43,46
N-acetyl β-D-glucosaminidase	44,46
Aryl sulphatase	45,46
β-glucosidase	46
Acid ribonuclease	42
Cathepsin D	47
Peroxisomal enzymes	
Catalase	13
Uricase	48
Amino acid oxidase	49
Palmityl-CoA reductase	50
Mitochondrial enzymes	
Succinic dehydrogenase	51
Cytochrome oxidase	52
Monoamine oxidase	53
Endoplasmic reticulum enzymes	
Glucose-6-phosphatase	9
NADPH cytochrome c reductase	54
Cytochrome P-450	55
Plasma membrane enzymes	
5'-Nucleotidase	56
Alkaline phosphatase	43
Alkaline phosphodiesterase	56

lysosomal acid hydrolases and peroximal catalase/peroxidases are uniquely distributed in lysosomes and peroxisomes, respectively, and thus represent fairly good markers, which may be used for locating these organelles and for assessing their purity in the liver and kidney. However, in some specific cells, e.g., polymorphonuclear leucocytes, some lysosomal populations (azurophil granules) may also contain peroxidative enzymes. The commonly employed mitochondrial marker enzymes, succinic dehydrogenase (SDH) and cytochrome oxidase are, in the case of liver and kidney, most suitable mitochondrial markers. Endoplasmic reticulum marker enzymes, glucose-6-phosphatase and NADPH cytochrome c reductase are adequate for assessing the degree of ER contamination in lysosomal preparations but precautions must be taken with glucose-6-phosphatase to inhibit the non-specific hydrolysis of glucose-6-phosphate by the highly enriched lysosomal acid phosphatase. If this is not done, by the inclusion of tartrate in the assay (see ref. 9), the apparently high glucose-6-phosphatase activity would suggest an erroneously high degree of ER contamination. Plasma membrane contamination is frequently estimated by the 5'-nucleotidase activity in the lysosomal preparations, but lysosomes also contain a 5'-nucleotidase which gives rise to an overestimate of contamination by plasma membrane.

The enzymes commonly used and suitable as marker enzymes in liver and kidney lysosomes and peroxisomes are listed in *Table 2*, together with references to the assay procedures. The manner in which these enzyme activities can be used to assess purification and contamination is discussed in Sections 3.3 and 3.4.

3.3 **Criteria of Purity and Recovery (Yield)**

The purity of a lysosome or peroxisome preparation is normally expressed as the relative specific activity or RSA. This is the specific activity of a typical marker enzyme in the preparation divided by the specific activity of the same enzyme in the homogenate. In order to arrive at a most representative purification, it is best and indeed strongly recommended, that a number of marker enzymes are assayed and their RSA values averaged. Thus the mean RSA of three acid hydrolases quoted for the method given in Section 2.4 is 73 with a range of 64 – 80 and for the method in Section 2.6 it is 66 with a range of 48 – 71. Whilst these RSA values are 'close' to the mean, there are distinct variations, with some enzymes, which, if measured alone, would indicate either a low or a high purification.

Liver lysosomes purified by free-flow electrophoresis (39) were, according to RSA values, extremely pure, but the wide variation of 40 – 238 for different hydrolases suggests that some populations of lysosomes were more highly purified than others. This means that the preparation is not representative of the whole lysosomal population.

Purity of kidney lysosomes was measured by Maunsbach (14) by quantitative morphology of the pelleted preparation and was expressed as 97.7% of the volume of the pellet. This method is, however, not suitable for routine assessments of purification, although it is useful to confirm visually that intact lysosomes are present in the preparation.

The yield or recovery of a preparation is normally expressed as the percentage of the total homogenate activity present in the particular preparation. The average yield of the lysosomal preparation described in Section 2.4 is about 11%, which, in view of the difficulties in purifying lysosomes, is a most respectable figure. From these figures and the purifications, it is not possible to determine how representative of the whole spectrum of liver lysosomes this 11% might be, but it would for most experimental purposes be acceptable. The important point is that this yield value should be used as an index of quality control and all preparations which differ widely from the established average yield should be rejected.

The recovery of liver lysosomes prepared on a large scale (see Section 2.6) is about 5%. This reflects the size properties of lysosomes and how they overlap with sedimentation of 'microsomes' and with mitochondria as is clear from the S-ρ diagram (*Figure 7*). In the rate zonal spin of the liver 'L' fraction, the width of the lysosomal region is chosen deliberately to exclude most of the ER on one side and mitochondria on the other, although the lysosomal band is considerably wider than the chosen region. Thus recovery in the final equilibrium spin may be significantly improved by taking a wider 'cut' of the lysosomes from the rate spin. However, this improved recovery can only be achieved at the expense of purification, since the larger proportion of mitochondria taken will significantly contaminate the final preparation.

3.4 **Assessment of Contamination**

Expressing purity as RSA only indicates the degree of enrichment over the original

Table 3. Assignment of Protein to the Individual Components Present in the Purified Liver Lysosomes.

	Method 2.6		Method 2.4
	% Protein × RSA (enzyme) in liver	% Protein attributed to organelle	% Protein quoted in ref. 8
Mitochondria	20×0.98 (SDH)	19.6	3.8
Peroxisomes	2.5×3.76 (Catalase)	9.4	0.3
ER	21×0.83 (G-6-P-ase)	17.4	4.9
Plasma membrane	2.6×0.54 (5′ AMP-ase)	1.4	19.7
Lysosomes	1×61 (Acid hydrolases)	61	68

homogenate. In order to establish what proportion of the sample is attributable to lysosomes and to other subcellular components it is necessary to apply the specific activities of a number of marker enzymes (for lysosomes and other components) to the 'assignment' formula. This biochemical assessment of purity is based on the approach of Leighton *et al.* (13) which assumes that the liver homogenate contains 20% mitochondrial (57), 2.5% peroxisomal, 21% endoplasmic reticulum (13), 2.6% plasma membrane (54) and 1% lysosomal (57,58) protein. The percentage protein contribution of the cellular organelles in the homogenate is multiplied by the RSA of the best marker enzyme for that organelle to give the percentage protein contribution of that organelle in the purified sample. For lysosomes the value is obviously numerically the same as RSA, since the liver contains 1% lysosomal protein. The relationship is however very useful for determining the relative proportions of the contaminants. From the purifications of a number of marker enzymes, the percentage of protein attributable to each organelle-type present in the purified lysosomes (Section 2.6) is presented in *Table 3*, together with the assignment values quoted by Wattiaux *et al.* (8), for the method given in Section 2.4 .

The assignment results of the two liver lysosomal methods described show rather different levels of contamination. However, in view of the inaccuracies in attributing protein contributions of specific organelles, the variations in RSA values for a series of marker enzymes and the difficulty in accurately measuring very low activities of other marker enzymes, these assignment calculations should only be used to give an indication of the relative proportion of contaminants.

3.5 Integrity of the Preparation

As well as determining the purity of the lysosomal preparation, by examining the relative specific activity of the lysosomal enzymes, it is also important to confirm that the preparation contains intact lysosomes. Whilst morphological examination by electron microscopy is useful, it does not provide a rapid result nor a quantitative assessment of lysosomal integrity.

If lysosomes are perfectly intact, the lysosomal enzymes are contained within them and are inaccessible to the substrates commonly employed for assaying these enzymes. In order to measure this activity, often referred to as 'latent', the acid hydrolases are measured in the presence of detergents (e.g., Triton X-100, digitonin, etc.) which completely disrupt all lysosomal membranes and thus release the latent or bound activity.

This latency may be conveniently used to assess the integrity of the lysosomal preparation. Latency may be determined by measuring the acid hydrolase activity in the whole lysosomal suspension and also the 'free' or soluble activity still present after removing the lysosomes by sedimentation; the difference is the latent activity. A more common method of determining latency is to assay the appropriate acid hydrolase in the presence of the detergent (this gives 'total' activity) and in the absence of detergent (this gives 'free' activity). For this purpose it is essential to use an enzyme whose activity can be measured reliably and rapidly. This is important since true 'free' activity can only be determined if no autolytic breakage of intact lysosomes occurs during incubation. Latency, representing bound activity may be conveniently expressed as a percentage i.e.,

$$\frac{\text{'Total'} - \text{'Free' activity}}{\text{'Total' activity}} \times 100$$

It should be noted that although theoretically there should be no free activity in a good preparation, in practice the latency value of the best preparations is always well below 100.

4. APPLICATION AND USE OF THE PREPARATIONS IN TOXICOLOGY

4.1 Effects of Exogenous Compounds on Lysosomes

In order to appreciate how small molecules are processed in lysosomes, it is useful to discuss briefly lysosomotropism. This term was coined by de Duve's group to describe how various types of compounds are accumulated in lysosomes (59). There are essentially four types of lysosomotropism.

(i)   Weakly basic compounds rapidly accumulate in lysosomes because the proton pump of the lysosomal membrane allows them to enter but not to exit.

(ii)   Permeant molecules may be modified, by the action of lysosomal enzymes, into forms which become impermeant.

(iii)   A wide range of small molecules which are readily bound to macromolecules may be delivered to lysosomes by the powerful process of receptor-mediated endocytosis. This is sometimes referred to as 'piggy back' endocytosis.

(iv)   Some small molecules may reach the lysosome directly by endocytosis.

The lysosomotropic properties of numerous compounds which may have therapeutic activity have led to the development of the important field of drug targeting to lysosomes. As pointed out by de Duve *et al.* (59), in their classic review of lysosomotropism, it is indeed possible to make most compounds lysosomotropic by chemically coupling to a suitable carrier. Such modifications may also be effectively employed to achieve tissue selectivity. Drug targeting and delivery, especially of compounds attached to macromolecules has been extensively reviewed by Gregoriadis *et al.* (60).

Since they do not possess the mixed function oxidases, which are found in the endoplasmic reticulum, lysosomes appear not to be involved in drug metabolism and would seem to be of little interest to toxicologists. However, small molecules which enter lysosomes may be exposed to a range of chemical and enzymic reactions. Firstly, compounds are subjected to an acid medium (pH 4 − 5) and to a large number (at least 60) of hydrolytic enzymes which may cause some breakdown. To quote just two examples:

esters of amino acids are cleaved by acid esterases, releasing the free amino acid, and compounds with peptide bonds will undoubtedly be broken down by one or more of the numerous peptidases. Whilst such reactions may cause some inactivation of drugs, it is considered unlikely that lysosomes are instrumental in the toxic mechanisms of small organic molecules. However, since many proteins are effective carriers of toxic metals and their compounds, it is likely that lysosomes may play an important role in the toxicity of metals which are delivered to lysosomes aboard carrier macromolecules.

It is important to note that the accumulation of large quantities of lysosomotropic bases in lysosomes leads to neutralisation of the intralysosomal pH, impairment of degradative capacity and a build up of undegraded and partially degraded macromolecules within lysosomes. The clinical effect of the accumulation of bases may thus be similar to that of some of the lysosomal storage diseases. Whilst basic compounds inhibit all lysosomal catabolism by altering the pH, some compounds may inhibit only some acid hydrolases. The plant toxin swainsonine, for example, is a potent inhibitor of mannosidases only.

The majority of published work cited under 'pharmacology and toxicology of lysosomes' deals with effects of compounds on the stability and integrity of the lysosomal membrane. This interest is understandable in view of the diversity of conditions which cause disruption of the lysosomal membrane and release of lysosomal enzymes and subsequently result in inflammation and cell death. It is believed that many anti-inflammatory agents act by stabilising the lysosomal membrane. Whilst some evidence of this stabilising effect is available from experiments utilising isolated lysosomal preparations, there is presently no definitive evidence to suggest that anti-inflammatory agents act in this manner *in vivo*.

Compounds which may cause rupture of the lysosomal membrane are clearly not suitable for therapeutic application unless this property is expressed only in tissues (e.g., tumours, eukaryotic parasites, etc.) which need to be eradicated.

4.2 Uses of Lysosomal Preparations

There are two particular approaches which are relevant to the use of isolated lysosomal preparations in toxicology. Firstly lysosomes from normal untreated tissue may be isolated and incubated with different compounds. This *in vitro*-type approach is useful for assessing the effect of the compound on the lysosomal membrane and also on the activities of lysosomal enzymes. Secondly, lysosomes may be isolated from experimental animals previously treated with appropriate compounds. This approach is more suited for studying uptake and lysosomal processing.

4.2.1 *Use of In Vitro Preparations*

Preparations of liver and kidney lysosomes, described in Sections 2.4 and 2.7, respectively, are ideally suited for examining the *in vitro* effects of various compounds on lysosomes. The preparations should be in the appropriate (isotonic) medium at a pH of $7-7.4$. It may be advantageous to include Mg^{2+} and ATP in this medium in order to maintain the lysosomal proton pump during the incubation period. It is not advisable to resuspend and incubate lysosomes at acid pH, since this will have the effect of neutralising the proton pump and thus disrupt this important functional component of the lysosomal membrane.

Such preparations may be employed to assess the effect of compounds on membrane stability by measuring the rate of rupture of the lysosomes and the release of acid hydrolases into the medium. Since lysosomes in suspension tend to autolyse during incubation it is important to measure this spontaneous breakage by using appropriate controls which contain no exogenous compound. Rupture of lysosomes in suspension and the subsequent release of lysosomal acid hydrolases is most conveniently express-ed as 'latency' (described in detail in Section 3.5). This method, representing the bound or inaccessible acid hydrolase activity, may be used to determine stabilising or labilis-ing effects of compounds being investigated on the lysosomal membrane. However, it is important to stress again that such measurements should not be taken as definitive evidence that the compound will exhibit the same effect *in vivo*. They should serve only as an indication that the compound may be investigated further by a more rigorous experimental approach (e.g., by examining the stability/lability of lysosomes isolated from tissues of experimental animals treated with the compound).

The use of *in vitro* lysosomal preparations is elegantly illustrated in the work describing the mechanisms of the lysosomal proton pump (see ref. 16). Although the compound employed, a potential-sensitive flourescent dye, is not of pharmacological relevance, the experimental approach employed is most relevant.

As well as using lysosomal preparations, purified lysosomal extracts may be utilised to assess their potential for hydrolysing new drugs. Such an approach was successfully employed to screen *in vitro* a variety of daunorubicin conjugates which were later shown to exhibit their anti-cancer therapeutic activity *in vivo* against L-1210 leukemia in the mouse (60).

4.2.2 *Lysosomal Preparations from Experimentally Treated Tissues*

Lysosomal preparations isolated after appropriate treatment of the experimental animal and analysed for their content of exogenous compound, offer a powerful technique for examining toxic mechanisms. It is useful in this approach to utilise a radiolabelled com-pound, but this may not be necessary if the compound or its metabolites can be readily measured by other methods (e.g., h.p.l.c.).

We have isolated liver lysosomes, by the method described in Section 2.5.1 after treatment of rats with gold thiomalate and have demonstrated that gold is indeed handled in liver lysosomes (61). The great advantage of this approach is that discrete profiles of different lysosomal hydrolases can be compared with the distribution profiles of the administered compound. With the aid of appropriately chosen marker enzymes (see Section 3.2) the profile of the compound may also be compared with the distribution of other cellular structures, in order to establish if they are also involved in the bind-ing, uptake, transport, metabolism, etc. of the compound. If such experiments are repeated at different time intervals after administration of the pharmacologically active compound, a time-course picture of how the compound is handled in lysosomes may be built up. The results can be particularly meaningful if during the time-course blood and bile samples are also analysed, thus supplementing the lysosomal fractionation data with liver pharmacokinetics of the compound. The examination of liver lysosomes by these methods should have particular application for compounds which are either excreted in bile or may cause impairment of biliary excretion or bile production and also for compounds which reach liver cells by the 'piggy back' mechanism aboard carrier proteins.

In our laboratory we have also employed this approach for studying the handling of different compounds in kidney lysosomes. Administration of ^{109}Cd metallothionein intravenously and the subsequent isolation of kidney lysosomes, by the method given in Section 2.8, has shown that ^{109}Cd reaches a peak between 30 and 90 min in large lysosomes (protein droplets) and disappears rapidly from these lysosomes, whilst remaining associated with smaller lysosomes for much longer periods (62). Whilst this type of experiment lends itself readily to investigating toxic metals, there is no reason why such approaches should not be used with a variety of nephrotoxins, especially since a number of these have been observed, by electron microscopic morphology, to affect the lysosomal populations, e.g., cyclosporin A, aminoglycosides, light hydrocarbons, radiocontrast materials, etc.

Another *in vivo* type approach for studying the effect of compounds on lysosomes is worth mentioning although it does not involve any isolation procedure. It is possible to investigate how compounds affect kidney lysosomes simply by measuring the release of acid hydrolases in urine. However, it should be remembered that the release of lysosomal enzymes into urine may be elevated or reduced by a number of factors other than effects of the administered compounds. For example, an increased rate of protein filtration (in the glomeruli) will cause a higher rate of endocytic reabsorption of the proteins in the proximal tubule and result in elevated urinary levels of lysosomal enzymes.

4.3 Use of Peroxisomal Preparations

As pointed out by Connock and Temple (5), peroxisomal preparations have not yet been systematically employed for studying the effects of a range of compounds which are now known to affect the normal peroxisomal function of fatty acid oxidation.

A wide range of peroxisomal proliferators are of particular interest, although it is fair to say that the great knowledge of peroxisomal properties has not been utilised to trace the association of small molecules with peroxisomes by systematic isolation procedures following administration of the appropriate compound to experimental animals. Whilst hypolipidaemic drugs, such as the clofibrates, elevate liver lipid levels and thus indirectly cause peroxisomal proliferation, other compounds such as the phthalates may affect proliferation directly by acting as analogues of the fatty acids. However, despite a great deal of work also being devoted to phthalates, there appear to be no reports on their association with or uptake into peroxisomes.

Proliferation of peroxisomes results in a large increase in hydrogen peroxide-producing enzymes but a relatively smaller increase in catalase, thus causing a net increase in hydrogen peroxide and the possibility of free radical build up. The favourite hypothesis for the non-mutagenic carcinogenesis attributed to various hypolipidaemic drugs, i.e., that it is related to free radicals generated in proliferated peroxisomes, has not been subjected to any rigorous experimental protocol with isolated preparations. For example, isolated preparations may help to understand the morphologically observed association of proliferating peroxisomes with rough ER *via* the 'loops' or 'hooks' (63) particularly with reference to the fundamental area of peroxisomal biogenesis, by a systematic approach to characterising populations of microperoxisomes (see ref. 5).

Structural and biochemical properties of peroxisomes and the effects of many compounds on peroxisomes are covered in great detail by Bock *et al.* (63). This reference is also an excellent source of ideas for application of peroxisomal preparations in toxicology.

5. REFERENCES

1. Holtzman,E. (1976) *Lysosomes: A Survey, Cell Biology Monographs,* Vol. **3**, published by Springer-Verlag, Vienna.
2. Davies,M. (1975) in *Lysosomes in Biology and Pathology,* Vol. **4**, Dingle,J.T. and Dean,R.T. (eds.), North Holland, Amsterdam, p. 305.
3. de Duve,C. (1969) *Proc. R. Soc. Lond., Ser. B.,* **173**, 71.
4. Leighton,F., Brandan,E., Lazo,O. and Bronfman,M. (1982) *Ann. N.Y. Acad. Sci.,* **386**, 62.
5. Connock,M.J. and Temple,N.J. (1983) *Int. J. Biochem.,* **15**, 125.
6. Hinton,R.H. and Dobrota,M. (1976) in *Laboratory Techniques in Biochemistry and Molecular Biology,* Vol. **6**, Work,T.S. and Work.E. (eds.), North Holland, Amsterdam, p. 1.
7. *Handbook of Chemistry and Physics,* published by Chemical Rubber Company.
8. Wattiaux,R., Wattiaux-de Coninck,S., Ronveaux-Dupal,M.F. and Dubois,F. (1978) *J. Cell Biol.,* **78**, 346.
9. Dobrota,M. and Hinton,R.H. (1980) *Anal. Biochem.,* **102**, 97.
10. Wattiaux,R. and Wattiaux-De Coninck,S. (1983) in *Iodinated Density Gradient Media — A Practical Approach,* Rickwood,D. (ed.), IRL Press, Oxford and Washington, D.C., p. 119.
11. Neat,C.E., Thomassen,M.S. and Osmundsen,H. (1981) *Biochem J.,* **196**, 149.
12. Van Veldhoven,P., Debeer,L.J. and Mannaerts,G.P. (1983) *Biochem. J.,* **210**, 685.
13. Leighton,F., Poole,B., Beaufay,H., Baudhuin,P., Coffey,J.W., Fowler,S. and de Duve,C. (1968) *J. Cell Biol.,* **37**, 482.
14. Maunsbach,A. (1966) *J. Ultrastruct. Res.,* **16**, 13.
15. Goldstone,A. and Koenig,H. (1972) *Life Sci.,* **11**, 511.
16. Harikumar,P. and Reeves,J.P. (1983) *J. Biol. Chem.,* **258**, 10403.
17. Andersen,K.J., Haga,H.J. and Dobrota,M. (1980), *Biochem. Soc. Trans.,* **8**, 597.
18. Kitano,R. and Morimoto,S. (1975) *Biochim. Biophys. Acta,* **411**, 113.
19. Small,G.M., Hocking,T.J., Sturdee,A.P., Burdett,K. and Connock.M.J. (1981) *Life Sci.,* **28**, 1875.
20. Smith.A.D. and Winkler,H. (1969) in *Lysosomes in Biology and Pathology,* Vol. **1**, Dingle,J.T. and Fell,H.B. (eds.), North Holland, Amsterdam, p. 155.
21. Peters,T.J. (1975) in *Lysosomes in Biology and Pathology,* Vol. **4**, Dingle,J.T. and Dean,R.T. (eds.), North Holland, Amsterdam, p. 47.
22. Vaes,G. and Jacques,P. (1965) *Biochem. J.,* **97**, 380.
23. Lisman,J.J.W., De Haan,J. and Overdijk,B. (1979) *Biochem. J.,* **178**, 79.
24. Karmali,A., Montague,D.J., Holloway,B.R. and Peters,T.J. (1984) *Cell Biochem. Function,* **2**, 155.
25. Rome,L.H., Garvin.A.J., Allietta,M.M. and Neufeld,E.F. (1979) *Cell,* **17**, 143.
26. Connock.M.J., Kirk,P.R. and Sturdee,A.P. (1974) *J. Cell Biol.,* **61**, 123.
27. Burge,M.L.E. and Hinton,R.H. (1971) in *Separations with Zonal Rotors,* Reid,E. (ed.), Wolfson Bioanalytical Centre, University of Surrey, p. S-5.1.
28. Mochizuki,Y., Hruban,Z., Morris,H.P., Slesers,A. and Vigil,E.L. (1971) *Cancer Res.,* **31**, 763.
29. Baggiolini,M., Hirsch,J.G. and de Duve,C. (1970) *J. Cell Biol.,* **45**, 586.
30. Hook,G.E.R. and Gilmore,L.B. (1982) *J. Biol. Chem.,* **257**, 9211.
31. Bowers,W.E. (1969) in *Lysosomes in Biology and Pathology,* Vol. **1**, Dingle,J.T. and Fell,H.B. (eds.), North Holland, Amsterdam, p. 167.
32. Brown,J.A. and Swank,R.T. (1983) *J. Biol. Chem.,* **258**, 15323.
33. Stauber,W.T. and Bird,J.W.C. (1974) *Biochim. Biophys. Acta,* **338**, 234.
34. Reeves,J.P., Decker,R.S., Crie,J.S. and Wildenthal,K. (1981) *Proc. Natl. Acad. Sci. USA,* **78**, 4426.
35. Gordon,J.L. (1975) in *Lysosomes in Biology and Pathology,* Vol. **4**, Dingle,J.T. and Dean,R.T. (eds.), North Holland, Amsterdam, p. 3.

36. Lazarus,L.S. and Hatcher,V.B. (1975) in *Lysosomes in Biology and Pathology*, Vol. **4**, Dingle,J.T. and Dean,R.T. (eds.), North Holland, Amsterdam, p. 111.
37. Miquelis,R. and Simon,C. (1974) *Biochem. J.*, **138**, 299.
38. Woessner,J.F. (1969) in *Lysosomes in Biology and Pathology*, Vol. **1**, Dingle,J.T. and Dean,R.T. (eds.), North Holland, Amsterdam, p. 299.
39. Stahn,R., Maier,K.P. and Hannig,K. (1970) *J. Cell Biol.*, **46**, 576.
40. Kamrath,F.J. and Dodt,G., Debuch,H. and Uhlenbruck,G. (1984) *Hoppe-Seyler's Z. Physiol. Chem.*, **365**, 539.
41. Morre,D.J., Cline,G.B., Coleman,R., Evans,W.H., Glaumann,H., Headon,D.R., Reid,E., Siebert,G. and Widnell,C.C. (1979) *Eur. J. Cell. Biol.*, **20**, 195.
42. Dobrota,M., Burge,M.L.E. and Hinton,R.H. (1979) *Eur. J. Cell. Biol.*, **19**, 139.
43. Hinton,R.H. and Norris,K.A. (1972) *Anal. Biochem.*, **48**, 247.
44. Hultberg,B. and Ockerman,P.A. (1972) *Clin. Chim. Acta*, **39**, 49.
45. Milson,D.W., Rose,F.A. and Dodgson,K.S. (1972) *Biochem. J.*, **128**, 331.
46. Barrett,A.J. and Heath,M.F. (1977) in *Lysosomes: A Laboratory Handbook*, second edition, Dingle,J.T. (ed.), Elsevier/North Holland, Amsterdam, p. 19.
47. Barrett,A.J. (1967) *Biochem. J.*, **104**, 601.
48. Beaufay,H., Bendall,D.S., Baudhuin,P., Wattiaux,R. and de Duve C. (1959) *Biochem. J.*, **73**, 628.
49. Baudhuin,P., Rahman-Li,Y., Sellinger,O.Z., Wattiaux,R., Jacques,P. and de Duve,C. (1964) *Biochem. J.*, **92**, 179.
50. Lazarow,P.B. and de Duve,C. (1976) *Proc. Natl. Acad. Sci. USA*, **73**, 2043.
51. Pennington,R.J. (1961) *Biochem. J.*, **80**, 649.
52. Appelmans,F., Wattiaux,R. and de Duve,C. (1955) *Biochem. J.*, **59**, 438.
53. Krajl,M. (1965) *Biochem. Pharmacol.*, **14**, 1684.
54. Beaufay,H., Amar-Costesec,A., Thines-Sempoux,D., Wibo,M., Robbi,M. and Berthet,J. (1974) *J. Cell Biol.*, **61**, 213.
55. Omura,T. and Sato,R. (1964) *J. Biol. Chem.*, **239**, 279.
56. Prospero,T.D., Burge,M.L.E., Norris,K.A., Hinton,R.H. and Reid,E., (1973) *Biochem. J.*, **132**, 449.
57. Beaufay,H. (1969) in *Lysosomes in Biology and Pathology*, Vol. **2**, Dingle,J.T. and Fell,H.B. (eds.), North Holland, Amsterdam, p. 516.
58. Blouin,A., Bolender,R.P. and Weibel,E.R. (1977) *J. Cell Biol.*, **72**, 441.
59. de Duve,C., De Basey,T., Poole,B., Trouet,A., Tulkens,P. and Van Hoof,F. (1974) *Biochem. Pharmacol.*, **93**, 2495.
60. Gregoriais,G., Senior,J. and Trouet,A. (1982) *Targeting of Drugs*, published by Plenum Press, New York/London.
61. Taylor,M., Dobrota,M., Taylor,A. and Hinton,R.H., unpublished results.
62. Dobrota,M., Bonner,F.W. and Carter,B.A. (1982) in *Nephrotoxicity: Assessment and Pathogenesis*, Bach,P.H., Bonner,F.W., Bridges,J.W., Look,E.A. (eds.), Wiley, Chichester, p. 320.
63. Bock,P., Kramar,R. and Pavelka, M., eds. (1980) *Peroxisomes and Related Particles in Animal Tissues, Cell Biology Monographs*, Vol. **7**, published by Springer-Verlag, Vienna.

INDEX

Published in the Practical Approach series

Mutagenicity testing
a practical approach

Mutagenicity testing

a practical approach

Edited by
S Venitt & J M Parry

Published in the
Practical Approach Series
Series editors: D Rickwood and B D Hames

IRL PRESS
Oxford · Washington DC

Edited by
S Venitt, Institute of Cancer Research,
and
J M Parry, University College of Swansea

A laboratory manual of genetic toxicology

Mutagenicity testing: a practical approach describes in great practical detail nine different tests for detecting mutagenic activity, chromosomal aberrations and induction of DNA damage and repair.

The book is written for researchers in universities and industry, post-graduates and undergraduates in toxicology, industrial hygiene, medicine and biochemistry. The editors introduce the aims and concepts of genetic toxicology and nine chapters give practical instructions for conducting tests in organisms ranging from bacteria to laboratory rodents.

Particular emphasis is given to well-validated tests required by national and international regulatory agencies. The text also offers guidance on the presentation and interpretation of results and includes key references and advice on laboratory safety.

CONTENTS

December 1984; 368 pp; 0 904147 72 X (softbound)

For details of price and ordering consult our current catalogue or contact:
IRL Press Ltd, PO Box 1, Eynsham, Oxford OX8 1JJ, UK
IRL Press Inc, PO Box Q, McLean, VA 22101, USA

◇ **IRL PRESS**
Oxford · Washington DC

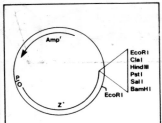

DNA cloning
(Volumes I and II)
a practical approach

Edited by D M Glover,
Imperial College of Science and
Technology, London

Published
in the
Practical
Approach
series

A STEP-BY-STEP GUIDE TO PROVEN NEW TECHNIQUES

Breakthroughs in the manipulation of DNA have already revolutionised biology; they are set to do the same for drug and food production. *DNA cloning* contains the background and detailed protocols for molecular biologists to perform these experiments with success. It supersedes previous manuals in describing recent developments with widespread applications that use *E coli* as the host organism.

Up-to-the-minute contributions cover the use of phage λ insertion vectors for cDNA cloning and the use of phage λ replacement vector systems to select recombinants for DNA cloning.

Two chapters evaluate *E coli* transformation and methods for *in vitro* mutagenesis of DNA cloning in other organisms including yeast, plant cells and Gram-negative and Gram-positive bacteria. Finally, the last three chapters of Volume II offer three different approaches to the introduction of cloned genes into animal cells.

Contents

Volume I: *June 1985; 204pp ;*
0 947946 18 7 *(softbound)*
Volume II: *June 1985; 260pp ;*
0 947946 19 5 *(softbound)*
Volumes I and II; *0 947946 20 9*

For details of price
and ordering consult
our current catalogue
or contact:

IRL Press Ltd,
Box 1, Eynsham,
Oxford OX8 1JJ, UK

IRL Press Inc,
PO Box Q,
McLean VA 22101,
USA

 **IRL PRESS**
Oxford · Washington DC